T0331823

Measure Theory and Filtering
Introduction and Applications

The estimation of noisily observed states from a sequence of data has traditionally incorporated ideas from Hilbert spaces and calculus-based probability theory. As conditional expectation is the key concept, the correct setting for filtering theory is that of a probability space. Graduate engineers, mathematicians, and those working in quantitative finance wishing to use filtering techniques will find in the first half of this book an accessible introduction to measure theory, stochastic calculus, and stochastic processes, with particular emphasis on martingales and Brownian motion. Exercises are included, solutions to which are available from www.cambridge.org. The book then provides an excellent user's guide to filtering: basic theory is followed by a thorough treatment of Kalman filtering, including recent results that exend the Kalman filter to provide parameter estimates. These ideas are then applied to problems arising in finance, genetics, and population modelling in three separate chapters, making this a comprehensive resource for both practitioners and researchers.

LAKHDAR AGGOUN is Associate Professor in the Department of Mathematics and Statistics at Sultan Qabos University, Oman.

ROBERT ELLIOTT is RBC Financial Group Professor of Finance at the University of Calgary, Canada.

Measure Theory and Filtering

Introduction and Applications

Lakhdar Aggoun

Department of Mathematics and Statistics, Sultan Qaboos University, Oman

Robert J. Elliott

Haskayne School of Business, University of Calgary

CAMBRIDGE
UNIVERSITY PRESS

CAMBRIDGE
UNIVERSITY PRESS

University Printing House, Cambridge CB2 8BS, United Kingdom

Cambridge University Press is part of the University of Cambridge.

It furthers the University's mission by disseminating knowledge in the pursuit of
education, learning and research at the highest international levels of excellence.

www.cambridge.org
Information on this title: www.cambridge.org/9780521838030

© Cambridge University Press 2004

First published 2004

A catalogue record for this publication is available from the British Library

Library of Congress Cataloguing in Publication data

Aggoun Lakhdar
Measure theory and filtering: introduction and applications/Lakhdar Aggoun, Robert J. Elliott.
p.   cm. – (Cambridge series on statistical and probabilistic mathematics)
Includes bibliographical references and index.
ISBN 0 521 83803 7 (hardback)   ·
1. Measure theory.   2. Kalman filtering.   I. Elliott, Robert J., 1940–   II. Title.   III. Series.
QA312 A34 2004
515'.42 – dc22   2004040397

ISBN 978-0-521-83803-0 Hardback
ISBN 978-1-107-41071-8 Paperback

Contents

Preface

Traditional courses for engineers in filtering and signal processing have been based on elementary linear algebra, Hilbert space theory and calculus. However, the key objective underlying such procedures is the (recursive) estimation of indirectly observed states given observed data. This means that one is discussing conditional expected values, given the observations. The correct setting for conditional expected value is in the context of measurable spaces equipped with a probability measure, and the initial object of this book is to provide an overview of required measure theory. Secondly, conditional expectation, as an inverse operation, is best formulated as a form of Bayes' Theorem. A mathematically pleasing presentation of Bayes' theorem is to consider processes as being initially defined under a "reference probability." This is an idealized probability under which all the observations are independent and identically distributed. The reference probability is a much nicer measure under which to work. A suitably defined change of measure then transforms the distribution of the observations to their real world form. This setting for the derivation of the estimation and filtering results enables more general results to be obtained in a transparent way.

The book commences with a leisurely and intuitive introduction to σ-fields and the results in measure theory that will be required.

The first chapter also discusses random variables, integration and conditional expectation.

Chapter 2 introduces stochastic processes, with particular emphasis on martingales and Brownian motion.

Stochastic calculus is developed in Chapter 3 and techniques related to changing probability measures are described in Chapter 4.

The change of measure method is the basic technique used in this book.

The second part of the book commences with a treatment of Kalman filtering in Chapter 5. Recent results, which extend the Kalman filter and enable parameter estimates to be obtained, are included. These results are applied to financial models in Chapter 6. The final two chapters give some filtering applications to genetics and population models.

The authors would like to express their gratitude to Professor Nadjib Bouzar of the Department of Mathematics and Computer Science, University of Indianapolis, for the incredible amount of time he spent reading through the whole manuscript and making many useful suggestions.

Robert Elliott would like to acknowledge the support of NSERC and the hospitality of the Department of Applied Mathematics at the University of Adelaide, South Australia.

Lakhdar Aggoun would like to acknowledge the support of the Department of Mathematics and Statistics, Sultan Qaboos University, Al-Khoud, Sultanate of Oman; the hospitality of the Department of Mathematical Sciences at the University of Alberta, Canada; and the Haskayne School of Business, University of Calgary, Calgary, Canada.

Part I
Theory

1

Basic probability concepts

1.1 Random experiments and probabilities

An experiment is random if its outcome cannot be predicted with certainty. A simple example is the throwing of a die. This experiment can result in any of six unpredictable outcomes 1, 2, 3, 4, 5, 6 which we list in what is usually called a *sample space* $\Omega = \{1, 2, 3, 4, 5, 6\} \triangleq \{\omega_1, \omega_2, \omega_3, \omega_4, \omega_5, \omega_6\}$. Another example is the amount of yearly rainfall in each of the next 10 years in Auckland. Each outcome here is an ordered set containing ten nonnegative real numbers (a vector in $\mathbb{R}^{10}_+$); however, one has to wait 10 years before observing the outcome ω.

Another example is the following.

Let X_t be the water level of a dam at time t. If we are interested in the behavior of X_t during an interval of time $[t_0, t_1]$ say, then it is necessary to consider simultaneously an uncountable family of X_ts, that is,

$$\Omega = \{0 \le X_t < \infty, \quad t_0 \le t \le t_1\}.$$

The "smallest" observable outcome ω of an experiment is called *simple*.

The set $\{1\}$ containing 1 resulting from a throw of a die is simple. The outcome "odd number" is not simple and it occurs if and only if the throw results in any of the three simple outcomes 1, 3, 5. If the throw results in a 5, say, then the same throw results also in "a number larger than 3" or "odd number." Sets containing outcomes are called *events*. The events "odd number" and "a number larger than 3" are not mutually exclusive, that is, both can happen simultaneously, so that we can define the event "odd number *and* a number larger than 3."

The event "odd number *and* even number" is clearly impossible or empty. It is called the *impossible event* and is denoted, in analogy with the empty set in set theory, by Ø. The event "odd number *or* even number" occurs no matter what is the event ω. It is Ω itself and is called the *certain event*.

In fact possible events of the experiment can be combined naturally using the set operations *union, intersection,* and *complementation*. This leads to the concept of field or algebra (σ-field (sigma-field) or σ-algebra, respectively) which is of fundamental importance in the theory of probability.

A nonempty class $\mathcal{F}$ of subsets of a nonempty set Ω is called a *field* or *algebra* if

1. $\Omega \in \mathcal{F}$,
2. $\mathcal{F}$ is closed under finite unions (or finite intersections),
3. $\mathcal{F}$ is closed under complementation.

It is a σ-*field* or (σ-*algebra*) if the stronger condition

2.′ $\mathcal{F}$ is closed under countable unions (or countable intersections)

holds.

If $\{\mathcal{F}\}$ is a σ-field the pair $(\Omega, \mathcal{F})$ is called a *measurable space*. The sets $B \in \mathcal{F}$ are called *events* and are said to be *measurable sets*.

For instance, the collection of finite unions of the half open intervals $(a, b]$, $(-\infty < a < b \leq +\infty)$ in $\mathbb{R}$ plus the empty set is a field but not a σ-field because it is not closed under infinite countable unions. The open interval $(0, 1) = \bigcup_{n=1}^{\infty}(0, 1 - 1/n]$ is not in this collection despite the fact it contains *each* interval $(0, 1 - 1/n]$. Neither does it contain the singletons $\{x\}$, even though $\{x\} = \bigcap_{n=1}^{\infty}(x - 1/n, x]$ and it does not contain many other useful sets. This suggests that the notion of σ-field is indeed needed. There exists a minimal σ-field denoted $\mathcal{B}(\mathbb{R})$ containing all half open intervals $(a, b]$. This is the *Borel* σ-field on the real line and it is the smallest σ-field containing the collection of open intervals and hence all intervals. It contains also:

1. all singletons $\{x\}$ since $\{x\} = \bigcap_{n=1}^{\infty} \left(x - \dfrac{1}{n}, x + \dfrac{1}{n} \right)$,
2. the set Q of all rational numbers because it is a countable union: $Q = \bigcup_{r \in Q}\{r\}$,
3. the complement of Q, which is the set of all irrational numbers,
4. all open sets since any open set $\mathcal{O} = \bigcup_n I_n$, where $\{I_n\}$ are disjoint intervals. To see this recall that since $\mathcal{O}$ is open, then for any $x \in \mathcal{O}$ there exits a maximal interval I_x containing x and contained in $\mathcal{O}$ and $I_x = \mathcal{O}$ if $\mathcal{O}$ is itself an interval. If $\mathcal{O}$ is not an interval then there is a collection of disjoint maximal intervals contained in $\mathcal{O}$, one for each $x \in \mathcal{O}$. Moreover, each of these intervals contains a rational number because of the density of Q. Let $\{r_n : n = 1, 2, \ldots\}$ be an enumeration of these rationals. Consequently, there is only at most a countable number of these intervals $I_1, I_2, \ldots$. Therefore, since each of these intervals is contained in $\mathcal{O}$, their union $\bigcup_n I_n \subset \mathcal{O}$. Conversely, for each $x \in \mathcal{O}$ there exits a maximal interval $I_{n(x)}$ containing x and contained in $\bigcup_n I_n$, that is, $\mathcal{O} \subset \bigcup_n I_n$. Consequently $\mathcal{O} = \bigcup_n I_n$.

Sets in $\mathcal{B}(\mathbb{R})$ are called *Borel sets*. Note that a topological space, unlike a measure space, is not closed under complementation. A word of caution here: even σ-fields are not in general closed under uncountable unions.

The largest possible σ-field on any set Ω is the *power class* 2^Ω containing all the subsets of Ω. However this σ-field is in general "too big" to be of any use in probability theory. At the other extreme we have the smallest σ-field consisting of Ω and the empty set $\emptyset$.

Given any collection C of subsets of Ω, the σ-field generated by C, denoted by $\sigma\{C\}$, is made up of the class of all countable unions, all countable intersections and all complements of the subsets in C and all countable unions, intersections and complements of these sets, and so on. For instance, if C contains one subset, F say, then $\sigma\{F\}$ consists of the subset

F itself, its complement $\bar{F}$ (also denoted F^c), their union $F \cup \bar{F}$ (which is always Ω) and their intersection $F \cap \bar{F}$ (which is always $\emptyset$).

The σ-field $\sigma\{C\}$ generated by a class of subsets C contains by definition C itself (as a subset); however, there are other σ-fields also containing C, one of them being 2^Ω (the largest one). The point here is that $\sigma\{C\}$ is the *smallest* σ-field containing C. In the set theory context "smallest" means that $\sigma\{C\}$ is in the intersection of all the σ-fields containing C. In summary:

$$C \subset \sigma\{C\} \subset \{\text{any } \sigma\text{-field containing } C\}.$$

It is left as an exercise to show that any σ-field is either finite or uncountably infinite.

Fields, or σ-fields, are convenient mathematical objects that express how much we know about the outcome ω of a random experiment. For instance, if $\Omega = \{1, 2, 3, 4, 5, 6\}$ we may not be able to observe ω but we may observe a "larger" event like "odd number"$= \{(1, 3, 5)\}$, so that our "observed" σ-field is smaller than the one generated by Ω. In fact it is equal to $\{(1, 3, 5), (2, 4, 6), \Omega, \emptyset\}$, which does not contain events like $\{(1, 3)\}$ or $\{6\}$.

When the sample space Ω is finite, it is enough to represent information through partitions of Ω into *atoms*, which are the smallest observable events. Since a field is just a collection of finite unions and complements of these atoms, it represents the same information as the partition. This is not true on infinite sample spaces as partitions and fields are not big enough to represent information in all practical situations.

Suppose that when the experiment of throwing a die is performed, an indirect observer of the outcome ω can only learn that the event $\{1, 2\}$ did or did not occur. So for this observer the (smallest) decidable events, or atoms, are in the field

$$\mathcal{F}_1 = \sigma\{\{1, 2\}, \{3, 4, 5, 6\}\} = \{\emptyset, \{1, 2, 3, 4, 5, 6\}, \{1, 2\}, \{3, 4, 5, 6\}\}.$$

Another observer with a better access to information might be able to observe the richer field

$$\mathcal{F}_2 = \sigma\{\{1, 2\}, \{3, 4\}, \{5, 6\}\},$$

which contains more atoms. The point here is that, given a set of outcomes Ω, it is possible to define many fields, or σ-fields, ranging from the coarsest (containing only Ω and the empty set $\emptyset$), to the finest (containing all the subsets of Ω).

A natural question is: what extra conditions will make a field into a σ-field? We have the following useful result.

A field is a σ-field if and only if it is closed under monotonic sequences of events, that is, it contains the limit of every monotonically increasing or decreasing sequence of events. (A sequence of events A_i, $i \in \mathbb{N}$, is monotonic increasing if $A_1 \subset A_2 \subset A_3 \ldots$).

Let the index parameter t be either a nonnegative integer or a nonnegative real number.

To keep track, to record, and to benefit from the flow of information accumulating in time and to give a mathematical meaning to the notions of *past*, *present* and *future* the concept of *filtration* is introduced. This is done by equipping the measurable space $(\Omega, \mathcal{F})$ with a nondecreasing family $\{\mathcal{F}_t, t \geq 0\}$ of "observable" sub-σ-fields of $\mathcal{F}$ such that $\mathcal{F}_t \subset \mathcal{F}_{t'}$ whenever $t \leq t'$. That is, as time flows, our information structures or σ-fields are becoming finer and finer.

We define $\mathcal{F}_\infty = \sigma(\bigcup_{t\geq0} \mathcal{F}_t) \stackrel{\triangle}{=} \bigvee_{t\geq0} \mathcal{F}_t$ where the symbol $\stackrel{\triangle}{=}$ stands for "by definition."

Example 1.1.1 Let $\Omega = \{\omega_1, \omega_2, \omega_3, \omega_4, \omega_5, \omega_6\}$. The σ-fields

$$\mathcal{F}_0 = \sigma\{\Omega, \emptyset\},$$
$$\mathcal{F}_1 = \sigma\{\{\omega_1, \omega_2, \omega_3\}, \{\omega_4, \omega_5, \omega_6\}\},$$
$$\mathcal{F}_2 = \sigma\{\{\omega_1, \omega_2\}, \{\omega_3\}, \{\omega_4, \omega_5, \omega_6\}\},$$
$$\mathcal{F}_3 = \sigma\{\{\omega_1\}, \{\omega_2\}, \{\omega_3\}, \{\omega_4, \omega_5, \omega_6\}\},$$

form a filtration since $\mathcal{F}_0 \subset \mathcal{F}_1 \subset \mathcal{F}_2 \subset \mathcal{F}_3$. However, the σ-fields

$$\mathcal{F}_0 = \sigma\{\Omega, \emptyset\},$$
$$\mathcal{F}_1 = \sigma\{\{\omega_1, \omega_2, \omega_3\}, \{\omega_4, \omega_5, \omega_6\}\},$$
$$\mathcal{F}_2 = \sigma\{\{\omega_1, \omega_4\}, \{\omega_2, \omega_5\}, \{\omega_3, \omega_6\}\},$$
$$\mathcal{F}_3 = \sigma\{\{\omega_1\}, \{\omega_2\}, \{\omega_3, \omega_4\}, \{\omega_5, \omega_6\}\},$$

do not form a filtration since, for instance, $\mathcal{F}_1 \not\subset \mathcal{F}_2$.   □

Example 1.1.2 Suppose Ω is the unit interval $(0, 1]$ and consider the following σ-fields:

$$\mathcal{F}_0 = \sigma\{\Omega, \emptyset\},$$
$$\mathcal{F}_1 = \sigma\{(0, \tfrac{1}{2}], (\tfrac{1}{2}, \tfrac{3}{4}], (\tfrac{3}{4}, 1]\},$$
$$\mathcal{F}_2 = \sigma\{(0, \tfrac{1}{4}], (\tfrac{1}{4}, \tfrac{1}{2}], (\tfrac{1}{2}, \tfrac{3}{4}], (\tfrac{3}{4}, 1]\},$$
$$\mathcal{F}_3 = \sigma\{(0, \tfrac{1}{8}], (\tfrac{1}{8}, \tfrac{2}{8}], \ldots, (\tfrac{7}{8}, 1]\}.$$

These form a filtration since $\mathcal{F}_0 \subset \mathcal{F}_1 \subset \mathcal{F}_2 \subset \mathcal{F}_3$.   □

When the time index $t \in \mathbb{R}^+$ we are led naturally to introduce the concepts of *right-continuity* and *left-continuity* of a filtration as a function of t.

A filtration $\{\mathcal{F}_t, \ t \geq 0\}$ is right-continuous if $\mathcal{F}_t$ contains events *immediately after t*, that is $\mathcal{F}_t = \bigcap_{\epsilon > 0} \mathcal{F}_{t+\epsilon}$. We may also say that a filtration $\{\mathcal{F}_t, \ t \geq 0\}$ is right-continuous if new information at time t arrives precisely at time t and not an instant after t.

It is left-continuous if $\{\mathcal{F}_t\}$ contains events *strictly prior* to t, that is $\mathcal{F}_t = \bigvee_{s < t} \mathcal{F}_s$.

Probability measures

Given a measurable space $(\Omega, \mathcal{F})$ a *probability measure* P is a countably additive function defined on events in $\mathcal{F}$ with values in $[0, 1]$. More precisely:

A set function $P: \mathcal{F} \to [0, 1]$, where $\mathcal{F}$ is either a field or a σ-field, is called a probability measure if

1. $P(\Omega) = 1$;
2. If B_k is a countable sequence of pairwise disjoint events in $\mathcal{F}$, then $P(\bigcup B_k) = \sum P(B_k)$. This is termed σ-additivity of P.

Of course, (1) and (2) imply that $P(\emptyset) = 0$. Also, if A is an event such that $P(A) = 0$ and B is any event contained in A then $P(B) = 0$.

It is easily seen that if Ω is finite we need only specify P on atoms of $\mathcal{F}$.

The triple $(\Omega, \mathcal{F}, P)$ is called a *probability space*.

Nonempty events which are unlikely to occur and to which a zero probability is assigned are called *negligible events* or *null* events.

A σ-field $\mathcal{F}$ is *P-complete* if all subsets of null events are also events. Of course, their probability is zero.

A filtration is *complete* if $\mathcal{F}_0$ is complete, i.e. all the null events are known at the initial time.

The mathematical object $(\Omega, \mathcal{F}, \mathcal{F}_t, P)$, where the filtration $\{\mathcal{F}_t, t \geq 0\}$ is right-continuous and complete, is sometimes called a *stochastic basis* or a *filtered probability space* .

The filtration $\{\mathcal{F}_t, t \geq 0\}$ is said to satisfy the "usual conditions" if it is right-continuous and complete.

For monotonic sequences of events we have the following result on *continuity* of probability measures.

Theorem 1.1.3 *Let $(\Omega, \mathcal{F}, P)$ be a probability space. If $\{A_n\}$ is an increasing sequence of events with limit A, then*

$$P(A_n) \uparrow P(A),$$

and if $\{B_n\}$ is a decreasing sequence of events with limit B, then

$$P(B_n) \downarrow P(B).$$

Proof To prove the first statement, visualize the sequence $\{A_n\}$ as a sequence of increasing concentric disks and then define the sequence of disjoint rings $\{R_n\}$ (except for R_1 which is the disk A_1):

$$R_1 = A_1, \quad R_2 = A_2 - A_1, \ldots, \quad R_n = A_n - A_{n-1}.$$

Note that

$$A_k = \cup_{n=1}^{k} R_n, \quad A = \cup_{n=1}^{\infty} A_n = \cup_{n=1}^{\infty} R_n,$$

so that by σ-additivity

$$P(A) = \sum_{n=1}^{\infty} P(R_n) = \lim_k \sum_{n=1}^{k} P(R_n) = \lim_k P(\cup_{n=1}^{k} R_n) = \lim_k P(A_k).$$

The proof of the second statement follows by considering the sequence of complementary events $\{\bar{B}_n\}$ which is increasing with limit $\bar{B}$, so that

$$1 - P(A_n) \uparrow 1 - P(A) \implies P(A_n) \downarrow P(A).$$

■

Example 1.1.4 Consider the experiment of tossing a fair coin infinitely many times and "observing" the outcomes of *all* tosses. Here each $\omega \in \Omega = (H, T)^{\infty}$ is a countably infinite sequence of "Heads" and "Tails". If we denote "Heads" and "Tails" by 0 and 1, each ω is a sequence of 0s and 1s and it can be shown that there are as many ωs as there are points in the interval $[0, 1)$!

Suppose we wish to estimate the probability of the event consisting of those ωs for which the proportion of heads converges to $1/2$. The so-called *Strong Law of Large Numbers* says that this probability is equal to one, i.e. the ωs for which the convergence to $1/2$ does not hold form a negligible set. However, this negligible set is rather huge, as can be imagined!   □

Example 1.1.5 In Example 1.1.4 let $F_{n,S}$ be the collection of infinite sequences of Hs and Ts with some restriction S put on the first n tosses. For instance, if $n = 3$,

$$S = \{HHT \ldots, HTH \ldots, THH \ldots\} \subset (H, T)^3,$$

$F_{3,S}$ is the collection of infinite sequences of Hs and Ts for which the first three entries contain exactly two Hs.

It is left as an exercise to show that the class
$\mathcal{F} = \{F_{n,S}, \; S \subset (H, T)^n, \; n \in \mathbb{N}\}$ is a field.   □

We now quote without proof from [4] the following result on extending a function P defined on sets in a field.

Theorem 1.1.6 *([4]) If P is a probability measure on a field $\mathcal{A}$, then it can be extended uniquely to the σ-field $\mathcal{F} = \sigma\{\mathcal{A}\}$ generated by $\mathcal{A}$, i.e. the restriction of the extension measure to the field $\mathcal{A}$ is P itself and by tradition they are both denoted by P.*

Let us return to the coin-tossing situation of Example 1.1.5.

Using the extension theorem (Theorem 1.1.6) one can construct a (unique) probability measure P called *product probability measure* on the space $((H, T)^\infty, \mathcal{F})$, starting from an initial probability $(p(H), p(T)) = (1/2, 1/2)$ by setting

$$P(F_{n,S}) = \sum_S \left(\frac{1}{2}\right)^n = \text{(number of infinite sequences in } S) \times \left(\frac{1}{2}\right)^n.$$

It is left as an exercise to show that P does not depend on the representations of sets in $\mathcal{F}$ and that it is countably additive. (See [4]).

An immediate generalization of the coin tossing experiment in Example 1.1.5 is to consider an infinite sequence of independent experiments, to which corresponds an infinite sequence of probability spaces $(\Omega_1, \mathcal{F}_1, P_1), (\Omega_2, \mathcal{F}_2, P_2), \ldots$. We are interested in the space $\Omega^{(\infty)} = \Omega_1 \times \Omega_2 \times \ldots$ of all infinite sequences $\omega = (\omega_1, \omega_2, \ldots)$. Events of interest are again cylinder sets, i.e. infinite sequences with restrictions put on the first n outcomes. The collection of all these cylinders form a field which generates a σ-field $\mathcal{F}$, often denoted $\mathcal{F}_1 \otimes \mathcal{F}_2 \otimes \ldots$. A probability measure P can be defined on cylinder sets then extended uniquely to $\mathcal{F}$ using the Extension Theorem 1.1.6.

In the coin-tossing experiment, an example of an event which is in $\mathcal{F}$ is the event F that a "Head" will occur. Clearly, $F = \bigcup_{k=1}^\infty F_k$, where F_k is the event that a "Head" occurs on the k-th trial and not before. Since each F_k is a cylinder set, $P(F_k)$ is well defined for each

$k \geq 1$. Moreover the F_ks are pairwise disjoint, hence

$$P(F) = \sum_{k=1}^{\infty} P(F_k) = \sum_{k=1}^{\infty} \frac{1}{2^k} = 1.$$

Note that this probability is still 1 regardless of the size of the probability of occurrence of a "Head", (as long as it is not 0).

Modeling with infinite sample spaces is not a mathematical fantasy. In many very simple minded problems infinite sequences of outcomes cannot be avoided. For example, "the first time a Head occurs" event cannot be described in a finite sample space model because the number of trials before it occurs cannot be bounded in advance.

In general, it is impossible to define a probability measure on all the subsets of an infinite sample space; that is, one cannot say any subset is an event. However, consider the following case.

Example 1.1.7 Suppose that Ω is countable and let $\mathcal{F}$ be the σ-field 2^Ω. Then it is not difficult to define a probability measure on $\mathcal{F}$. Choose P such that

$$0 \leq P(\{\omega\}) \leq 1 \text{ and } P(\{\Omega\}) = \sum_{\omega \in \Omega} P(\omega) = 1,$$

and for any $F \in \mathcal{F}$, define $P(F) = \sum_{\omega \in F} P(\omega)$.

Let $\{F_n\}_{n \in \mathbb{N}}$ be a sequence of disjoint sets in $\mathcal{F}$ and let ω_n, denote the simple events in F_n. Since we have an infinite series of nonnegative numbers,

$$P(\bigcup_n F_n) = \sum_{n,m} P(\omega_{n,m}) = \sum_n \sum_m P(\omega_{n,m}) = \sum_n P(F_n).$$

$\square$

1.2 Conditional probabilities and independence

Given a probability space $(\Omega, \mathcal{F}, P)$ and some event B with $P(B) \neq 0$, we define a new *posterior* probability measure as follows. If A is any event we define the probability of A given B as

$$P(A \mid B) = \frac{P(A \text{ and } B)}{P(B)} = \frac{P(A \cap B)}{P(B)},$$

provided $P(B) > 0$. Otherwise $P(A \mid B)$ is left undefined.

What we mean by "given event B" is that *we know* that event B has occurred, that is we know that $\omega \in B$, so that we no longer assign the same probabilities given by P to events but assign new, or updated, probabilities given by the probability measure $P(. \mid B)$. Any event which is mutually exclusive with B has probability zero under $P(. \mid B)$ and the new probability space is now $(B, \mathcal{F} \cap B, P(. \mid B))$.

If our observation is limited to knowing whether event B has occurred or not we may as well define $P(. \mid \overline{B})$, where $\overline{B}$ is the complement of B within Ω. Prior to knowing where the outcome ω is we define the, now random, quantity:

$$P(. \mid B \text{ or } \overline{B})(\omega) = P(. \mid \sigma\{B\})(\omega) \overset{\triangle}{=} P(. \mid B)I_B(\omega) + P(. \mid \overline{B})I_{\overline{B}}(\omega).$$

This definition extends in an obvious way to a σ-field $\mathcal{G}$ generated by a finite or countable partition $\{B_1, B_2, \dots\}$ of Ω and the random variable $P(. \mid \mathcal{G})(\omega)$ is called the *conditional probability given* $\mathcal{G}$. The random function $P(. \mid \mathcal{G})(\omega)$ whose values on the atoms B_i are ordinary conditional probabilities $P(. \mid B_i) = \dfrac{P(. \cap B_i)}{P(B_i)}$ is not defined if $P(B_i) = 0$. In this case we have a family of functions $P(. \mid \mathcal{G})(\omega)$, one for each possible arbitrary value assigned to the undefined $P(. \mid B_i)$. Usually, one *version* is chosen and different versions differ only on a set of probability 0.

Example 1.2.1   Phone calls arrive at a switchboard between 8:00 a.m. and 12:00 p.m. according to the following probability distribution:

1. $P(k \text{ calls within an interval of length } l) = e^{-l} \dfrac{l^k}{k!}$;
2. If I_1 and I_2 are disjoint intervals,

$$P((k_1 \text{ calls within } I_1) \cap (k_2 \text{ calls within } I_2))$$
$$= P(k_1 \text{ calls within } I_1) P(k_2 \text{ calls within } I_2),$$

that is, events occurring within disjoint time intervals are independent.

Suppose that the operator wants to know the probability that 0 calls arrive between 8:00 and 9:00 given that the total number of calls from 8:00 a.m. to 12:00 p.m., N_{8-12}, is known. From past experience, the operator assumes that this number is near 30 calls, say. Hence

$$P(0 \text{ calls within } [8, 9) \mid 30 \text{ calls within } [8, 12])$$
$$= \frac{P((0 \text{ calls within } [8, 9)) \cap (30 \text{ calls within } [9, 12]))}{P(30 \text{ calls within } [8, 12])}$$
$$= \frac{P(0 \text{ calls within } [8, 9)) P(30 \text{ calls within } [9, 12])}{P(30 \text{ calls within } [8, 12])} = \left(\frac{3}{4}\right)^{30},$$

which can be written as

$$P(0 \text{ calls within } [8, 9) \mid N_{8-12} = N) = \left(\frac{3}{4}\right)^N. \tag{1.2.1}$$

$\square$

Remarks 1.2.2   Consider again Example 1.2.1.

1. The events $F_i = \{\omega : N_{8-12}(\omega) = i\}$, $i = 0, 1, \dots$ form a partition of Ω and are atoms of the σ-field generated by observing only N_{8-12}, so we may write:

$$P(0 \text{ calls within } [8, 9) \mid F_i, \ i \in \mathbb{N})(\omega)$$
$$= P(0 \text{ calls within } [8, 9) \mid \sigma\{F_i, \ i \in \mathbb{N}\})(\omega)$$
$$= \sum_i^\infty \left(\frac{3}{4}\right)^i I_{F_i}(\omega).$$

2. Observe that since each event $F \in \sigma\{F_i, \ i \in \mathbb{N}\}$ is a union of some $F_{i_1}, F_{i_2}, \dots$, and since we know, at the end of the experiment, which F_j contains ω, then we know

whether or not ω lies in F, that is whether F or the complement of F has occurred. In this sense, $\sigma\{F_i,\ i \in \mathbb{N}\}$ is indeed *all* we can answer about the experiment from what we know. □

The likelihood of occurrence of any event A could be affected by the realization of B. Roughly speaking if the "proportion" of A within B is the same as the "proportion" of A within Ω then it is intuitively clear that $P(A \mid B) = P(A \mid \Omega) = P(A)$. Knowing that B has occurred does not change the prior probability $P(A)$. In that case we say that events A and B are *independent*. Therefore two events A and B are *independent* if and only if $P(A \cap B) = P(A)P(B)$.

Two σ-fields $\mathcal{F}_1$ and $\mathcal{F}_2$ are independent if and only if $P(A_1 \cap A_2) = P(A_1)P(A_2)$ for all $A_1 \in \mathcal{F}_1$, $A_2 \in \mathcal{F}_2$.

If events A and B are independent so are $\sigma\{A\}$ and $\sigma\{B\}$ because the impossible event $\emptyset$ is independent of everything else including itself, and so is Ω. Also A and B^c, A^c and B, A^c and B^c are independent. We can say a bit more, if $P(E) = 0$ or $P(E) = 1$ then the event E is independent of any other event including E itself, which seems intuitively clear.

Mutually exclusive events with positive probabilities provide a good example of dependent events.

Example 1.2.3 In the die throwing experiment the σ-fields

$$\mathcal{F}_1 = \sigma\{\{1, 2\}, \{3, 4, 5, 6\}\},$$

and

$$\mathcal{F}_2 = \sigma\{\{1, 2\}, \{3, 4\}, \{5, 6\}\},$$

are not independent since if we know, for instance, that ω has landed in $\{5, 6\}$ (or equivalently $\{5, 6\}$ has occurred) in $\mathcal{F}_2$ then we also know that the event $\{3, 4, 5, 6\}$ in $\mathcal{F}_1$ has occurred. This fact can be checked by direct calculation using the definition. However, the σ-fields

$$\mathcal{F}_3 = \sigma\{\{1, 2, 3\}, \{4, 5, 6\}\},$$

and

$$\mathcal{F}_4 = \sigma\{\{1, 4\}, \{2, 5\}, \{3, 6\}\},$$

are independent. The occurrence of any event in any of $\mathcal{F}_3$ or $\mathcal{F}_4$ does not provide any nontrivial information about the occurrence of any (nontrivial) event in the other field. □

Another fundamental concept of probability theory is *conditional independence*. Events A and C are said to be conditionally independent given event B if $P(A \cap C \mid B) = P(A \mid B)P(C \mid B)$, $P(B) > 0$.

The following example shows that it is not always easy to decide, under a probability measure, if conditional independence holds or not between events.

Example 1.2.4 Consider the following two events:

A_1="person 1 is going to watch a football game next weekend,"
A_2="person 2, with no relation at all with person 1, is going to watch a football game next weekend."

There is no reason to doubt the independence of A_1 and A_2 in our model. However consider now the event $B =$ "next weekend weather is good." Suppose that

$$P(A_1 \mid B) = .90, \quad P(A_2 \mid B) = .95, \quad P(A_1 \mid \overline{B}) = .40,$$
$$P(A_2 \mid \overline{B}) = .30, \quad P(B) = .75 \quad \text{and} \quad P(\overline{B}) = .25.$$

Using this information it can be checked that $P(A_1 \cap A_2) \neq P(A_1)P(A_2)$. The reason is that event B has "linked" events A_1 and A_2 in the sense that if we knew that A_1 has occurred the probability of B should be high, resulting in the probability of A_2 increasing.   □

The independence concept extends to arbitrary families of events. A family of events $\{A_\alpha, \alpha \in I\}$ is said to be a family of independent events if and only if any finite subfamily is independent, i.e., for any finite subset of indices $\{i_1, i_2, \ldots, i_k\} \subset I$,

$$P(A_{i_1} \cap A_{i_2} \cap \cdots \cap A_{i_k}) = P(A_{i_1})P(A_{i_2}) \ldots P(A_{i_k}).$$

A family of σ-fields $\{\mathcal{F}_\alpha, \alpha \in I\}$ is said to be a family of independent σ-fields if and only if any finite subfamily $\{\mathcal{F}_{i_1}, \mathcal{F}_{i_2}, \ldots, \mathcal{F}_{i_k}\}$ is independent; that is, if and only if any collection of events of the form $\{A_{i_1} \in \mathcal{F}_{i_1}, A_{i_2} \in \mathcal{F}_{i_2}, \ldots, A_{i_k} \in \mathcal{F}_{i_k}\}$ is independent.

An extremely powerful and standard tool in proving properties which are true with probability one is the Borel–Cantelli Lemma. This lemma concerns sequences of events.

Let $\{A_n\}$ be a monotone *decreasing* sequence of events, i.e.

$$A_1 \supset A_2 \supset \cdots \supset A_n \supset A_{n+1} \supset \cdots,$$

then by definition

$$\lim_{n \to \infty} A_n = \bigcup_{n=1}^{\infty} A_n.$$

Let $\{B_n\}$ be a monotone *increasing* sequence of events, i.e.

$$B_1 \subset B_2 \subset \cdots \subset B_n \subset B_{n+1} \subset \cdots,$$

then by definition

$$\lim_{n \to \infty} B_n = \bigcap_{n=1}^{\infty} B_n.$$

Let $\{C_n\}$ be an arbitrary sequence of events. Define

$$A_n = \sup_{k \geq n} C_k \overset{\Delta}{=} \bigcup_{k=n}^{\infty} C_k,$$

and

$$B_n = \inf_{k \geq n} C_k \overset{\Delta}{=} \bigcap_{k=n}^{\infty} C_k.$$

Event A_n occurs if and only if at least one of the events $C_n, C_{n+1}, \ldots$ occurs and event B_n occurs if and only if all the C_n occur simultaneously except for a finite number.

By construction, A_n and B_n are monotone. A_n is decreasing and B_n is increasing so that:

$$A = \lim_{n \to \infty} A_n = \bigcap_{n=1}^{\infty} A_n = \bigcap_{n=1}^{\infty} \bigcup_{k=n}^{\infty} C_k,$$

and

$$B = \lim_{n \to \infty} B_n = \bigcup_{n=1}^{\infty} B_n = \bigcup_{n=1}^{\infty} \bigcap_{k=n}^{\infty} C_k.$$

Event $A = \bigcap_{n=1}^{\infty} \bigcup_{k=n}^{\infty} C_k \overset{\triangle}{=} \limsup C_n$ occurs if and only if infinitely many C_n occur, or C_n occurs infinitely often (C_n i.o.). To see this suppose that ω belongs to an infinite number of C_ns; then for every n, $\omega \in \bigcup_{k=n}^{\infty} C_k$. Therefore, $\omega \in \bigcap_{n=1}^{\infty} \bigcup_{k=n}^{\infty} C_k$. Conversely, if ω belongs to only a finite number of C_ns, then there is some n_0 such that $\omega \notin \bigcup_{k=n_0}^{\infty} C_k$. Since $\bigcap_{n=1}^{\infty} \bigcup_{k=n}^{\infty} C_k \subset \bigcup_{k=n_0}^{\infty} C_k$, this shows that $\omega \notin \bigcap_{n=1}^{\infty} \bigcup_{k=n}^{\infty} C_k$ if ω belongs to only a finite number of C_ns.

Event $B = \bigcup_{n=1}^{\infty} \bigcap_{k=n}^{\infty} C_k \overset{\triangle}{=} \liminf C_n$ occurs if and only if all but a finite number of C_n occur.

Clearly $\liminf C_n \subset \limsup C_n$.

Consider the following simple example of sequences of intervals in $\mathbb{R}$.

Example 1.2.5 Let A and B be any subsets of Ω and define the sequences $C_{2n} = A$ and $C_{2n+1} = B$. Then:

$$\limsup C_n = A \cup B, \quad \liminf C_n = A \cap B.$$

$\square$

Example 1.2.6 Let

$$C_k = \{(x, y) \in \mathbb{R}^2 : \quad 0 \leq x < k, \, 0 \leq y < \frac{1}{k}\},$$

then

$$A_n = \sup_{k \geq n} C_k = \bigcup_{k=n}^{\infty} C_k = \{x, y \in \mathbb{R}^2 : \quad 0 \leq x < \infty, \, 0 \leq y < \frac{1}{n}\},$$

and

$$B_n = \inf_{k \geq n} C_k = \bigcap_{k=n}^{\infty} C_k = \{x, y \in \mathbb{R}^2 : \quad 0 \leq x < n, \, y = 0\}.$$

A_n and B_n are monotone and decreasing and increasing respectively so that:

$$A = \lim_{n \to \infty} A_n = \bigcap_{n=1}^{\infty} A_n = \limsup C_n = \{x, y \in \mathbb{R}^2 : \quad 0 \leq x < \infty, \, y = 0\},$$

and

$$B = \lim_{n \to \infty} B_n = \bigcup_{n=1}^{\infty} B_n = \liminf C_n = \{x, y \in \mathbb{R}^2 : \quad 0 \leq x < \infty, \, y = 0\}.$$

$\square$

Lemma 1.2.7 *(Borel–Cantelli). Let $(\Omega, \mathcal{F}, P)$ be a probability space.*

1. *For an arbitrary sequence of events $\{C_n\}$, $\sum_{n=1}^{\infty} P(C_n) < \infty$ implies $P(\limsup C_n) = 0$.*
2. *If $\{C_n\}$ is a sequence of independent events, $\sum_{n=1}^{\infty} P(C_n) = \infty$ implies $P(\limsup C_n) = 1$.*

Proof

1.
$$\limsup C_n = \bigcap_{n=1}^{\infty} \bigcup_{k=n}^{\infty} C_k \subset \bigcup_{k=n}^{\infty} C_k,$$

which implies $P(\limsup C_n) \leq \sum_{k=n}^{\infty} P(C_k) \to 0$ as $n \to \infty$.

2. Consider the complementary event of $\limsup C_n$ which is $\bigcup_{n=1}^{\infty} \bigcap_{k=n}^{\infty} \bar{A}_k$. Now

$$P(\bigcap_{k=n}^{n+m} \bar{A}_k) = \prod_{k=n}^{n+m} (1 - P(C_k)) \leq \exp\left\{ -\sum_{k=n}^{n+m} P(C_k) \right\} \to 0,$$

for all n as $m \to \infty$ because of the divergence of the series $\sum_{n=1}^{\infty} P(C_n)$. ■

1.3 Random variables

Definition 1.3.1 *If $(\Omega, \mathcal{F})$ and $(E, \mathcal{E})$ are measurable spaces a map $X : \Omega \to E$ is measurable if $X^{-1}(B) \in \mathcal{F}$ for all $B \in \mathcal{E}$.*

Definition 1.3.2 *A measurable real valued function $X : (\Omega, \mathcal{F}, P) \to (\mathbb{R}, \mathcal{B}(\mathbb{R}))$ is called a random variable.*

It is left as an exercise to show that if $\{\omega : X(\omega) \leq x\} = \{\omega : X(\omega) \in (-\infty, x]\} \in \mathcal{F}$ for all real $x \in \mathbb{R}$ then X is a random variable.

For $C \in \Omega$ define $I_C(\omega)$, (also denoted $\chi_C(\omega)$ or simply $I(C)$), the *indicator function* of the set C, as follows:

$$I_C(\omega) = \begin{cases} 1 \text{ if } \omega \in C, \\ 0 \text{ otherwise.} \end{cases}$$

Example 1.3.3 Let Ω be the unit interval $(0, 1]$ and on it are given the following σ-fields:

$$\mathcal{F}_1 = \sigma\{(0, \tfrac{1}{2}], (\tfrac{1}{2}, \tfrac{3}{4}], (\tfrac{3}{4}, 1]\},$$
$$\mathcal{F}_2 = \sigma\{(0, \tfrac{1}{4}], (\tfrac{1}{4}, \tfrac{1}{2}], (\tfrac{1}{2}, \tfrac{3}{4}], (\tfrac{3}{4}, 1]\},$$
$$\mathcal{F}_3 = \sigma\{(0, \tfrac{1}{8}], (\tfrac{1}{8}, \tfrac{2}{8}], \ldots, (\tfrac{7}{8}, 1]\}.$$

Consider the mapping

$$X(\omega) = x_1 I_{(0, \frac{1}{4}]}(\omega) + x_2 I_{(\frac{1}{4}, \frac{1}{2}]}(\omega) + x_3 I_{(\frac{1}{2}, \frac{3}{4}]}(\omega) + x_4 I_{(\frac{3}{4}, 1]}(\omega).$$

It is an easy exercise to check that the inverse image X^{-1} of any interval in $\mathbb{R}$ is in $\mathcal{F}_2$ and in $\mathcal{F}_3$ ($\mathcal{F}_2 \subset \mathcal{F}_3$). $\mathcal{F}_2$ is *coarser* than $\mathcal{F}_3$ because the atoms (smallest sets) of $\mathcal{F}_2$ are unions of the smaller atoms of $\mathcal{F}_3$. So if we know in which atom of $\mathcal{F}_3$ the outcome ω is we can determine in which atom it is of $\mathcal{F}_2$. However, if we know that ω is in the $\mathcal{F}_2$ atom $(0, \tfrac{1}{4}]$, say, then it could be in either of $\mathcal{F}_3$ atoms $(0, \frac{1}{8}]$ or $(\frac{1}{8}, \frac{2}{8}]$. X is not a random variable with respect to $\mathcal{F}_1$ since $X^{-1}(\{x_1\}) = (0, \frac{1}{4}]$ is not an atom (and, a fortiori, it is not an event) of $\mathcal{F}_1$. To put it another way, knowing for instance that $\omega \in (0, \frac{1}{2}]$ leaves us undecided about which value X has taken; that is to say X is not $\mathcal{F}_1$-measurable. □

Note that in the above example $\mathcal{F}_2$ is the smallest σ-field with respect to which X is measurable and it coincides with the class of *all* inverse images of X. For this reason it is called the σ-field generated by X and is denoted $\sigma(X)$. We have $\mathcal{F}_1 \subset \sigma(X) \subset \mathcal{F}_3$.

For more general cases where X takes its values in some topological space E, the Borel σ-field $\mathcal{B}$ on E is the smallest σ-field generated by the open sets of E.

In general it is not possible to assign probabilities to all subsets of Ω and, therefore, we cannot treat any function X as a random variable since X^{-1} might not be an event and so its probability is not defined. However, it is not an easy task to come up with an example of a function which is not a random variable!

In the finite state space set up we have:

X is $\mathcal{F}$-measurable if and only if X is a constant function when restricted to any of the atoms of $\mathcal{F}$. To see this suppose first that $F \in \mathcal{F}$ is an atom and that X takes values a and b on F with $a < b$. Let $\alpha = (a+b)/2$, then $\{\omega \in \Omega : X(\omega) \leq \alpha\} \in \mathcal{F}$ is a nonempty proper subset of F, a contradiction. For the converse, let $\{F_1, \ldots, F_p\}$ be the collection of atoms of $\mathcal{F}$ and suppose that $X(\omega) = \alpha_i$ for $\omega \in F_i, i = 1, 2, \ldots, p$. Then

$$\{\omega \in \Omega : X(\omega) \leq \alpha\} = \bigcup_{\alpha_i \leq \alpha} F_i \in \mathcal{F},$$

that is, X is $\mathcal{F}$-measurable.

It is interesting to note here that we can express a random variable in two ways:

$$X(\omega) = \sum_{F_i} \alpha_i I_{F_i}(\omega) = \sum_{\alpha \in \mathbb{R}} \alpha I_{X^{-1}(\alpha)}(\omega).$$

Definition 1.3.4 *X is a simple function on $(\Omega, \mathcal{F})$ if there exists a partition of $\Omega = \bigcup A_i$, $A_i \cap A_j = \emptyset$ for $i \neq j$ and*

$$X(\omega) = \sum_{i=1}^{k} x_i I_{A_i}(\omega). \tag{1.3.1}$$

Given a finite σ-field $\mathcal{A} = \sigma\{A_1, \ldots, A_N\}$, where we assume without loss of generality that $A_1, \ldots, A_N$ form a partition and a random variable X assuming finitely many (distinct) values, $x_1, \ldots, x_k$, and with inverse images $X^{-1}(\{x_i\})$ contained in $\mathcal{A}$ (so $k \leq N$) we can write uniquely $X = \sum_{l=1}^{N} y_l I_{A_l}$. To see this note that the inverse image of any point x_i in the range of X is a union of atoms of $\mathcal{A}$, $A_{i_1}, \ldots, A_{i_{s(x_i)}}$, say, that is

$$X = \sum_{i=1}^{k} x_i I_{[\cup_s A_{i_s}]} = \sum_{i=1}^{k} x_i \left(\sum_s I_{A_{i_s}} \right) = \sum_{l=1}^{N} y_l I_{A_l}.$$

The mapping in Example 1.3.3 is a simple random variable.

The following result is fundamental in the theory of integration.

Theorem 1.3.5 *If X is a positive random variable (possibly with infinite values), there exists an increasing sequence of simple random variables $X_1, X_2, \ldots$ converging monotonically to X for each ω such that $X_n(\omega) \leq X(\omega)$ for all n and all ω.*

Proof   If X is nonnegative, define

$$X_n(\omega) = \sum_{k=1}^{n2^n} \frac{k-1}{2^n} I_{[\frac{k-1}{2^n}, \frac{k}{2^n})}(\omega).$$

Clearly, $X_n(\omega)$ converges to $X(\omega)$ for all ω.   ∎

As a corollary, notice that a general random variable X can be represented as the difference of two nonnegative random variables: $X = XI_{X \geq 0} - (-X)I_{X < 0} \stackrel{\triangle}{=} X^+ - X^-$.

Let us notice that, unlike spaces of continuous functions, the spaces of measurable functions are closed under pointwise limits, i.e. the spaces of measurable functions are more stable.

In applications we often work with functions of random variables. Suppose $f : \mathbb{R} \to \mathbb{R}$ is a function, then $Y(.) = f(X(.))$ is also a function from Ω to $\mathbb{R}$. Y will be a random variable if

$$Y^{-1}(B) = [f(X)]^{-1}(B) = X^{-1}(f^{-1}(B)) \in \mathcal{F},$$

for each Borel set B; which holds whenever $f^{-1}(B)$ is a Borel set. By definition the random variable Y generates events which are in $\sigma(X)$; in other words $\sigma(Y) \subset \sigma(X)$. In this case we say that Y is $\sigma(X)$-measurable. Conversely, we have the important result:

Theorem 1.3.6 *If Y is $\sigma(X)$-measurable then there exists a measurable function f such that $Y(.) = f(X(.))$.*

Proof   Start with simple functions. Let $A \in \sigma(X)$ and consider the random variable

$$Y(\omega) = I_A(\omega) = \begin{cases} 1 \text{ if } \omega \in A, \\ 0 \text{ otherwise.} \end{cases}$$

Since $A \in \sigma(X)$ there exists a Borel set B such that $X^{-1}(B) = A$ and clearly $Y(\omega) = I_A(\omega) = I_B(X(\omega)) \stackrel{\triangle}{=} f(X(\omega))$. Hence the result is true for simple functions. Using Theorem 1.3.5 and limiting arguments we can establish the result for general random variables which are $\sigma(X)$-measurable.

A more precise argument is (see [36]): let

$$\Phi = \{Y : Y \text{ is } \sigma(X)\text{-measurable}\},$$

and

$$\Phi_X = \{Z : Z \text{ is } \sigma(X)\text{-measurable and } Z = f(X), \ f \text{ is a Borel function}\}.$$

We first note that $\Phi \subseteq \Phi_X$. Now for any $F \in \sigma(X)$, there is a Borel set $B \in \mathcal{B}(\mathbb{R})$ such that $F = \{\omega : X(\omega) \in B\}$ and we see that $I_F(\omega) = I_B(X(\omega)) \in \Phi_X$. Hence every simple $\sigma(X)$-measurable random variable is in Φ_X. However, if Y is an arbitrary, $\sigma(X)$-measurable random variable, Theorem 1.3.5 guarantees the existence of a sequence of simple random variables $Y_1, Y_2, \ldots$ which converges pointwise to Y. By the previous step, for each n, there is a Borel function f_n such that $Y_n(\omega) = f_n(X(\omega)) \in \Phi_X$. Then $f_n(X(\omega))$ converges to $Y(\omega)$ pointwise. Define the Borel function:

$$f(x) = \begin{cases} \lim_n f_n(x) \text{ if the limit exists,} \\ 0 \text{ otherwise.} \end{cases}$$

Clearly $Y(\omega) = \lim_n f_n(X(\omega)) = f(X(\omega))$ for all $\omega \in \Omega$, which gives the result. ∎

Example 1.3.7 Let X be a real valued function on Ω with finite range. Suppose that the range of X is $\{x_1, \ldots, x_r\}$. Then we know that $\sigma(X)$ is the σ-algebra generated by the atoms $\{X^{-1}(x_j); j = 1, \ldots, r\}$. Suppose that Y is $\sigma(X)$-measurable. Then Y must be constant on the atoms of $\sigma(X)$. That is, there exist constants $y_1, \ldots, y_r$ such that

$$Y(\omega) = \sum_j y_j I_{X^{-1}(x_j)}(\omega) = \sum_j g(x_j) I_{X^{-1}(x_j)}(\omega) = g(X(\omega)).$$

where, by definition, $g(x_j) = y_j$. □

Remark 1.3.8 The result of Example 1.3.7 can be extended as follows. Let $X_1, \ldots, X_n$ be n random variables on $(\Omega, \mathcal{F}, P)$. Write $\sigma(X_1, \ldots, X_n)$ for the σ-algebra generated by these random variables. The atoms of this σ-algebra are all sets of the form

$$F_1 \cap F_2 \cap \cdots \cap F_n,$$

where F_i is an atom of $\sigma(X_i)$ for $i = 1, 2, \ldots, n$. Then a function Y on Ω is $\sigma(X_1, \ldots, X_n)$-measurable if and only if $Y = g(X_1, \ldots, X_n)$ for some function $g : \mathbb{R}^n \to \mathbb{R}$. Thus the use of σ-algebra is a powerful way of expressing dependence of one random variable on a (possibly infinite) family of others. □

It should be pointed out that the class of Borel functions is rich enough for most practical purposes. It contains continuous functions, piecewise continuous functions, etc. It is closed under pointwise limit operations, linear combinations, products, compositions, etc.

A real random variable X induces a probability measure on the Borel sets of $\mathcal{B}(\mathbb{R})$, called the *probability distribution* of X, defined by

$$P_X(B) = P(X \in B),$$

for any Borel set B.

If there exists a nonnegative function f such that for any Borel set B

$$P_X(B) = \int_B f(x)\mathrm{d}x, \quad (\Rightarrow \int_{\mathbb{R}} f(x)\mathrm{d}x = 1),$$

we say that P_X is *absolutely continuous* and f is the *probability density function* of X.

The *probability distribution function* of a real valued random variable is defined on $\mathbb{R}$ by the formula:

$$F(x) = P(X \leq x).$$

The probability distribution function has the following properties:

1. $0 \leq F(x) \leq 1$ for all $x \in \mathbb{R}$.
2. F is nondecreasing and right-continuous with left limits at each $x \in \mathbb{R}$.

Conversely, given a distribution function $F(x)$ on the real line, there exists a probability space $(\Omega = \mathbb{R}, \mathcal{F} = \mathcal{B}(\mathbb{R}), P)$ and a random variable $X(\omega) = \omega$ on it such that

$$P(X \leq x) = F(x).$$

P is defined on intervals as $P((a, b]) = F(b) - F(a)$ and extended uniquely to a probability measure on $\mathcal{B}(\mathbb{R})$. (See Theorem 1, page 152 in [36]). The measure P constructed from the function F is usually called the Lebesgue–Stieltjes probability measure corresponding to the distribution F.

A similar situation exists for random vectors $(X_1, \ldots, X_n)$ and probability distributions on $(\mathbb{R}^n, \mathcal{B}(\mathbb{R}^n))$. Given a probability measure P on $(\mathbb{R}^n, \mathcal{B}(\mathbb{R}^n))$, write:

$$F_n(x_1, \ldots, x_n) = P((-\infty, x_1] \times \cdots \times (-\infty, x_n]).$$

An absolutely continuous random vector is one for which there exists a nonnegative *probability density function* $f_n(.) : \mathbb{R}^n \to \mathbb{R}$ such that:

$$\int_{-\infty}^{+\infty} \cdots \int_{-\infty}^{+\infty} f_n(u_1, \ldots, u_n) du_1 \ldots du_n = 1,$$

and

$$F_n(x_1, \ldots, x_n) = \int_{-\infty}^{x_1} \cdots \int_{-\infty}^{x_n} f_n(u_1, \ldots, u_n) du_1 \ldots du_n.$$

Conversely we have

Theorem 1.3.9 *(Theorem 2, page 160 in Shiryayev [36]) Let F_n be a distribution function on $\mathbb{R}^n$. Then there is a unique probability measure P on $(\mathbb{R}^n, \mathcal{B}(\mathbb{R}^n))$ such that*

$$P((a_1, b_1] \times \cdots \times (a_n, b_n]) = \Delta_{a_1 b_1} \ldots \Delta_{a_n b_n} F_n(x_1, \ldots, x_n),$$

where

$$\Delta_{a_i b_i} F_n(x_1, \ldots, x_n) = F_n(x_1, \ldots, x_{i-1}, b_i, x_{i+1}, \ldots, x_n)$$
$$- F_n(x_1, \ldots, x_{i-1}, a_i, x_{i+1}, \ldots, x_n).$$

The remarkable fact is that a similar construction for probability measures also works for the space $(\mathbb{R}^\infty, \mathcal{B}(\mathbb{R}^\infty))$, where $\mathbb{R}^\infty = \{(x_1, x_2, \ldots) \in \mathbb{R} \times \mathbb{R} \times \ldots\}$ and the Borel σ-field of $\mathbb{R}^\infty$ $\mathcal{B}(\mathbb{R}^\infty)$ is the smallest σ-field containing all the *cylinder* sets

$$\{x = (x_1, x_2, \ldots) \in \mathbb{R}^\infty : x_1 \in I_1, x_2 \in I_2, \ldots, x_n \in I_n\}, \quad n \geq 1,$$

where each I_i is an interval of the form $(a_i, b_i]$. In other words, a cylinder set is a set of infinite sequences with restrictions placed on a finite number of coordinates.

The next theorem is Kolmogorov's Theorem on the Extension of Measures in $(\mathbb{R}^\infty, \mathcal{B}(\mathbb{R}^\infty))$. See [36] page 163.

Theorem 1.3.10 *Suppose that P_1, P_2, ... are probability measures on the spaces $(\mathbb{R}, \mathcal{B}(\mathbb{R}))$, $(\mathbb{R}^2, \mathcal{B}(\mathbb{R}^2))$, ... such that for $n \geq 1$ and $B \in \mathcal{B}(\mathbb{R}^n))$*

$$P_{n+1}(B \times \mathbb{R}) = P_n(B).$$

then there is a unique probability measure P on $(\mathbb{R}^\infty, \mathcal{B}(\mathbb{R}^\infty))$ such that the restriction of P to $\mathcal{B}(\mathbb{R}^n))$ is P_n for all $n \geq 1$.

Properties of random variables are usually not required to hold for every ω. They may fail to hold on sets of measure zero, that is negligible subsets of Ω. In this case they are said to hold *almost surely* (*a.s.*) or *almost everywhere* (*a.e.*) with respect to some probability measure P. σ-fields are P-completed by including in them all subsets of the P-negligible subsets of Ω.

Example 1.3.11 A sequence of random variables $\{X_k\}$ is said to converge almost surely to a random variable X ($X_k \overset{a.s.}{\to} X$) if

$$P[\omega : \lim_{k \to \infty} X_k(\omega) \text{ exists and is finite}] = 1.$$

That is, $\{X_k(\omega)\}$ converges pointwise to $X(\omega)$, except possibly on a negligible subset of Ω. □

Independence of random variables is expressed in terms of the σ-fields they generate.

Definition 1.3.12 *Suppose $X_1, \ldots, X_n$ are random variables on a probability space $(\Omega, \mathcal{F}, P)$. Then $X_1, \ldots, X_n$ are independent if for any $k \leq n$,*

$$P(X_1 \in B_1, \ldots, X_k \in B_k) = \prod_{i=1}^{k} P(X_i \in B_i),$$

for all Borel sets $(B_1, \ldots, B_k)$.

An arbitrary class of random variables is independent if and only if any finite subfamily is independent.

Theorem 1.3.13 *Suppose $X_1, \ldots, X_n$ are random variables on a probability space $(\Omega, \mathcal{F}, P)$. Then $X_1, \ldots, X_n$ are independent if and only if*

$$P(X_1 \leq x_1, \ldots, X_n \leq x_n) = \prod_{i=1}^{n} P(X_i \leq x_i), \quad (x_1, \ldots, x_n) \in \mathbb{R}^n.$$

Proof   See [36] page 179.   ■

Definition 1.3.14 *Let $(\Omega, \mathcal{F})$ be a measurable space. A map $\mu : \mathcal{F} \to [0, \infty]$ is called a measure on $(\Omega, \mathcal{F})$ if μ is countably additive, that is if $\{B_k\}$ is a countable sequence of pairwise disjoint elements of $\mathcal{F}$, then $\mu(\bigcup B_k) = \sum \mu(B_k)$. The triple $(\Omega, \mathcal{F}, \mu)$ is then called a measure space.*

We say that μ is finite if $\mu(\Omega) < \infty$ and it is σ-finite if there exists a partition of Ω into disjoint subsets $\{S_k\}$ such that $\bigcup_{n=1}^{\infty} S_k = \Omega$ and $\mu(S_k) < \infty$, for all $k \geq 1$.

When we move from intervals, which are essential in the definition of Riemann integral, to more complicated subsets of $\mathbb{R}$ whose "lengths" or measures are not obvious we use Lebesgue measure and integration theory. Roughly speaking, the Lebesgue measure of a set F (denoted $\lambda(F)$) is the infimum of the sums of lengths of intervals which cover the set F. If this infimum exists we say that F is measurable. More formally, consider the field $\mathcal{A}$ generated by finite unions of intervals of the form $(a, b]$ and define the set function $\lambda_0(\cup(a_i, b_i]) = \sum_{i=1}^{n}(b_i - a_i)$, $a_1 \leq b_1 \ldots a_n \leq b_n$. Then λ_0 is well defined, σ-finite and countably additive. Hence there exists a unique measure λ on the Borel σ-field $\mathcal{B}(\mathbb{R})$ [34]. This measure is called the *Lebesgue measure* on $(\mathbb{R}, \mathcal{B}(\mathbb{R}))$.

Given a finite probability space $(\Omega, \mathcal{F}, P\}$ and a simple random variable

$$X(\omega) = \sum_{\ell=1}^{n} x_\ell I_{A_\ell}(\omega),$$

the *expected value* or *mean* or *integral* with respect to the measure P is defined by the formula:

$$E[X] = \sum_{\ell=1}^{n} x_\ell P(A_\ell).$$

For a nonnegative random variables X there exists, by Theorem 1.3.5, a sequence of simple random variables $X_1, X_2, \ldots$ increasing to X pointwise, and we define its Lebesgue integral or expectation by:

$$E[X] = \lim_n E[X_n] \overset{\triangle}{=} \int X(\omega)dP(\omega).$$

Furthermore, it can be shown that the limit is independent of the increasing sequence of functions.

If the above limit is finite, X is said to be integrable. If both its positive and negative parts, $(X^+ = X I_{\{X \geq 0\}}, X^- = -X I_{\{X < 0\}})$ are integrable, X is said to be *integrable* and we define

$$\int X(\omega)dP(\omega) = \int X^+(\omega)dP(\omega) - \int X^-(\omega)dP(\omega).$$

Note that

$$E[X] = \int_{\Omega} X(\omega)dP(\omega) = \int_{\mathbb{R}} x\,dF_X(x).$$

Here $F_X(x) = P(X \leq x)$ is the usual probability distribution function of X and dF_X is the Lebesgue–Stieltjes probability measure corresponding to the probability distribution function F.

Theorem 1.3.15 (*Monotone Convergence Theorem.*) *If X_n is a sequence of nonnegative random variables increasing almost surely to X, then $\int X_n dP$ increases to $\int X dP$.*

Proof Let $X_{n,k}$ be a sequence of simple functions increasing pointwise to X_k as $n \to \infty$. Put $Z_n = \max_{1 \le k \le n} X_{n,k}$. Then Z_n is increasing and $Z_n \le \max_{1 \le k \le n} X_k = X_n$. Clearly,

$$X_k = \lim_n X_{n,k} \le \lim_n Z_n = Z \le \lim_n X_n = X,$$

for all $k \ge 1$, that is to say $X = Z$. However, the random variables Z_n are simple and increasing to Z, so that

$$\lim_n \int X_n dP \le \int X dP = \int Z dP = \lim_n \int Z_n dP \le \lim_n \int X_n dP.$$

and hence $\lim_n \int X_n dP = \int X dP$. ∎

An important consequence of the Monotone Convergence Theorem is *Fatou's Lemma*.

Theorem 1.3.16 *(Fatou's Lemma.) If X_n is a sequence of nonnegative random variables then*

$$\int \liminf_n X_n dP \le \liminf_n \int X_n dP,$$

where $\liminf_n X_n \overset{\triangle}{=} \lim_n \inf_{m \ge n} X_m$.

Proof Let $\liminf_n X_n = \lim_n \inf_{m \ge n} X_m = \lim_n Z_n$. Then Z_n increases to $\liminf_n X_n$ so by the Monotone Convergence Theorem (Theorem 1.3.15)

$$\int \liminf_n X_n dP = \int \lim_n Z_n dP = \lim_n \int Z_n dP = \liminf_n \int Z_n dP$$

$$\le \liminf_n \int X_n dP,$$

which establishes the result. ∎

Theorem 1.3.17 *(Lebesgue's Dominated Convergence Theorem.) Suppose $\{X_n\}$ is a sequence of random variables such that $|X_n| \le Y$ almost surely where Y is an integrable random variable. If X_n converges to X a.s., then X_n and X are integrable, $\int X_n dP$ converges to $\int X dP$, and $\int |X_n - X| dP \to 0$ as $n \to \infty$.*

Proof Using the hypothesis and Fatou's Lemma,

$$\int X dP = \int \liminf_n X_n dP \le \liminf_n \int X_n dP$$

$$\le \limsup_n \int X_n dP \le \int \limsup_n X_n dP = \int X dP.$$

Since Y is integrable and $|X| \le Y$, X is also integrable. Now, note that $|X_n - X| \le 2Y$ so the same argument can be used to prove the second part of the theorem. ∎

A special case of Lebesgue's Dominated Convergence Theorem is the *Bounded Convergence Theorem* where Y is replaced by some positive constant.

If X and Y are independent random variables with finite means, then their product is a random variable with finite mean and

$$E[XY] = E[X]E[Y]. \tag{1.3.2}$$

To see this consider first independent events A and B and their indicator functions: $X = I_A$, $Y = I_B$. Then

$$E[XY] = E[I_A I_B] = E[I_{A \cap B}] = P(A \cap B) = P(A)P(B).$$

If $X = \sum_{i=1}^{m} a_i I_{A_i}$, $Y = \sum_{j=1}^{n} b_j I_{B_j}$ are independent, we may assume that the A_i (resp. B_i) form a partition of Ω and $a_i \neq a_j$ if $i \neq j$ (resp. $b_i \neq b_j$ if $i \neq j$). Then

$$E[XY] = \int_\Omega \sum_{i=1}^{m} a_i I_{A_i}(\omega) \sum_{j=1}^{n} b_j I_{B_j}(\omega) dP(\omega)$$

$$= \sum_{i=1}^{m} \sum_{j=1}^{n} a_i b_j \int_\Omega I_{A_i}(\omega) dP(\omega) \int_\Omega I_{B_j}(\omega) dP(\omega)$$

$$= \int_\Omega X(\omega) dP(\omega) \int_\Omega Y(\omega) dP(\omega)$$

$$= E[X]E[Y].$$

When X and Y are integrable, they are limits of simple functions X_n, Y_n for which $E[X_n Y_n] = E[X_n]E[Y_n]$ is true by the above argument. Taking limits, as we are allowed to by Theorem 1.3.17, finishes the proof.

We also have the *change of variable formula in Lebesgue Integral* :

Theorem 1.3.18 *Let X be a real random variable and B a Borel set. Then:*

$$\int_B g(x) dF_X(x) = \int_{X^{-1}(B)} g(X(\omega)) dP(\omega).$$

Here g is a Borel function and when $B = \mathbb{R}$

$$\int_\Omega g(X(\omega)) dP(\omega) = \int_{\mathbb{R}} g(x) dF_X(x).$$

(If Ω is countable, the integral is replaced by a summation

$$E[g(X)] = \sum_{\omega_i} g(X(\omega_i)) P(\omega_i) = \sum_{x \in \mathbb{R}} g(x) P(X^{-1}(x)) = \sum_{x \in \mathbb{R}} g(x) \Delta F_X(x).$$

Here $\Delta F_X(x) = F_X(x) - F_X(x-))$.

Proof Let B and C be two Borel sets and $g(x) = I_C(x)$. Then

$$\int_B g(x) dF_X(x) = P(X^{-1}(B) \cap X^{-1}(C)) = \int_{X^{-1}(B)} g(X(\omega)) dP(\omega).$$

Hence, the result is true for nonnegative simple functions, and by the Monotone Convergence Theorem 1.3.15 it is true for all nonnegative random variables. In the general case we need only represent g as the difference of two nonnegative functions: $g = g^+ - g^-$. ∎

Definition 1.3.19 *Given two σ-fields $\mathcal{F}_1$ and $\mathcal{F}_2$ the* product σ-field *of $\mathcal{F}_1$ and $\mathcal{F}_2$, denoted $\mathcal{F}_1 \otimes \mathcal{F}_2$, is the smallest σ-field containing all "rectangles" $F_1 \times F_2$, $F_1 \in \mathcal{F}_1$, $F_2 \in \mathcal{F}_2$.*

Definition 1.3.20 *Let $(\Omega_1, \mathcal{F}_1, \mu_1)$, $(\Omega_2, \mathcal{F}_2, \mu_2)$ be two measure spaces. The* direct product *of $(\Omega_1, \mathcal{F}_1, \mu_1)$ and $(\Omega_2, \mathcal{F}_2, \mu_2)$ is defined as the measure space $(\Omega = \Omega_1 \times \Omega_2, \mathcal{F} = \mathcal{F}_1 \otimes \mathcal{F}_2, \mu = \mu_1 \times \mu_2)$ such that:*

$$\mu_1 \times \mu_2(F_1 \times F_2) = \mu_1(F_1)\mu_2(F_2), \quad F_1 \in \mathcal{F}_1, F_2 \in \mathcal{F}_2.$$

The following theorem is on the reduction of a (Lebesgue) double integral to an iterated integral.

Theorem 1.3.21 *(Fubini Theorem). Suppose $X(\omega_1, \omega_2)$ is an $\mathcal{F}_1 \otimes \mathcal{F}_2$-measurable function which is integrable with respect to $\mu_1 \times \mu_2$. Then*

1. $\displaystyle\int_{\Omega_1} X d\mu_1(\omega_1)$ *is defined, $\mathcal{F}_2$-measurable and finite μ_2-a.e.*

2. $\displaystyle\int_{\Omega_2} X d\mu_2(\omega_2)$ *is defined, $\mathcal{F}_1$-measurable and finite μ_1-a.e.*

3. $\displaystyle\int_{\Omega_1 \times \Omega_2} X d(\mu_1 \times \mu_2) = \int_{\Omega_1} \left[\int_{\Omega_2} X d\mu_2 \right] d\mu_1 = \int_{\Omega_2} \left[\int_{\Omega_1} X d\mu_1 \right] d\mu_2.$

Proof See [36] page 198. ∎

For $C \in \mathcal{F}$, we write

$$\int_C X dP = \int I_C X dP = E[I_C X].$$

If $X = I_C(.)$ we see that

$$E(X) = E[I_C] = \int_C dP = P(C),$$

and we have the following important special cases.

$$P(X \in B) = E[I_B(X(.))],$$

$$P(X \in B, Y \in C) = E[I_B(X(.))I_C(Y(.))].$$

Remark 1.3.22 When integrating random variables we are implicitly assuming measurability of the integrand with respect to the σ-field and measure at hand. However, consider the following example. □

Example 1.3.23 Let Ω be the unit interval $(0, 1]$ and on Ω consider the σ-field

$$\mathcal{F} = \sigma\{(0, \tfrac{1}{2}], (\tfrac{1}{2}, \tfrac{3}{4}], (\tfrac{3}{4}, 1]\}.$$

Define the measure

$$\lambda((0, \tfrac{1}{2}]) = \tfrac{1}{2}, \quad \lambda((\tfrac{1}{2}, \tfrac{3}{4}]) = \tfrac{1}{4}, \quad \lambda((\tfrac{3}{4}, 1]) = \tfrac{1}{4},$$

Consider the mapping

$$X(\omega) = x_1 I_{(0,\frac{1}{4}]}(\omega) + x_2 I_{(\frac{1}{4},\frac{1}{2}]}(\omega) + x_3 I_{(\frac{1}{2},\frac{3}{4}]}(\omega) + x_4 I_{(\frac{3}{4},1]}(\omega).$$

There is no natural definition of the integral of X with respect to the given measure λ on the field $\mathcal{F}$, because λ is not defined on the sets $(0, \frac{1}{4}]$ and $(\frac{1}{4}, \frac{1}{2}]$. The problem here is that as far as $\mathcal{F}$ is concerned, the event $(0, \frac{1}{2}]$ is an indivisible atom of the measure space. That is to say, X is not $\mathcal{F}$-measurable.

Thus, integration theory in this context is forced to consider only those integrands X with $x_1 = x_2$, and in this case the integral is as defined earlier for simple functions:

$$\int_\Omega X \mathrm{d}\lambda = x_1 \lambda((0, \tfrac{1}{2}]) + x_3 \lambda((\tfrac{1}{2}, \tfrac{3}{4}]) + x_4 \lambda((\tfrac{3}{4}, 1]).$$

$\square$

Remark 1.3.24 Given any nonnegative random variable X on a probability space $(\Omega, \mathcal{F}, P)$ with finite mean $E(X) = 1$, one can define another (probability) measure $\overline{P}$ on $\mathcal{F}$ by setting for $F \in \mathcal{F}$:

$$\overline{P}(F) = \int_F X \mathrm{d}P.$$

Clearly, if $P(F) = 0$, then $\overline{P}(F) = 0$ for any $F \in \mathcal{F}$ and we say that $\overline{P}$ is *absolutely continuous* with respect to P ($\overline{P} \ll P$). However, the remarkable fact that the converse is true is given by the following theorem. (See [36].)

$\square$

Theorem 1.3.25 *(Radon–Nikodym). Let $(\Omega, \mathcal{F})$ be a measurable space, μ a σ-finite measure, and $\overline{\mu}$ a signed measure (i.e. $\overline{\mu} = \overline{\mu}_1 - \overline{\mu}_2$, where at least one of the measures $\overline{\mu}_1$ and $\overline{\mu}_2$ is finite) such that for each $F \in \mathcal{F}$, $\mu(F) = 0$ implies $\overline{\mu}(F) = 0$. We write $\overline{\mu} \ll \mu$. Then there exists an $\mathcal{F}$-measurable function Λ with values in the extended real line $[-\infty, +\infty]$, such that*

$$\overline{\mu}(C) = \int_C \Lambda(\omega) \mathrm{d}\mu(\omega),$$

for all $C \in \mathcal{F}$. The function Λ is unique up to sets of μ-measure zero; if h is another $\mathcal{F}$-measurable function such that $\overline{\mu}(C) = \int_C h(\omega) \mathrm{d}\mu(\omega)$ for all $C \in \mathcal{F}$, then $\mu\{\omega : \Lambda(\omega) \neq h(\omega)\} = 0$. If $\overline{\mu}$ is a positive measure, then Λ has its values in $[0, +\infty]$. We write $\left.\dfrac{\mathrm{d}\overline{\mu}}{\mathrm{d}\mu}\right|_{\mathcal{F}} = \Lambda$.

Remark 1.3.26 In the case of probability measures the Radon–Nikodym Theorem reads as follows. If P and $\overline{P}$ are two probability measures on $(\Omega, \mathcal{F})$ such that for each $B \in \mathcal{F}$, $P(B) = 0$ implies $\overline{P}(B) = 0$ ($\overline{P} \ll P$), then there exists a nonnegative random variable Λ, such that $\overline{P}(C) = \int_C \Lambda \mathrm{d}P$ for all $C \in \mathcal{F}$. We write $\left.\dfrac{\mathrm{d}\overline{P}}{\mathrm{d}P}\right|_{\mathcal{F}} = \Lambda$.

Taking $C = \Omega$ we see in particular that

$$\overline{P}(\Omega) = 1 = \int_\Omega \Lambda \mathrm{d}P = E[\Lambda],$$

so that $\overline{P}$ is a probability measure if and only if Λ is nonnegative and $E[\Lambda] = 1$.

Λ is called the density of $\overline{P}$, with respect to P, or the Radon–Nikodym derivative of $\overline{P}$, with respect to P. □

Example 1.3.27 Suppose $(\Omega, \mathcal{F}, P)$ is a probability space, where Ω is a finite set containing N outcomes and $p_i = P(\{\omega_i\})$.

Suppose Λ is a positive, real valued function on Ω such that $\Lambda(\omega_i) = \dfrac{\overline{p}_i}{p_i}$, where $\overline{p}_i = \overline{P}(\omega_i)$ is some new probability we would like to assign to outcome ω_i.

Since

$$1 = \sum \overline{p}_i = \sum \Lambda(\omega_i) p_i,$$

the expected value of Λ under probability measure P is equal to 1 and the random variable Λ is the Radon–Nikodym derivative of $\overline{P}$ with respect to P.

If X is a random variable on Ω then:

$$\overline{E}[X] = \sum X(\omega_i)\overline{p}_i = \sum X(\omega_i)\Lambda(\omega_i)p_i = E[\Lambda X].$$

Hence $\overline{E}[X] = E[\Lambda X]$. Conversely, if $\overline{p}_i \neq 0$ for all i, $E[X] = \overline{E}[\Lambda^{-1}X]$. □

Example 1.3.28 Let f be a continuous function defined on $[0, \infty)$ and $F(x) = \int_0^x f(u)du = \int_0^x f(u)d\lambda(u)$, where λ is the Lebesgue measure. Consider the signed measure μ (i.e. $\mu = \mu_1 - \mu_2$, where at least one of the measures μ_1 and μ_2 is finite) defined on $([0, \infty), \mathcal{B}([0, \infty)))$ by setting $\mu([0, x]) = \int_0^x f(u)du$. We know that the derivative of F is given by the limit:

$$\lim_{h \to 0} \frac{F(x+h) - F(x)}{h} = \lim_{h \to 0} \frac{\mu([x, x+h])}{\lambda([x, x+h])} = \frac{d\mu(x)}{d\lambda(x)} = f(x),$$

and we may think of f as the Radon–Nikodym derivative of the measure μ induced by the integral with respect to Lebesgue measure λ. □

Example 1.3.29 Let $(\Omega, \mathcal{F}, P)$ be a probability space on which are defined the random variables $Y_1, Y_2, \ldots, Y_n$. Let $\mathcal{F}_n = \sigma\{Y_1, Y_2, \ldots, Y_n\}$. Let $\overline{P}$ be another probability measure on $\mathcal{F}$. Suppose that under P and $\overline{P}$ the random vector $\{Y_1, Y_2, \ldots, Y_n\}$ has densities $f_n(.)$ and $\overline{f}_n(.)$ respectively, with respect to n-dimensional Lebesgue measure. Then the Radon–Nikodym derivative

$$\left. \frac{d\overline{P}}{dP} \right|_{\mathcal{F}_n} = \frac{\overline{f}_n(Y_1, Y_2, \ldots, Y_n)}{f_n(Y_1, Y_2, \ldots, Y_n)} \tag{1.3.3}$$

is the likelihood ratio of the two probability measures in the presence of a sample of observations $\{Y_1, Y_2, \ldots, Y_n\}$. The higher (1.3.3) is the stronger is the evidence against $f_n(.)$. □

Now we state a few useful and standard theorems in the theory of integration.

Definition 1.3.30 *Suppose X is a square integrable random variable, that is $E[X^2] < \infty$, then*

$$\text{Var}(X) = E[(X - E[X])^2] = E[X^2] - E[X]^2,$$

is called the variance *of X. The square root of the variance $\sqrt{\text{Var}(X)} \overset{\triangle}{=} \sigma_X$ is called the* standard deviation *of the random variable X.*

Definition 1.3.31 *If X and Y are random variables with finite means μ_X and μ_Y and finite variances σ_X^2 and σ_Y^2, the* covariance *of X, Y is given by*

$$\text{Cov}(X, Y) = E[(X - \mu_X)(Y - \mu_Y)] = E[XY] - \mu_X \mu_Y,$$

and the correlation coefficient *of X, Y is given by*

$$\rho(X, Y) = \frac{\text{Cov}(X, Y)}{\sigma_X \sigma_Y} = \text{Cov}(\overline{X}, \overline{Y}),$$

where $\overline{X} = \dfrac{X - \mu_x}{\sigma_X}$ and $\overline{Y} = \dfrac{Y - \mu_Y}{\sigma_Y}$ are the normalized *versions of X and Y.*

In view of (1.3.2) independence of X and Y implies $\text{Cov}(X, Y) = 0 = \rho(X, Y)$. However, the converse is not true. In fact $\text{Cov}(X, Y)$ can vanish even if Y is a (nonlinear) function of X. The covariance, or the correlation coefficient, are related only with the linear dependence of X and Y.

Lemma 1.3.32 *(Cauchy–Schwarz Inequality). If X and Y are square integrable random variables,*

$$(E[XY])^2 \leq E[X^2]E[Y^2].$$

Proof   Let $R(t) = E[X + tY]^2 = E[X^2] + t^2 E[Y^2] + 2t E[XY] \geq 0$. Then for the quadratic function of t, $R(t)$, to be nonnegative its discriminant must satisfy $(E[XY])^2 - E[X^2]E[Y^2] \leq 0$, which is the result.   ∎

From the Cauchy–Schwarz Inequality 1.3.32 the integrability of XY is guaranteed by the square integrability of X and Y.

Another useful inequality is

Lemma 1.3.33 *(Chebyshev–Markov inequality). Let $(\Omega, \mathcal{F}, \mu)$ be a measurable space. For every measurable real valued function on Ω, and every pair of real numbers $p > 0$, $\alpha > 0$,*

$$\mu(\{\omega : |f(\omega)| \geq \alpha\}) \leq \frac{1}{\alpha^p} \int_\Omega |f(\omega)|^p d\mu(\omega).$$

Proof   Let $F_\alpha = \{\omega : |f(\omega)| \geq \alpha\}$. Then

$$\int_\Omega |f(\omega)|^p d\mu(\omega) \geq \int_{F_\alpha} |f(\omega)|^p d\mu(\omega) \geq \alpha^p \int_{F_\alpha} d\mu = \alpha^p \mu(F_\alpha).$$

∎

In addition to almost sure convergence, which was defined in Example 1.3.11, we have the following types of convergence.

First recall that $L^p(\Omega, \mathcal{F}, P)$, $p \geq 1$, is the space of random variables with finite absolute p-th moments, that is, $E[|X|^p] < \infty$.

$\{X_k\}$ converges to X in L^p ($X_k \overset{L^p}{\to} X$), $(0 < p < \infty)$, if

$$E[|X_k|^p] < \infty, \qquad E[|X|^p] < \infty,$$

and $E[|X_k - X|^p] \to 0$ ($k \to \infty$).

$\{X_k\}$ converges to X *in probability* or *in measure* ($X_k \overset{P}{\to} X$) if for each $\epsilon > 0$ the sequence

$$P[|X_k - X| > \epsilon] \to 0 \quad (k \to \infty).$$

Let $F_n(x) = P[X_n \leq x]$, $F(x) = P[X \leq x]$. X_n converges *in distribution* to X ($X_n \overset{D}{\to} X$) if

$$\int_{\mathbb{R}} g(x)\, \mathrm{d} F_n(x) \to \int_{\mathbb{R}} g(x)\, \mathrm{d} F(x),$$

for every real valued, continuous bounded function g defined on $\mathbb{R}$. A necessary and sufficient condition for that is:

$$F_n(x) \to F(x),$$

at every continuity point x of F [7].

These convergence concepts are in the following relationship to each other.

$(X_k \overset{a.s.}{\to} X) \Rightarrow (X_k \overset{P}{\to} X) \Rightarrow (X_n \overset{D}{\to} X).$

A useful concept is the *uniform integrability* of a family of random variables which permits the interchange of limits and expectations.

Definition 1.3.34 *A sequence $\{X_n\}$ of random variables is said to be uniformly integrable if*

$$\sup_n E[|X_n| I_{\{|X_n| > A\}}] \to 0, \qquad (A \to \infty). \tag{1.3.4}$$

A family $\{X_t\}$, $t \geq 0$ of random variables is said to be uniformly integrable if

$$\sup_t E[|X_t| I_{\{|X_t| > A\}}] \to 0, \qquad (A \to \infty). \tag{1.3.5}$$

Example 1.3.35 If $\mathcal{L}$ is bounded in $L^p(\Omega, \mathcal{F}, P)$ for some $p > 1$, then $\mathcal{L}$ is uniformly integrable.

Proof Choose A so large that $E[|X|^p] < A$ for all $X \in \mathcal{L}$. For fixed $X \in \mathcal{L}$, let $Y = |X| I_{\{|X| > K\}}$. Then $Y(\omega) \geq K I_{\{|X| > K\}} > 0$ for all $\omega \in \Omega$. Since $p > 1$, $\dfrac{Y^{p-1}}{K^{p-1}} \geq I_{\{|X| > K\}}$, and

$$K^{1-p} Y^p = \frac{Y^{p-1}}{K^{p-1}} Y \geq Y I_{\{|X| > K\}} = Y.$$

Thus

$$E[Y] \leq K^{1-p} E[Y^p] \leq K^{1-p} E[|X|^p] \leq K^{1-p} A,$$

which goes to 0 when $K \to \infty$, from which the result follows. ■

□

The following result is a somewhat stronger version of Fatou's Lemma 1.3.16.

Theorem 1.3.36 *Let $\{X_n\}$ be a uniformly integrable family of random variables. Then*

$$E[\liminf X_n] \leq \liminf E[X_n].$$

Proof   The proof is left as an exercise   ∎

Corollary 1.3.37 *Let $\{X_n\}$ be a uniformly integrable family of random variables such that $X_n \to X$ (a.s.), then*

$$E|X_n| < \infty, \quad E(X_n) \to E(X), \quad \text{and } E|X_n - X| \to 0.$$

The following deep result (Shiryayev [36]) gives a necessary and sufficient condition for taking limits under the expectation sign.

Theorem 1.3.38 *Let $0 \leq X_n \to X$ and $E(X_n) < \infty$. Then*

$$E(X_n) \to E(X) \iff \text{the family } \{X_n\} \text{ is uniformly integrable}.$$

Proof   The sufficiency part follows from Theorem 1.3.36. To prove the necessity, note that if x is not a point of positive probability for the distribution of the random variable X then $X_n I_{\{X_n < x\}} \to X I_{\{X < x\}}$ and the family $\{X_n I_{\{X_n < x\}}\}$ is uniformly integrable. Hence, by the sufficiency part of the theorem, we have

$$E[X_n I_{\{X_n < x\}}] \to E[X I_{\{X < x\}}], \text{ and } E[X_n I_{\{X_n \geq x\}}] \to E[X I_{\{X \geq x\}}].$$

So given $\epsilon > 0$ there exists N_ϵ such that $E[X_n I_{\{X_n \geq x\}}] \leq E[X I_{\{X \geq x\}}] + \frac{\epsilon}{2}$ and also, since $E[X]$ is finite, there exists, for the same ϵ, an x_0 (which is not a point of discontinuity of the distribution of X) such that $E[X I_{\{X \geq x_0\}}] < \frac{\epsilon}{2}$. Then if n is large enough,

$$E[X_n I_{\{X_n \geq x_0\}}] \leq E[X I_{\{X \geq x_0\}}] + \frac{\epsilon}{2} \leq \epsilon.$$

Since $E[X_n]$ is finite for all n, choose x_1 so large that $E[X_n I_{\{X_n \geq x_1\}}] < \epsilon$ for all $n \leq N_\epsilon$. Then we can conclude that $\sup_n E[X_n I_{\{X_n \geq x_1\}}] < \epsilon$, which shows the uniform integrability of $\{X_n\}$.   ∎

Remark 1.3.39 If we delete n in (1.3.4),

$$E[|X| I_{\{|X| > A\}}] \to 0, \quad (A \to \infty),$$

is the requirement for X to be integrable.   □

1.4 Conditional expectations

We often have insights into the occurrence of events in Ω gained through observations of realizations of related random variables. Conditional expectations incorporate this information by conditioning on events or random variables, i.e. on the σ-fields generated by these random variables.

Let $X = \sum_i x_i I_{A_i}$ be a simple random variable on a probability space $(\Omega, \mathcal{F}, P)$. What is the expected value of X given some event B having positive probability $P(B)$? Under the posterior probability measure $P(. \mid B)$ this is

$$E[X \mid B] = \sum x_i P(X = x_i \mid B)$$

$$= \frac{1}{P(B)} \sum x_i P(\{X = x_i\} \cap B) = \frac{1}{P(B)} E[X I_B].$$

$E[X I_B]$ is the probability weighted sum of the values taken on by X in the event B. We divide the weighted sum by $P(B)$ to obtain the weighted average.

We could write as a definition:

$$E[X \mid B] = \frac{E[X I_B]}{E[I_B]} = \frac{E[X I_B]}{P(B)}.$$

Let $X = I_C$ and $Y = I_B$. The σ-field $\sigma(Y)$ is generated by the atoms B and $\overline{B}$. To see this, consider any Borel set $\mathcal{B}$:

$$Y^{-1}\{\mathcal{B}\} = \begin{cases} \emptyset & \text{if } \{0, 1\} \notin \mathcal{B}, \\ \overline{B} & \text{if } 0 \in \mathcal{B}, 1 \notin \mathcal{B}, \\ B & \text{if } 1 \in \mathcal{B}, 0 \notin \mathcal{B}, \\ \Omega & \text{if } \{0, 1\} \in \mathcal{B}. \end{cases}$$

Hence $\sigma(Y) = \{\Omega, B, \overline{B}, \emptyset\}$.

Define

$$E[X \mid Y] = E[X \mid \sigma(Y)] = E[X \mid \text{atoms of } \sigma(Y)] = E[X \mid B, \overline{B}].$$

Or,

$$E[I_C \mid B, \overline{B}](\omega) = P(C \mid B, \overline{B})(\omega) \overset{\triangle}{=} P(C \mid B)I_B(\omega) + P(C \mid \overline{B})I_{\overline{B}}(\omega).$$

Hence $E[X \mid Y]$ is a function constant on the atoms of $\sigma(Y)$. That is

$$E[X \mid Y] \text{ is } \sigma(Y)\text{-measurable.}$$

Since $E[X \mid Y]$ is a random variable its mean is:

$$E[E[X \mid Y]] = E[P(C \mid B)I_B(\omega) + P(C \mid \overline{B})I_{\overline{B}}(\omega)]$$
$$= P(C \cap B) + P(C \cap \overline{B}) = P(C) = E[X].$$

If X is an integrable random variable and $Y = \sum_i y_i I_{B_i}$ is a simple random variable, we write

$$E[X \mid Y] \overset{\triangle}{=} E[X \mid \sigma(Y)] = \sum \frac{E[X I_{B_i}]}{P(B_i)} I_{B_i}(\omega).$$

Hence $E[X \mid Y]$ is $\sigma(Y)$-measurable and

$$E[E[X \mid Y]] = \sum E[X I_{B_i}] = E[X].$$

The expected value of $E[X \mid Y]$ is the same as the expected value of X.

Let $X \in L^1$ ($E|X| < \infty$) be a (nonnegative for simplicity) random variable on a probability space $(\Omega, \mathcal{F}, P)$ and $\mathcal{G}$ be a sub-σ-field of $\mathcal{F}$. The probability space $(\Omega, \mathcal{G}, P)$ is a coarsening of the original one and X is, in general, not measurable with respect to $\mathcal{G}$. We seek now a $\mathcal{G}$-measurable random variable, which we denote temporarily by $X_{\mathcal{G}}$, that assumes, on average, the same values as X. That is, we seek an integrable random variable $X_{\mathcal{G}}$ such that $X_{\mathcal{G}}$ is $\mathcal{G}$-measurable and

$$\int_A X_{\mathcal{G}} dP = \int_A X dP, \quad \text{for all } A \in \mathcal{G}.$$

Now the set function $Q(A) = \int_A X dP$ is a measure absolutely continuous with respect to P, so that the Radon–Nikodym Theorem 1.3.25 guarantees the existence of a $\mathcal{G}$-measurable random variable suggestively denoted by $E(X \mid \mathcal{G})$, which is uniquely determined except on an event of probability zero, such that

$$\int_A X dP = \int_A E[X \mid \mathcal{G}] dP,$$

for all $A \in \mathcal{G}$. We say that $X_{\mathcal{G}}$ is a version of $E(X \mid \mathcal{G})$. For a general integrable random variable X we define $E[X \mid \mathcal{G}]$ as $E[X^+ \mid \mathcal{G}] - E[X^- \mid \mathcal{G}]$.

Remark 1.4.1 Let $(\Omega, \mathcal{F}, P)$ be given, and suppose X is an L^2 random variable (measurable with respect to $\mathcal{F}$). Let $\mathcal{G}$ be a sub-σ-algebra of $\mathcal{F}$, that is, $\mathcal{G}$ is less informative than $\mathcal{F}$. A natural question is: by observing only $\mathcal{G}$ how much can we learn about X? Or, among all random variables which are $\mathcal{G}$-measurable which one gives us the best information (in the mean square sense) about the random variable X? It turned out that $E[X \mid \mathcal{G}]$ is the closest ($\mathcal{G}$-measurable) random variable to X. This is seen by considering, for any $\mathcal{G}$-measurable random variable,

$$Z = X - E[X \mid \mathcal{G}].$$

Then:

$$E[(Z - Y)^2] = E[(X - E[X \mid \mathcal{G}])^2 + Y^2 + 2Y(X - E[X \mid \mathcal{G}])]$$
$$= E[E[(X - E[X \mid \mathcal{G}])^2 \mid \mathcal{G}]] + E[Y^2].$$

□

This is minimized when $Y = 0$ a.s.

Example 1.4.2 Let $\Omega = (0, 1]$, $X(\omega) = \omega$, P be Lebesgue measure and consider the σ-field

$$G = \sigma\{(0, \tfrac{1}{4}], (\tfrac{1}{4}, \tfrac{1}{2}], (\tfrac{1}{2}, \tfrac{3}{4}], (\tfrac{3}{4}, 1]\} \stackrel{\triangle}{=} \sigma\{A_1, A_2, A_3, A_4\}.$$

$E[X \mid G]$ must be constant on the atoms of G so that

$$E[X \mid G](\omega) = \sum x_i I_{A_i}(\omega).$$

where $x_i = \dfrac{E[X I_{A_i}]}{P(A_i)}$.

Clearly $P(A_i) = \dfrac{1}{4}$ and $E[X I_{A_i}] = \displaystyle\int_{A_i} x dx.$

Hence

$$E[X \mid G](\omega) = \frac{1}{8} I_{A_1}(\omega) + \frac{2}{8} I_{A_2}(\omega) + \frac{5}{8} I_{A_3}(\omega) + \frac{7}{8} I_{A_4}(\omega),$$

which is a G-measurable random variable. □

Example 1.4.3 Let X_1, X_2 and X_3 be three independent, identically distributed (i.i.d.) random variables such that

$$P(X_i = 1) = p = 1 - P(X_i = 0) = 1 - q.$$

Let $S = X_1 + X_2 + X_3$. Suppose that we observe X_1 and X_2 and we wish to find the (conditional) probability that $S = 2$ given X_1 and X_2. The σ-field generated by the (vector) random variable (X_1, X_2) is generated by the atoms $\{A_{ij}\}$, $i, j = 0, 1$, where $A_{ij} = [\omega : X_1(\omega) = i, X_2(\omega) = j]$.

$$
\begin{aligned}
P(S = 2 \mid X_1, X_2)(\omega) &= P(S = 2 \mid \sigma\{X_1, X_2\})(\omega) \\
&= \sum_{i,j=0,1} P(S = 2 \mid A_{ij}) I_{A_{ij}}(\omega) \\
&= \sum_{i,j=0,1} \frac{P(S = 2 \cap A_{ij})}{P(A_{ij})} I_{A_{ij}}(\omega) \\
&= \sum_{i,j=0,1} \frac{P(i + j + X_3 = 2) P(A_{ij})}{P(A_{ij})} I_{A_{ij}}(\omega) \\
&= \sum_{i,j=0,1} P(X_3 = 2 - i - j) I_{A_{ij}}(\omega) \\
&= P(X_3 = 0) I_{A_{11}} + P(X_3 = 1) I_{\{A_{10} \cup A_{01}\}} \\
&= q I_{A_{11}}(\omega) + p I_{\{A_{10} \cup A_{01}\}}(\omega).
\end{aligned}
$$

The expected value of the ($\sigma\{X_1, X_2\}$-measurable) random variable $P(S = 2 \mid X_1, X_2)$ is

$$E[q I_{A_{11}}(\omega) + p I_{\{A_{10} \cup A_{01}\}}(\omega)] = q P(A_{11}) + p[P(A_{01}) + P(A_{01})] = P(S = 2).$$

□

Example 1.4.4 Let $f \in L^1[0, 1]$, i.e. the Lebesgue integral $\int_{[0,1)} |f(x)| dx$ exists and is finite. Let $\mathcal{F}_n = \sigma\{[\frac{j}{2^n}, \frac{j+1}{2^n}), \quad j = 0, \ldots, 2^n - 1\}$. Then

$$E[f \mid \mathcal{F}_n](\omega) = \sum_{j=0}^{2^n-1} \frac{\int_{j2^{-n}}^{(j+1)2^{-n}} f(x) dx}{2^{-n}} I_{[j2^{-n}, (j+1)2^{-n})}(\omega).$$

□

Theorem 1.4.5 *If X is real $\mathcal{F}$-measurable random variable and if $\int_A X dP = 0$ for all $A \in \mathcal{F}$, then $X = 0$ a.s.*

Proof Suppose $X \geq 0$ and $\int_A X \mathrm{d}P = 0$ for all $A \in \mathcal{F}$. Write $A_n = \{\omega : X(\omega) \geq \frac{1}{n}\}$.

$$\int_{A_n} X \mathrm{d}P \geq \frac{1}{n} P(A_n) \geq 0.$$

But $\int_{A_n} X \mathrm{d}P = 0$ so $P(A_n) = 0$ for all n. Therefore,

$$P(\{X > 0\}) = P(\bigcup A_n) \leq \sum P(A_n) = 0.$$

For a general random variable X, recall that $X = X^+ - X^-$, where both X^+ and X^- are nonnegative. ∎

The following is a list of classical results on conditional expectation:

1. $E(X \mid \mathcal{A})$ is unique (a.s.)
   Proof Let $X_1 = E(X \mid \mathcal{A})$ and X_2 be an $\mathcal{A}$-measurable random variable such that

$$\int_A X_2 \mathrm{d}P = \int_A X \mathrm{d}P,$$

for all $A \in \mathcal{A}$ and let $\Omega_0 = \{\omega : X_1 > X_2\} \in \mathcal{A}$. Hence

$$\int_{\Omega_0} X_1 \mathrm{d}P = \int_{\Omega_0} E(X \mid \mathcal{A}) = \int_{\Omega_0} X \mathrm{d}P,$$

and

$$\int_{\Omega_0} X_2 \mathrm{d}P = \int_{\Omega_0} X \mathrm{d}P,$$

so that

$$\int_{\Omega_0} X_1 \mathrm{d}P = \int_{\Omega_0} X_2 \mathrm{d}P,$$

or

$$\int_{\Omega_0} (X_1 - X_2) \mathrm{d}P = 0.$$

Using Theorem 1.4.5 $X_1 = X_2$ a.s. ∎

2. If $\mathcal{A}_1$ and $\mathcal{A}_2$ are two sub-σ-fields of $\mathcal{F}$ such that $\mathcal{A}_1 \subset \mathcal{A}_2$, then

$$E(E(X \mid \mathcal{A}_1) \mid \mathcal{A}_2) = E(E(X \mid \mathcal{A}_2) \mid \mathcal{A}_1) = E(X \mid \mathcal{A}_1). \tag{1.4.1}$$

   Proof Clearly $E(E(X \mid \mathcal{A}_1) \mid \mathcal{A}_2) = E(E(X \mid \mathcal{A}_2) \mid \mathcal{A}_1)$. Now $E(E(X \mid \mathcal{A}_2) \mid \mathcal{A}_1)$ is $\mathcal{A}_1$-measurable and for $A \in \mathcal{A}_1$,

$$\int_A E(E(X \mid \mathcal{A}_2) \mid \mathcal{A}_1)) \mathrm{d}P = \int_A E(X \mid \mathcal{A}_2) \mathrm{d}P$$

$$= \int_A X \mathrm{d}P = \int_A E(X \mid \mathcal{A}_1) \mathrm{d}P.$$

Hence $E(E(X \mid \mathcal{A}_2) \mid \mathcal{A}_1) = E(X \mid \mathcal{A}_1)$ a.s. ∎

3. If $X, Y, XY \in L^1$ and Y is $\mathcal{A}$-measurable then

$$E[XY \mid \mathcal{A}] = YE[X \mid \mathcal{A}]. \qquad (1.4.2)$$

Proof   It is sufficient to prove the result when X and Y are positive. If $Y = I_A$, $A \in \mathcal{A}$, then for every $B \in \mathcal{A}$

$$\int_B XY \mathrm{d}P = \int_{A \cap B} X \mathrm{d}P = \int_{A \cap B} E[X \mid \mathcal{A}] \mathrm{d}P$$
$$= \int_B I_A E[X \mid \mathcal{A}] \mathrm{d}P = \int_B YE[X \mid \mathcal{A}] \mathrm{d}P.$$

That is, $E[XY \mid \mathcal{A}] = YE[X \mid \mathcal{A}]$, if Y is an indicator function. It follows that the result is true for simple functions of sets in $\mathcal{A}$ and therefore for a limit of bounded increasing sequence of such functions converging to Y. ∎

4. If X is independent of the σ-field $\mathcal{A}$, then

$$E(X \mid \mathcal{A}) = E(X). \qquad (1.4.3)$$

Proof   First note that $E(X)$ is $\mathcal{A}$-measurable. Now, for $A \in \mathcal{A}$ we have to show that

$$\int_A E(X \mid \mathcal{A}) \mathrm{d}P = \int_A E(X) \mathrm{d}P.$$

However, the left hand side is equal to $E[I_A X]$ and the right hand side is equal to $E[I_A]E[X]$, and their equality follows from the definition of independence of random variables. ∎

5. Conditional expectation is a projection operation, and so

$$E[E[X \mid \mathcal{A}] \mid \mathcal{A}] = E[X \mid \mathcal{A}]. \qquad (1.4.4)$$

Example 1.4.6 Consider the joint distribution function $F(x_1, x_2)$ of two real valued random variables X_1, X_2 and the probability measure P on the two-dimensional Borel sets generated by the distribution function $F(x_1, x_2)$. Suppose that P is absolutely continuous with respect to two-dimensional Lebesgue measure. Then, by the Radon–Nikodym theorem, there exists a nonnegative density function $f(x_1, x_2)$ such that for any Borel set B:

$$P(B) = \iint I_B(x_1, x_2) f(x_1, x_2) \mathrm{d}x_1 \mathrm{d}x_2.$$

If $f(x_1, x_2) > 0$ everywhere,

$$P(B \mid X_2 = x_2) = \frac{\int_{\{x_1 : (x_1, x_2) \in B\}} f(x_1, x_2) \mathrm{d}x_1}{\int_{-\infty}^{+\infty} f(x_1, x_2) \mathrm{d}x_1},$$

from which we can deduce that $\dfrac{f(x_1, x_2)}{\int_{-\infty}^{+\infty} f(x_1, x_2) \mathrm{d}x_1}$ is the density function of the conditional probability measure $P(. \mid X_2 = x_2)$. □

Example 1.4.7 Let X_1 and X_2 be two random variables with a normal joint distribution. Then their probability density function has the form

$$\phi(x_1, x_2) = \frac{1}{2\pi\sigma_1\sigma_2\sqrt{1-\rho^2}} \exp\left\{-\frac{1}{2(1-\rho^2)}\left[\bar{x}_1^2 - 2\rho\bar{x}_1\bar{x}_2 + \bar{x}_2^2\right]\right\},$$

where $0 \leq \rho < 1$ and $\bar{x}_i = \dfrac{x_i - \mu_i}{\sigma_i}$, $i = 1, 2$. The conditional density of X_1 given $X_2 = x_2$ is a normal density with mean $\mu_1 + \rho\dfrac{\sigma_1}{\sigma_2}(x_2 - \mu_2)$ and variance $\mathrm{Var}(X_1 \mid X_2 = x_2) = (1 - \rho^2)\sigma_1^2 < \sigma_1^2 = \mathrm{Var}(X_1)$. To see this, recall that, by definition, the conditional density of X_1 given X_2 is given by

$$\phi(x_1 \mid x_2) = \frac{\phi(x_1, x_2)}{\int_{\mathbb{R}} \phi(x_1, x_2)dx_1}$$

$$= \frac{\dfrac{1}{2\pi\sigma_1\sigma_2\sqrt{1-\rho^2}} \exp\left\{-\dfrac{1}{2(1-\rho^2)}\left[\bar{x}_1^2 - 2\rho\bar{x}_1\bar{x}_2 + \bar{x}_2^2\right]\right\}}{\dfrac{1}{2\pi\sigma_2} \exp\left\{-\dfrac{1}{2}\bar{x}_2^2\right\}}$$

$$= \frac{1}{2\pi\sigma_1\sqrt{1-\rho^2}} \exp\left\{-\frac{1}{2(1-\rho^2)}\left[\bar{x}_1^2 - 2\rho\bar{x}_1\bar{x}_2 + \rho^2\bar{x}_2^2\right]\right\}$$

$$= \frac{1}{2\pi\sigma_1\sqrt{1-\rho^2}} \exp\left\{-\frac{1}{2(1-\rho^2)}[\bar{x}_1 - \rho\bar{x}_2]^2\right\}$$

$$= \frac{1}{2\pi\sigma_1\sqrt{1-\rho^2}}$$

$$\times \exp\left\{-\frac{1}{2\sigma_1^2(1-\rho^2)}\left[x_1 - \left(\mu_1 + \rho\frac{\sigma_1}{\sigma_2}(x_2 - \mu_2)\right)\right]^2\right\},$$

and the result follows.

Thus by conditioning on X_2 we have gained some statistical information about X_1 which resulted in a reduction in the variability of X_1. $\qquad\square$

1.5 Problems

1. Let $\{\mathcal{F}_i\}_{i \in I}$ be a family of σ-fields on Ω. Prove that $\bigcap_{i \in I} \mathcal{F}_i$ is a σ-field.
2. Let A and B be two events. Express by means of the indicator functions of A and B

$$I_{A \cup B}, \quad I_{A \cap B}, \quad I_{A-B}, \quad I_{B-A}, \quad I_{(A-B) \cup (B-A)},$$

where $A - B = A \cap \bar{B}$.

3. Let $\Omega = \mathbb{R}$ and define the sequences $C_{2n} = [-1, 2 + \dfrac{1}{2n})$ and $C_{2n+1} = [-2 - \dfrac{1}{2n+1}, 1)$. Show that

$$\limsup C_n = [-2, -2], \quad \liminf C_n = [-1, 1].$$

4. Let $\Omega = (\omega_1, \omega_2, \omega_3, \omega_4)$ and $P(\omega_1) = \dfrac{1}{12}$, $P(\omega_2) = \dfrac{1}{6}$, $P(\omega_3) = \dfrac{1}{3}$, and $P(\omega_4) = \dfrac{5}{12}$. Let

$$A_n = \begin{cases} \{\omega_1, \omega_3\} & \text{if } n \text{ is odd,} \\ \{\omega_2, \omega_4\} & \text{if } n \text{ is even.} \end{cases}$$

Find $P(\limsup A_n)$, $P(\liminf A_n)$, $\limsup P(A_n)$, and $\liminf P(A_n)$ and compare.

5. Give a proof to Theorem 1.3.36.

6. Show that a σ-field is either finite or uncountably infinite.

7. Show that if X is a random variable, then $\sigma\{|X|\} \subseteq \sigma\{X\}$.

8. Show that the set $\mathcal{B}_0$ of countable unions of open intervals in $\mathbb{R}$ is not closed under complementation and hence is not a σ-field. (Hint: enumerate the rational numbers and choose, for each one of them, an open interval containing it. Now show that the complement of the union of all these open intervals is not in $\mathcal{B}_0$.)

9. Show that the class of finite unions of intervals of the form $(-\infty, a], (b, c]$, and (d, ∞) is a field but not a σ-field.

10. Show that a sequence of random variables $\{X_n\}$ converges (a.s.) to X if and only if $\forall\ \epsilon > 0 \lim_{m \to \infty} P[|X_n - X| \le \epsilon\ \ \forall n \ge m] = 1$.

11. Show that if $\{X_k\}$ converges (a.s.) to X then $\{X_k\}$ converges to X in probability but the converse is false.

12. Consider the probability space $(\mathbb{N}, \mathcal{F}, P)$, where $\mathbb{N}$ is the set of natural numbers, $\mathcal{F}$ is the collection of all the subsets of $\mathbb{N}$ and $P(\{k\}) = \dfrac{1}{2^k}$. Let $X_k(\omega) = I_{[\omega=k]}$. Discuss the convergence (a.s.) and in probability of X_k and show that on this particular space they are equivalent.

13. Let $\{X_n\}$ be a sequence of random variables with

$$P[X_n = 2^n] = P[X_n = -2^n] = \frac{1}{2^n},$$

$$P[X_n = 0] = 1 - \frac{1}{2^{n-1}}.$$

Show that $\{X_n\}$ converges (a.s.) to 0 but $E|X_n|^p$ does not converge to 0.

14. Let $\{X_n\}$ be a sequence of random variables with

$$P[X_n = n^{1/2^p}] = \frac{1}{n},$$

$$P[X_n = 0] = 1 - \frac{1}{n}.$$

Show that $\{X_n\}$ does not converge (a.s.) to 0 but $E|X_n|^p$ converges to 0.

15. Suppose Q is another probability measure on $(\Omega, \mathcal{F})$ such that $P(A) = 0$ implies $Q(A) = 0$ ($Q \ll P$). Show that P-a.s. convergence implies Q-a.s. convergence.

16. Prove that if $\mathcal{F}_1$ and $\mathcal{F}_2$ are independent sub-σ-fields and $\mathcal{F}_3$ is coarser than $\mathcal{F}_1$, then $\mathcal{F}_3$ and $\mathcal{F}_2$ are independent.

17. Let $\Omega = (\omega_1, \omega_2, \omega_3, \omega_4, \omega_5, \omega_6)$, $P(\omega_i) = p_i \ne \dfrac{1}{6}$ and the sub-σ-fields

$$\mathcal{F}_1 = \sigma\{\{\omega_1, \omega_2\}, \{\omega_3, \omega_4, \omega_5, \omega_6\}\},$$
$$\mathcal{F}_2 = \sigma\{\{\omega_1, \omega_2\}, \{\omega_3, \omega_4\}, \{\omega_5, \omega_6\}\}.$$

Show that $\mathcal{F}_1$ and $\mathcal{F}_2$ are not independent. What can be said about the sub-σ-fields

$$\mathcal{F}_3 = \sigma\{\{\omega_1, \omega_2\}, \{\omega_3\}, \{\omega_4, \omega_5, \omega_6\}\},$$

and

$$\mathcal{F}_5 = \sigma\{\{\omega_1, \omega_4\}, \{\omega_2, \omega_5\}, \{\omega_3, \omega_6\}\}?$$

18. Let $\Omega = \{(i, j) : i, j = 1, \ldots, 6\}$ and $P(\{i, j\}) = 1/36$. Define the quantity

$$X(\omega) = \sum_{k=0}^{\infty} k I_{\{(i,j):i+j=k\}}.$$

Is X a random variable? Find $P_X(x) = P(X = x)$, calculate $E[X]$ and describe $\sigma(X)$, the σ-field generated by X.

19. For the function X defined in the previous exercise, describe the random variable $P(A \mid X)$, where $A = \{(i, j) : i \text{ odd}, j \text{ even}\}$ and find its expected value $E[P(A \mid X)]$.

20. Let Ω be the unit interval $(0, 1]$ and on it be given the following σ-fields:

$$\mathcal{F}_1 = \sigma\{(0, \tfrac{1}{2}], (\tfrac{1}{2}, \tfrac{3}{4}], (\tfrac{3}{4}, 1]\},$$
$$\mathcal{F}_2 = \sigma\{(0, \tfrac{1}{4}], (\tfrac{1}{4}, \tfrac{1}{2}], (\tfrac{1}{2}, \tfrac{3}{4}], (\tfrac{3}{4}, 1]\},$$
$$\mathcal{F}_3 = \sigma\{(0, \tfrac{1}{8}], (\tfrac{1}{8}, \tfrac{2}{8}], \ldots, (\tfrac{7}{8}, 1]\}.$$

Consider the mapping

$$X(\omega) = x_1 I_{(0, \frac{1}{4}]}(\omega) + x_2 I_{(\frac{1}{4}, \frac{1}{2}]}(\omega) + x_3 I_{(\frac{1}{2}, \frac{3}{4}]}(\omega) + x_4 I_{(\frac{3}{4}, 1]}(\omega).$$

Find $E[X \mid \mathcal{F}_1]$, $E[X \mid \mathcal{F}_2]$, and $E[X \mid \mathcal{F}_3]$.

21. Let Ω be the unit interval and $((0, 1], P)$ be the Lebesgue-measurable space and consider the following sub-σ-fields:

$$\mathcal{F}_1 = \sigma\{(0, \tfrac{1}{2}], (\tfrac{1}{2}, \tfrac{3}{4}], (\tfrac{3}{4}, 1]\},$$
$$\mathcal{F}_2 = \sigma\{(0, \tfrac{1}{4}], (\tfrac{1}{4}, \tfrac{1}{2}], (\tfrac{1}{2}, \tfrac{3}{4}], (\tfrac{3}{4}, 1]\}.$$

Consider the mapping

$$X(\omega) = \omega.$$

Find $E[E[X \mid \mathcal{F}_1] \mid \mathcal{F}_2]$, $E[E[X \mid \mathcal{F}_2] \mid \mathcal{F}_1]$ and compare.

22. Consider the probability measure P on the real line such that:

$$P(0) = p, \quad P((0, 1)) = q, \quad p + q = 1,$$

and the random variables defined on $\Omega = \mathbb{R}$,

$$X_1(x) = 1 + x, \quad X_2(x) = 0 I_{\{x \le 0\}} + (1 + x) I_{\{0 < x < 1\}} + 2 I_{\{x \ge 1\}},$$
$$X_3(x) = \sum_{k=-\infty}^{+\infty} (1 + x + k) I_{\{k \le x \le k+1\}}.$$

Is there any P-a.s. equality between X_1, X_2 and X_3?

23. Let X_1, X_2 and X_3 be three independent, identically distributed (i.i.d.) random variables such that $P(X_i = 1) = p = 1 - P(X_i = 0) = 1 - q$. Find $P(X_1 + X_2 + X_3 = s \mid X_1, X_2)$.

24. Let X_1, X_2 and X_3 be three random variables with *multinomial distribution* with parameters p_1, p_2, p_3, n, that is

$$P(X_1 = n_1, X_2 = n_2, X_3 = n_3) = \frac{n! p_1^{n_1} p_2^{n_2} p_3^{n_3}}{n_1! n_2! n_3!},$$

where n_1, n_2 and n_3 are nonnegative integers such that $n_1 + n_2 + n_3 = n$. Show that if n is a random variable with Poisson distribution with parameter λ then the three random X_1, X_2 X_3 become mutually independent with Poisson distributions.

25. On $\Omega = [0, 1]$ and P being Lebesgue measure show that

$$X = x_1 I_{(0,\frac{1}{2}]} + x_2 I_{(\frac{1}{2},1]} \text{ and } Y = y_1 I_{(0,\frac{1}{4}]\cup(\frac{3}{4},1]} + y_2 I_{(\frac{1}{4},\frac{3}{4}]}$$

are independent.

26. Show that (see Example 1.4.4)

$$E[f \mid \mathcal{F}_n] = \sum_{j=0}^{2^n-1} \frac{\int_{j2^{-n}}^{(j+1)2^{-n}} f(x)dx}{2^{-n}} I_{[j2^{-n},(j+1)2^{-n})}$$

converges a.s. and in L^1 to f as $n \to \infty$.
In particular, if $f = I_E$ for some Borel set E, then

$$\sum_{j=0}^{2^n-1} \frac{\lambda(E \cap [j2^{-n}, (j+1)2^{-n}))}{2^{-n}} I_{[j2^{-n},(j+1)2^{-n})}(x) \overset{\text{a.s.}}{\to} I_E(x),$$

$x \in [0, 1]$. Here $\lambda(.)$ is the Lebesgue measure.

Stochastic processes

2.1 Definitions and general results

A *stochastic process* is a mathematical model for any phenomenon evolving or varying in time (or over some index set), subject to random influences. Examples include the price of a commodity observed through time, the fluctuating water level behind a dam or the distribution of shades in a noisy image observed over a region of $\mathbb{R}^2$. Suppose $(\Omega, \mathcal{F})$ is a measurable space. We shall define a stochastic process to be a mapping $X_{(\text{index})}(\omega)$ from $\Omega \times \{\text{index space}\}$ into a second measurable space $(E, \mathcal{E})$, called the *state space*, or the range space. Alternatively, we can consider a stochastic process as a family $\{X_t\}$ $t \in \{\text{index space}\}$ of random variables all defined on a measurable space $(\Omega, \mathcal{F})$.

For a fixed simple outcome ω, $X_{(.)}(\omega)$ is a function describing *one* possible trajectory, or sample path, followed by the process. If the time index is frozen at t, say, then we have a random variable $X_t(.)$, i.e. an $\mathcal{F}$-measurable function of ω.

When the time index t is continuous, measurability, continuity, etc. in t are considered.

A continuous-time stochastic process $\{X_t\}$ is said to have *independent increments* if for all $t_0 < t_1 < t_2 < \cdots < t_n$, the random variables $X_{t_1} - X_{t_0}, X_{t_2} - X_{t_1}, \ldots, X_{t_n} - X_{t_{n-1}}$ are independent. If for all s, $X_{t+s} - X_t$ has the same distribution for all t, $\{X_t\}$ is said to possess *stationary increments*.

Sometimes, a stochastic process is interpreted as just a single random variable taking values in a space of functions, that is, with each ω is associated a function. In analogy with real random variables, the state space is then endowed with a Borel σ-field (generated by the open sets of an underlying topology).

Example 2.1.1 Let

$$\Omega = \{\omega_1, \omega_2, \ldots\},$$

and let the time index n be finite $0 \le n \le N$. A stochastic process X in this setting is a two-dimensional array or matrix such that:

$$
X = \begin{array}{|c|c|c|}
\hline
X_1(\omega_1) & X_1(\omega_2) & \cdots \\
\hline
X_2(\omega_1) & X_2(\omega_2) & \cdots \\
\hline
\cdots & \cdots & \cdots \\
\hline
X_N(\omega_1) & X_N(\omega_2) & \cdots \\
\hline
\end{array}
$$

Each row represents a random variable and each column is a sample path or a realization of the stochastic process X. If the time index is unbounded, each sample path is given by an infinite sequence. □

Example 2.1.2 Let $N = 4$ in the previous example and suppose that X is given by the following array.

2	3	5	7	11	3	2.3	1
-1	1	5.7	$\sqrt{2}$	3	6	83	19
11	7	70	3	2	-5	2	21
5	3	2	1	0	1	2	3

The sample space of $\{X_n\}$ is $\mathbb{R}^4$ and the stochastic process can be thought of as a mapping (in fact a random variable)

$$\omega_i \to X(\omega_i) = (X_1(\omega_i), \ldots, X_4(\omega_i)) = (x_1^i, x_2^i, x_3^i, x_4^i) \overset{\triangle}{=} x^i \in \mathbb{R}^4.$$

The random variable X induces a probability measure P_X on the Borel σ-field $\mathcal{B}(\mathbb{R}^4)$ in the usual way, i.e., for any $B \in \mathcal{B}(\mathbb{R}^4)$,

$$P_X(B) \overset{\triangle}{=} P[\omega : X(\omega) \in B] = P(X^{-1}(B)).$$

For instance,

$$B_1 = \{x \in \mathbb{R}^4 : 3 \le x_1 \le 5, 2 \le x_2 \le 7\}$$

contains a single trajectory (column 6 in the table) so that $P_X(B_1) = P(\omega_6)$.

$$B_2 = \{x \in \mathbb{R}^4 : \max_{1 \le n \le 4} x_n \le 7\}$$

contains four trajectories (column 2, column 3, column 4 and column 6 in the table) so that $P_X(B_2) = P(\omega_2, \omega_3, \omega_4, \omega_6)$. □

Example 2.1.3 Let $\Omega = \{\omega_1, \omega_2, \ldots\}$ and P be a probability measure on $(\Omega, \mathcal{F})$. Suppose that the time index set is the set of positive integers. A real valued stochastic process X in this setting is a two-dimensional infinite array such that:

$$X = \begin{array}{|c|c|c|} \hline X_1(\omega_1) & X_1(\omega_2) & \cdots \\ \hline X_2(\omega_1) & X_2(\omega_2) & \cdots \\ \hline \cdots & \cdots & \cdots \\ \hline \end{array}.$$

Here the sample space is

$$\mathbb{R}^\infty = \{(x_1, x_2, \ldots) \in \mathbb{R} \times \mathbb{R} \times \ldots\}.$$

Note that the Borel σ-field $\mathcal{B}(\mathbb{R}^\infty)$ coincides with the smallest σ-field containing the open sets in $\mathbb{R}^\infty$ in the metric $\rho_\infty(x^1, x^2) = \sum_k 2^{-k} \dfrac{|x_k^1 - x_k^2|}{1 + |x_k^1 - x_k^2|}$ ([36]).

Now think of the stochastic process X as an $\mathbb{R}^\infty$ valued random variable

$$\omega_i \to X(\omega_i) = (X_1(\omega_i), X_2(\omega_i), \dots) = (x_1^i, x_2^i, \dots) \overset{\triangle}{=} x^i \in \mathbb{R}^\infty.$$

The random variable X induces a probability measure P_X on the σ-field $\mathcal{B}(\mathbb{R}^\infty)$. For instance, if

$$A = \{x \in \mathbb{R}^\infty : \sup x_n > a\} \in \mathcal{B}(\mathbb{R}^\infty),$$

then the set A consists of all sequences with some of their entries larger than a and $P_X(A) = P(\omega : X(\omega) \in A)$.   □

Example 2.1.4 (The Single Jump Process) Consider a stochastic process $\{X_t\}, t \geq 0$, which takes its values in some measurable space $\{E, \mathcal{E}\}$ and which remains at its initial value $z_0 \in E$ until a random time T, when it jumps to a random position Z. A sample path of the process is

$$X_t(\omega) = \begin{cases} z_0 \text{ if } t < T(\omega), \\ Z(\omega) \text{ if } t \geq T(\omega). \end{cases}$$

The underlying probability space can be taken to be

$$\Omega = [0, \infty] \times E,$$

with the σ-field $\mathcal{B} \times \mathcal{E}$. A probability measure P is given on $(\Omega, \mathcal{B} \times \mathcal{E})$ and we suppose

$$P([\infty, 0] \times \{z_0\}) = 0 = P(\{0\} \times E),$$

so that the probabilities of a zero jump and a jump at time zero are zero.
    Write

$$F_t = P[T > t, Z \in E],$$
$$c = \inf\{t : F_t = 0\}.$$

F_t is right-continuous and monotonic decreasing, so there are only countably many points of discontinuity $\{u\} = D$ where $\Delta F_u = F_u - F_{u-} \neq 0$. At points in D, there are positive probabilities that X jumps. Note that the more probability mass there is at a point u, the more predictable is the jump at that point.
    Formally define a function Λ by setting:

$$d\Lambda(t) = P(T \in]t - dt, t], Z \in E \mid T > t - dt).$$

Then Λ is the probability that the jump occurs in the interval $]t - dt, t]$, given it has not happened at $t - dt$. Roughly speaking we have

$$d\Lambda(t) = P(T \in]t - dt, t] \mid T > t - dt)$$
$$= \frac{P(T \in]t - dt, t])}{F_{t-dt}}$$
$$= \frac{1 - F_t - (1 - F_{t-dt})}{F_{t-dt}}$$
$$= \frac{-(F_t - F_{t-dt})}{F_{t-dt}}$$
$$= \frac{-(F_t - F_{t-})}{F_{t-}}$$
$$= \frac{-dF_t}{F_{t-}}.$$

Define

$$\Lambda(t) = -\int_{]0,t[} \frac{dF_s}{F_{s-}}. \tag{2.1.1}$$

For instance, if T is exponentially distributed with parameter θ we have

$$\Lambda(t) = -\int_{]0,t[} \frac{d\exp(-\theta s)}{\exp(-\theta s)} = \theta.$$

Write

$$F_t^A = P[T > t, Z \in A],$$

then clearly the measure on $(\mathbb{R}^+, \mathcal{B}(\mathbb{R}^+))$ given by F_t^A is absolutely continuous with respect to that given by F_t, so that there is a Radon–Nikodym derivative $\lambda(A, s)$ such that

$$F_t^A - F_0^A = \int_{]0,t[} \lambda(A, s) dF_s. \tag{2.1.2}$$

The pair (λ, Λ) is the *Lévy system* for the jump process. Roughly, $\lambda(dx, s)$ is the conditional distribution of the jump position Z given the jump happens at time s. ☐

Let X_t be a continuous time stochastic process. That is, the time index belongs to some interval of the real line, say, $t \in [0, \infty)$. If we are interested in the behavior of X_t during an interval of time $[t_0, t_1]$ it is necessary to consider simultaneously an uncountable family of X_ts $\{X_t, t_0 \le t \le t_1\}$. This results in a technical problem because of the uncountability of the index parameter t. Recall that σ-fields are, by definition, closed under countable operations only and that statements like $\{X_t \ge x, t_0 \le t \le t_1\} = \bigcap_{t_0 \le t \le t_1} \{X_t \ge x\}$ are not events! However, for most practical situations this difficulty is bypassed by replacing uncountable index sets by countable dense subsets without losing any significant information. In general, these arguments are based on the *separability* of a continuous time stochastic process. This is possible, for example, if the stochastic process X is almost surely continuous (see Definition 2.1.6).

Let $X = \{X_t : t \geq 0\}$ and $Y = \{Y_t : t \geq 0\}$ be two stochastic processes defined on the same probability space $(\Omega, \mathcal{F}, P)$. Because of the presence of ω, the functions $X_t(\omega)$ and $Y_t(\omega)$ can be compared in different ways.

Definition 2.1.5

1. *X and Y are called* indistinguishable *if*

$$P(\{\omega : X_t(\omega) = Y_t(\omega), t \geq 0\}) = 1.$$

2. *Y is a* modification *of X if for every $t \geq 0$, we have*

$$P(\{\omega : X_t(\omega) = Y_t(\omega)\}) = 1.$$

3. *X and Y have the same law or probability distribution if and only if all their finite dimensional probability distributions coincide, that is, if and only if for any sequence of times $0 \leq t_1 \leq \cdots \leq t_n$ the joint probability distributions of $(X_{t_1}, \ldots, X_{t_n})$ and $(Y_{t_1}, \ldots, Y_{t_n})$ coincide.*

Note that the first property is much stronger than the other two. The null sets in the second and third properties may depend on t.

Recall that there are different definitions of *limit* for sequences of random variables. So to each definition corresponds a type of continuity of a real valued time index process.

Definition 2.1.6

1. $\{X_t\}$ *is* continuous in probability *if for every t and $\epsilon > 0$,*

$$\lim_{h \to 0} P[|X_{t+h} - X_t| > \epsilon] = 0.$$

2. $\{X_t\}$ *is* continuous in L^p *if for every t,*

$$\lim_{h \to 0} E[|X_{t+h} - X_t|^p] = 0.$$

3. $\{X_t\}$ *is* continuous almost surely *(a.s.) if for every t,*

$$P[\lim_{h \to 0} X_{t+h} = X_t] = 1.$$

4. $\{X_t\}$ *is* right continuous *if for almost every ω the map $t \to X_t(\omega)$ is right continuous. That is,*

$$\lim_{s \downarrow t} X_s = X_t \ a.s.$$

If in addition

$$\lim_{s \uparrow t} X_s = X_{t-} \ exists \ a.s.,$$

$\{X_t\}$ *is* right continuous with left limits (rcll or corlol or càdlàg).

However, none of the above notions is strong enough to differentiate, for instance, between a process for which almost all sample paths are continuous for every t, and a process for which almost all sample paths have a countable number of discontinuities, when the two processes have the same finite dimensional distributions. A much stronger criterion for continuity is *sample paths continuity* which requires continuity for *all* ts simultaneously! In other words, for almost all ω the function $X_{(.)}(\omega)$ is continuous in the usual sense. Unfortunately, the definition of a stochastic process in terms of its finite dimensional distributions does not help here since we are faced with whole intervals containing uncountable numbers of ts. Fortunately, for most useful processes in applications, continuous versions (sample path continuous), or right-continuous versions, can be constructed.

If a stochastic process with index set $[0, \infty)$ is continuous its sample space can be identified with $C[0, \infty)$, the space of all real valued continuous functions. A metric on this space is

$$\rho(x, y) = \sum_k 2^{-k} \frac{\sup_{0 \le t \le k} |x(t) - y(t)|}{1 + \sup_{0 \le t \le k} |x(t) - y(t)|},$$

for $x, y \in C[0, \infty)$. (See [36].)

Let $\mathcal{B}(C)$ be the smallest σ-field containing the open sets of the topology induced by ρ on $C[0, \infty)$, the Borel σ-field. Then ([36]) the same σ-field $\mathcal{B}(C)$ is generated by the cylinder sets of $C[0, \infty)$ which have the form

$$\{x \in C[0, \infty) : x_{t_1} \in I_1, \, x_{t_2} \in I_2, \, \ldots, \, x_{t_n} \in I_n\},$$

where each I_i is an interval of the form $(a_i, b_i]$. In other words, a cylinder set is a set of functions with restrictions put on a finite number of coordinates, or, in the language of Shiryayev ([36]), it is the set of functions that, at times $t_1, \ldots, t_n$, "get through the windows" $I_1, \ldots, I_n$ and at other times have arbitrary values.

An example of a Borel set from $\mathcal{B}(C)$ is

$$A = \{x : \sup x_t > a, t \ge 0\}.$$

Remark 2.1.7 Note that the set given by A depends on the behavior of functions on an uncountable set of points and would not be in the σ-field $\mathcal{B}(C)$ if $C[0, \infty)$ were replaced by the much larger space $\mathbb{R}^{[0, \infty)}$ (see Theorem 3, page 146 of [36]). In this latter space every Borel set is determined by restrictions imposed on the functions x, on an at most countable set of points $t_1, t_2, \ldots$. $\square$

Suppose the index parameter t is either a nonnegative integer or a nonnegative real number. The σ-fields $\mathcal{F}_t^X = \sigma\{X_u, u \le t\}$ are the smallest ones with respect to which the random variables $X_u, u \le t$, are measurable, and are naturally associated with any stochastic process $\{X_t\}$. $\mathcal{F}_t^X$ is sometimes called the *natural filtration* associated with the stochastic process $\{X_t\}$.

The σ-field $\mathcal{F}_t^X$ contains all the events which by time t are known to have occurred or not by observing X up to time t.

Often it is convenient to consider larger σ-fields than $\mathcal{F}_t^X$. For instance, $\{\mathcal{F}_t = \sigma\{X_u, Y_u; u \le t\}$ where $\{Y_t\}$ is another stochastic process.

Definition 2.1.8 *The stochastic process X is* adapted to the filtration $\{\mathcal{F}_t, \, t \ge 0\}$ *if for each $t \ge 0$ X_t is a $\mathcal{F}_t$-measurable random variable.*

Clearly X is adapted to $\mathcal{F}_t^X$. A function f is $\mathcal{F}_t^X$-measurable if the value of $f(\omega)$ can be decided by observing the history of X up to time t (and nowhere else). This follows from the multivariate version of Theorem 1.3.6. For instance, $f(\omega) = X_{t^2}(\omega)$ is $\mathcal{F}_t^X$-measurable for $0 < t < 1$ but it is not $\mathcal{F}_t^X$-measurable for $t \ge 1$.

As a function of two variables (t, ω), a stochastic process should be measurable with respect to both variables to allow a minimum of "good behavior".

Definition 2.1.9 *A stochastic process* $\{X_t\}$ *with* $t \in [0, \infty)$ *on a probability space* $\{\Omega, \mathcal{F}, P\}$ *is measurable if, for all Borel sets B in the Borel* σ*-field* $\mathcal{B}(\mathbb{R}^d)$,

$$\{(\omega, t) : X_t(\omega) \in B\} \in \mathcal{F} \otimes \mathcal{B}([0, \infty)).$$

If the probability space $\{\Omega, \mathcal{F}, P\}$ is equipped with a filtration $\{\mathcal{F}_t\}$ then a much stronger statement of measurability which relates measurability in t and ω with the filtration $\{\mathcal{F}_t\}$ is progressive measurability.

Definition 2.1.10 *A stochastic process* $\{X_t\}$ *on a filtered probability space* $\{\Omega, \mathcal{F}, \mathcal{F}_t, P\}$ *is progressively measurable if, for any* $t \in [0, \infty)$ *and for any set B in the Borel* σ*-field* $\mathcal{B}(\mathbb{R}^d)$,

$$\{(\omega, s) : s \leq t, X_s(\omega) \in B\} \in \mathcal{F}_t \otimes \mathcal{B}([0, t]).$$

Here $\mathcal{B}([0, t])$ *is the* σ*-field of Borel sets on the interval* $[0, t]$.

A measurable process need not be progressively measurable since $\sigma(X_t)$ may contain events not in $\mathcal{F}_t$.

Lemma 2.1.11 *If X is a progressively measurable stochastic process, then X is adapted.*

Proof   The map $\omega \to (s, \omega)$ from $\Omega \to [0, t] \times \Omega$ is $\mathcal{F}_t$-measurable. The map $(s, \omega) \to X_s(\omega)$ from $[0, t] \times \Omega$ to the state space of X is $\mathcal{F}_t$-measurable. By composition of the two maps the result follows.   ∎

Theorem 2.1.12 *If the stochastic process* $\{X_t : t \geq 0\}$ *on the filtered probability space* $\{\Omega, \mathcal{F}, \mathcal{F}_t, P\}$ *is measurable and adapted, then it has a progressively measurable modification.*

Proof   See [28] page 68.   ∎

Typically, in a description of a random process, the measure space and the probability measure on it are not given. One simply describes the family of joint distribution functions of every finite collection of random variables of the process. A basic question is whether there is a stochastic process with such a family of joint distribution functions. The following theorem ([36] page 244), due to Kolmogorov, guarantees us that this is the case if the joint distribution functions satisfy a set of natural consistency conditions.

Theorem 2.1.13 *(Kolmogorov Consistency Theorem) For all* $t_1, \ldots, t_k$, $k \in \mathbb{N}$, *in the time index T, let* $P_{t_1,\ldots,t_k}$ *be probability measures on* $(\mathbb{R}^k, \mathcal{B}(\mathbb{R}^k))$ *such that*

$$P_{t_{\sigma(1)},\ldots,t_{\sigma(k)}}(F_1 \times \cdots \times F_k) = P_{t_1,\ldots,t_k}(F_{\sigma^{-1}(1)} \times \cdots \times F_{\sigma^{-1}(k)}).$$

for all permutations σ *on* $\{1, 2, \ldots, k\}$ *and*

$$P_{t_1,\ldots,t_k}(F_1 \times \cdots \times F_k) = P_{t_1,\ldots,t_k,t_{k+1},\ldots,t_{k+m}}(F_1 \times \cdots \times F_k \times \mathbb{R}^n \times \cdots \times \mathbb{R}^n),$$

for all $m \in \mathbb{N}$, and the set on the right hand side has a total of $k + m$ factors. Then there is a unique probability measure P on the space $(\mathbb{R}^T, \mathcal{B}(\mathbb{R}^T))$ such that the restriction of P to any cylinder set $B_n = \{x \in \mathbb{R}^T : x_{t_1} \in I_1, x_{t_2} \in I_2, \ldots, x_{t_n} \in I_n\}$ is $P_{t_1,\ldots,t_n}$, that is

$$P(B_n) = P_{t_1,\ldots,t_n}(B_n).$$

Proof   See [36] page 167.   ∎

Theorem 2.1.14 (*Kolmogorov's Existence Theorem*). *For all $\tau_1, \ldots, \tau_k$, $k \in \mathbb{N}$ and τ in the time index let $P_{\tau_1,\ldots,\tau_k}$ be probability measures on $\mathbb{R}^{nk}$ such that*

$$P_{\tau_{\sigma(1)},\ldots,\tau_{\sigma(k)}}(F_1 \times \cdots \times F_k) = P_{\tau_1,\ldots,\tau_k}(F_{\sigma^{-1}(1)} \times \cdots \times F_{\sigma^{-1}(k)}),$$

for all permutations σ on $\{1, 2, \ldots, k\}$ and

$$P_{\tau_1,\ldots,\tau_k}(F_1 \times \cdots \times F_k) = P_{\tau_1,\ldots,\tau_k,\tau_{k+1},\ldots,\tau_{k+m}}(F_1 \times \cdots \times F_k \times \mathbb{R}^n \times \cdots \times \mathbb{R}^n),$$

for all $m \in \mathbb{N}$, and the set on the right hand side has a total of $k + m$ factors. Then there exist a probability space $(\Omega, \mathcal{F}, P)$ and a stochastic process $\{X_\tau\}$ on Ω into $\mathbb{R}^n$ such that

$$P_{\tau_1,\ldots,\tau_k}(F_1 \times \cdots \times F_k) = P[X_{\tau_1} \in F_1, \ldots, X_{\tau_k} \in F_k],$$

for all τ_i in the time set, $k \in \mathbb{N}$ and all Borel sets F_i.

Proof   The proof follows essentially from Theorems 1.3.9, 1.3.10 and 2.1.13. See [36] page 247.   ∎

Definition 2.1.15 *Suppose X is a stochastic process whose index set is the positive integers Z^+. Suppose $\mathcal{F}_n$ is a filtration. Then $\{X_n\}$ is predictable if X_n is $\mathcal{F}_{n-1}$-measurable, that is, $X_n(\omega)$ is known from observing events in $\mathcal{F}_{n-1}$ at time $n - 1$.*

In continuous time, without loss of generality, we shall take the time index set to be $[0, \infty)$.

In the continuous time case, roughly speaking, a stochastic process $\{X_t\}$ is predictable if knowledge about the behavior of the process is left-continuous, that is, X_t is $\mathcal{F}_{t-}$-measurable. Stated differently, for processes which are continuous on the left one may predict their value at each point by their values at preceding points. A Poisson process (see Section 2.10) is not predictable (its sample paths are right-continuous) otherwise we would be able to predict a jump time immediately before it jumps. More precisely, a stochastic process is predictable if it is measurable with respect to the σ-field on $\Omega \times [0, \infty)$ generated by the family of all left-continuous adapted stochastic processes.

A stochastic process X with continuous time parameter is *optional* if it is measurable with respect to the σ-field on $\Omega \times [0, \infty)$ generated by the family of all right-continuous, adapted stochastic processes which have left limits.

Definition 2.1.16 *A measurable stochastic process $\{X_t\}$ with values in $[0, \infty)$, is called an increasing process if almost every sample path $X(\omega)$ is right-continuous and increasing.*

Theorem 2.1.17 *Suppose $\{X_t\}$ is an increasing process. Then X_t has a unique decomposition as $X_t^c + X_t^d$, where $\{X_t^c\}$ is an increasing continuous process, and $\{X_t^d\}$ is an increasing purely discontinuous process, that is, $\{X_t^d\}$ is the sum of the jumps of $\{X_t\}$.*

If $\{X_t\}$ is predictable $\{X_t^d\}$ is predictable. If $\{X_t\}$ is adapted $\{X_t^c\}$ is predictable.

Proof   See [11] page 69.   ∎

2.2 Stopping times

One of the most important questions in the study of stochastic processes is the study of when a process hits a certain level or enters a certain region in its state space for the first time. Since for each possible trajectory, or realization ω, there is a hitting time (finite or infinite), the *hitting time* is a random variable taking values in the index, or time, space of the stochastic process.

Let $\mathbb{N}_\infty = \{1, 2, 3, \ldots, \infty\}$ and $\mathcal{F}_\infty = \sigma\{\cup_{n=1}^\infty \mathcal{F}_n\} \overset{\triangle}{=} \bigvee_{n=1}^\infty \mathcal{F}_n$.

A random variable α taking values in $\mathbb{N}_\infty$ is a *stopping time* (or *optional* or *Markov time*) with respect to a filtration $\{\mathcal{F}_n\}$ if for all $n \in \mathbb{N}_\infty$ we have $\{\omega : \alpha(\omega) \le n\} \in \mathcal{F}_n$. An equivalent definition in discrete time is to require $\{\omega : \alpha(\omega) = n\} \in \mathcal{F}_n$.

The concept of stopping time is directly related to the concept of the flow of information through time, that is, the filtration. The event $\{\omega : \alpha(\omega) \le n\}$ is $\mathcal{F}_n$-measurable, that is, measurable with respect to the information available up to time n. This means a stopping time is a nonanticipative function, whereas a general random variable may anticipate the future.

Example 2.2.1 Let $\{X_n, \mathcal{F}_n\}$ be an adapted process (i.e. $\{\mathcal{F}_n\}$ is a filtration and X_n is $\mathcal{F}_n$-measurable for all n). Suppose A is a measurable set of the state space of X. Then the random time

$$\alpha = \min\{k : X_k \in A\}$$

is a stopping time since

$$\{\alpha \le n\} = \bigcup_{k=1}^n \{X_k \in A\} \in \mathcal{F}_n.$$

□

If α is a stopping time with respect to a filtration $\mathcal{F}_n$ so is $\alpha + m$, $m \in \mathbb{N}$. However, $\alpha - m$, $m \in \mathbb{N}$ is not a stopping time since the event $\{\alpha - m = n\} = \{\alpha = n + m\}$ is not in $\mathcal{F}_n$; it is in $\mathcal{F}_{n+m}$ and hence anticipates the future.

In order to measure the information accumulated up to a stopping time we should define the σ-field $\mathcal{F}_\alpha$ of events prior to a stopping time α. Suppose that some event B is part of this information. This means that if $\alpha \le n$ we should be able to tell whether or not B has occurred. However, $\{\alpha \le n\} \in \mathcal{F}_n$ so that we should have $B \cap \{\alpha \le n\} \in \mathcal{F}_n$ and $B^c \cap \{\alpha \le n\} \in \mathcal{F}_n$. We, therefore, define:

$$\mathcal{F}_\alpha = \{A \in \mathcal{F}_\infty : A \cap \{\omega : \alpha(\omega) \le n\} \in \mathcal{F}_n \quad \forall n \ge 0\}.$$

The next examples should help to clarify this concept.

Example 2.2.2 Let $\Omega = \{\omega_i; i = 1, \ldots, 8\}$ and the time index $T = \{1, 2, 3\}$. Consider the following filtration:

$$\mathcal{F}_1 = \sigma\{\{\omega_1, \omega_2, \omega_3, \omega_4, \omega_5, \omega_6\}, \{\omega_7, \omega_8\}\},$$
$$\mathcal{F}_2 = \sigma\{\{\omega_1, \omega_2\}, \{\omega_3, \omega_4\}, \{\omega_5, \omega_6\}, \{\omega_7, \omega_8\}\},$$
$$\mathcal{F}_3 = \sigma\{\{\omega_1\}, \{\omega_2\}, \{\omega_3\}, \{\omega_4\}, \{\omega_5\}, \{\omega_6\}, \{\omega_7\}, \{\omega_8\}\}.$$

Now define the random variable

$$\alpha(\omega_1) = \alpha(\omega_2) = \alpha(\omega_5) = \alpha(\omega_6) = 2,$$
$$\alpha(\omega_3) = \alpha(\omega_4) = \alpha(\omega_7) = \alpha(\omega_8) = 3,$$

so that

$$\{\alpha = 0\} = \emptyset, \quad \{\alpha = 1\} = \emptyset,$$
$$\{\alpha = 2\} = \{\omega_1, \omega_2, \omega_5, \omega_6\},$$
$$\{\alpha = 3\} = \{\omega_3, \omega_4, \omega_7, \omega_8\},$$

and α is a stopping time.

Now $\mathcal{F}_\alpha = \{$ all events $A \in \mathcal{F}_\infty (= \mathcal{F}_3)$ such that for some n the event A is a subset of the event $\{\omega : \alpha(\omega) \le n\}$ $\}$. In our situation

$$\mathcal{F}_\alpha = \sigma\{\{\omega_1, \omega_2\}, \{\omega_5, \omega_6\}, \{\omega_3\}, \{\omega_4\}, \{\omega_7\}, \{\omega_8\}\}.$$

Note that the first two simple events of $\mathcal{F}_\alpha$, $\{\omega_1, \omega_2\}, \{\omega_5, \omega_6\}$, are in $\mathcal{F}_2$ and the rest are in $\mathcal{F}_3$ as they should be. Also, note that $\mathcal{F}_\alpha$ *is not* the σ-field generated by the random variable α. However, a closer look shows that α is $\mathcal{F}_\alpha$-measurable. If, for instance, the outcome is ω_1 then $\alpha = 2$ and $\alpha^{-1}(2) = \{\alpha = 2\} = \{\omega_1, \omega_2, \omega_5, \omega_6\}$ is an atom of the σ-field generated by the random variable α but not an atom of $\mathcal{F}_\alpha$. □

Example 2.2.3 Consider again the experiment of tossing a fair coin infinitely many times. Each ω is an infinite sequence of heads and tails and

$$\Omega = \{H, T\}^{\mathbb{N}}.$$

Define the filtration:

$\mathcal{F}_1 = \sigma\{\{\omega$ starting with $H\}, \{\omega$ starting with $T\}\},$
$\mathcal{F}_2 = \sigma\{\{\omega$ starting with $HH\}, \{\omega$ starting with $HT\}, \{\omega$ starting with $TH\},$
$\qquad \{\omega$ starting with $TT\}\}, \ldots,$
$\mathcal{F}_n = \sigma\{\{\omega$ starting with n fixed letters$\}\}$

Suppose that we win one dollar each time "Heads" comes up and lose one otherwise. Let $S_0 = 0$ and S_n be our fortune after the n-th toss. Define the random variable $\alpha = \inf\{n : S_n > 0\}$, which is the first time our winnings exceed our losses. Clearly, α is a stopping time with respect to the filtration $\mathcal{F}_n$.

Here

$$\mathcal{F}_\alpha = \sigma\{\{\omega \text{ starting with } H\}, \{\omega \text{ starting with } THH\},$$
$$\{\omega \text{ starting with } THTHH\}, \{\omega \text{ starting with } TTHHH\}, \ldots\}.$$

and

$\alpha(\omega$ starting with $H) = 1$,
$\alpha(\omega$ starting with $THH) = 3$,
$\alpha(\omega$ starting with $THTHH) = \alpha(\omega$ starting with $TTHHH) = 5$.

If $\omega = THTHH\ldots$, then the information at time $\alpha(THTHH\ldots) = 5$ is in $\mathcal{F}_5$ and is given by the event composed of all the smaller events starting with $THTHH$ and is an atom of $\mathcal{F}_\alpha$. However $\{\alpha = 5\} = \{\{THTHH\ldots\}, \{TTHHH\ldots\}\}$ which is not an atom of $\mathcal{F}_\alpha$.   □

If $\alpha \leq \beta$ are two stopping times then $\mathcal{F}_\alpha \subset \mathcal{F}_\beta$, because if $A \in \mathcal{F}_\alpha$,

$$A \bigcap \{\beta \leq n\} = (A \bigcap \{\alpha \leq n\}) \bigcap \{\beta \leq n\} \in \mathcal{F}_n \qquad (2.2.1)$$

for all n. From this result we see that if $\{\alpha_n\}$ is an increasing sequence of stopping times, the sequence $\{\mathcal{F}_{\alpha_n}\}$ is a filtration.

Example 2.2.4 Let $\Omega = \{\omega_i, i = 1, \ldots, 8\}$ and the time index $T = \{1, 2, 3, 4\}$. Consider the following filtration:

$$\mathcal{F}_1 = \sigma\{\{\omega_i, i = 1, \ldots, 6\}, \{\omega_7, \omega_8\}\},$$
$$\mathcal{F}_2 = \sigma\{\{\omega_1, \omega_2, \omega_3\}, \{\omega_4, \omega_5, \omega_6\}, \{\omega_7, \omega_8\}\},$$
$$\mathcal{F}_3 = \sigma\{\{\omega_1, \omega_2\}, \{\omega_3\}, \{\omega_4\}, \{\omega_5, \omega_6\}, \{\omega_7, \omega_8\}\},$$
$$\mathcal{F}_4 = \sigma\{\{\omega_1\}, \{\omega_2\}, \{\omega_3\}, \{\omega_4\}, \{\omega_5\}, \{\omega_6\}, \{\omega_7\}, \{\omega_8\}\}.$$

Now define the stopping times α_1 and α_2:

$$\alpha_1(\omega_1) = \alpha_1(\omega_2) = \alpha_1(\omega_3) = \alpha_1(\omega_4) = \alpha_1(\omega_5) = \alpha_1(\omega_6) = 2,$$
$$\alpha_1(\omega_7) = \alpha_1(\omega_8) = 3,$$
$$\alpha_2(\omega_1) = \alpha_2(\omega_2) = \alpha_2(\omega_3) = 2, \quad \alpha_2(\omega_5) = \alpha_2(\omega_6) = 3,$$
$$\alpha_2(\omega_4) = \alpha_2(\omega_7) = \alpha_2(\omega_8) = 4,$$

so that $\alpha_1 \leq \alpha_2$ and $\mathcal{F}_{\alpha_1} \subset \mathcal{F}_{\alpha_2}$, where

$$\mathcal{F}_{\alpha_1} = \sigma\{\{\omega_1, \omega_2, \omega_3\}, \{\omega_4, \omega_5, \omega_6\}, \{\omega_7, \omega_8\}\},$$
$$\mathcal{F}_{\alpha_2} = \sigma\{\{\omega_1, \omega_2, \omega_3\}, \{\omega_4\}, \{\omega_5, \omega_6\}, \{\omega_7\}, \{\omega_8\}\}.$$

□

For any Borel set B,

$$\{\omega : X_{\alpha(\omega)}(\omega) \in B\} = \bigcup_{n=0}^{\infty}\{X_n(\omega) \in B, \alpha(\omega) = n\} \in \mathcal{F},$$

that is, X_α is a random variable.

If X_∞ has been defined and $X_\infty \in \mathcal{F}_\infty = \bigvee_n \mathcal{F}_n$, then we define $X_\alpha(\omega) = X_{\alpha(\omega)}(\omega)$, i.e. $X_\alpha = \sum_{n \in \mathbb{N}_\infty} X_n I_{\{\alpha = n\}} \in \mathcal{F}_\alpha$, that is, X_α is $\mathcal{F}_\alpha$-measurable.

In the continuous time situation, definitions are more involved and the time parameter t plays a much more important role since continuity, limits etc. enter the scene. Let $\{\mathcal{F}_t\}$, $t \in [0, \infty)$ be a filtration. A nonnegative random variable α is called a stopping time with respect to the filtration $\mathcal{F}_t$ if for all $t \geq 0$ we have $\{\omega : \alpha(\omega) \leq t\} \in \mathcal{F}_t$.

A nonnegative random variable α is an *optional time* with respect to the filtration $\mathcal{F}_t$ if for all $t \geq 0$ we have $\{\omega : \alpha(\omega) < t\} \in \mathcal{F}_t$.

Every stopping time is optional, and the two concepts coincide if the filtration is right-continuous since $\{\omega : \alpha(\omega) \leq t\} \in \mathcal{F}_{t+\epsilon}$ for every $\epsilon > 0$, and hence $\{\omega : \alpha(\omega) \leq t\} \in \bigcap_{\epsilon > 0} \mathcal{F}_{t+\epsilon} = \mathcal{F}_{t+} = \mathcal{F}_t$ provided that $\mathcal{F}_t$ is right-continuous.

Example 2.2.5 Suppose $\{X_t, t \geq 0\}$ is continuous and adapted to the filtration $\{\mathcal{F}_t, t \geq 0\}$.

1. Consider $\alpha(\omega) = \inf\{t, X_t(\omega) = b\}$, the first time the process X hits level $b \in \mathbb{R}$ (*first passage time to a level $b \in \mathbb{R}$*). Then α is a stopping time since

$$\{\alpha \leq t\} = \bigcap_{n \in \mathbb{N}} \bigcup_{\{r \in Q, r \leq t\}} \{|X_r - b| \leq \frac{1}{n}\} \in \mathcal{F}_t.$$

2. Consider $\alpha(\omega) = \inf\{t, |X_t(\omega)| \geq 1\}$, the first time the process X leaves the interval $[-1, +1]$. Then α is a stopping time.
3. Consider $\alpha(\omega) = \inf\{t, \Delta X_t(\omega) > 1\}$ which is the first time the jump $\Delta X_t = X_t - X_{t-}$ exceeds 1. Then α is a stopping time. $\qquad\qquad$ □

Similarly to the discrete time case, the σ-field of events prior to a stopping time α is defined by

$$\mathcal{F}_\alpha = \{A \in \mathcal{F}_\infty : A \cap \{\omega : \alpha(\omega) \leq t\} \in \mathcal{F}_t, \quad \forall t \geq 0\}. \qquad (2.2.2)$$

Any stopping time α is $\mathcal{F}_\alpha$-measurable as, for $s \leq t$,

$$\{\omega : \alpha(\omega) \leq s\} \cap \{\omega : \alpha(\omega) \leq t\} = \{\omega : \alpha(\omega) \leq \min(t, s)\} \in \mathcal{F}_{\min(t,s)} \subset \mathcal{F}_t. \qquad (2.2.3)$$

Hence $\{\omega : \alpha(\omega) \leq s\} \in \mathcal{F}_\alpha$.

If α_1, α_2 are stopping times, then $\min(\alpha_1, \alpha_2)$, $\max(\alpha_1, \alpha_2)$ and $\alpha_1 + \alpha_2$ are stopping times as:

1. $\{\min(\alpha_1, \alpha_2) \leq t\} = \{\alpha_1 \leq t\} \cup \{\alpha_2 \leq t\} \in \mathcal{F}_t$,
2. $\{\max(\alpha_1, \alpha_2) \leq t\} = \{\alpha_1 \leq t\} \cap \{\alpha_2 \leq t\} \in \mathcal{F}_t$,
3. $\{\alpha_1 + \alpha_2 \leq t\} = \{\alpha_1 = 0, \alpha_2 = t\} \cup \{\alpha_2 = 0, \alpha_1 = t\}$
$$\cup (\bigcup_{p,q \in Q, p+q \leq t} (\{\alpha_1 \leq p\} \cap \{\alpha_2 \leq q\})),$$

where Q is the set of rational numbers.
4. If $\{\alpha_n\}$ is a sequence of stopping times then $\sup \alpha_n$ is a stopping time since $\{\sup \alpha_n \leq t\} = \bigcap_n \{\alpha_n \leq t\} \in \mathcal{F}_t$.
5. If α_1, α_2 are stopping times such that $\alpha_1 \leq \alpha_2$ then $\mathcal{F}_{\alpha_1} \subset \mathcal{F}_{\alpha_2}$.

Perhaps one of the most important applications of the concept of stopping time is the so-called *strong Markov property*.

A stochastic process $\{X_t\}$ is a *Markov process* if

$$E[f(X_{t+s}) \mid \mathcal{F}_t^X] = E[f(X_{t+s}) \mid X_t], \quad (P\text{-a.s.}) \tag{2.2.4}$$

where f is any bounded measurable function and $\mathcal{F}_t^X = \sigma\{X_u, u \le t\}$. Equation (2.2.4) is termed the *Markov property*.

A natural generalization of the Markov property is the *strong Markov property*, where the "present" time t in (2.2.4) is replaced by a stopping time and the "future" time $t + s$ is replaced by another later stopping time. That is, if α and β are stopping times and $\alpha \le \beta$,

$$E[X_\beta \mid \mathcal{F}_\alpha] = X_\alpha \text{ a.s.}$$

In other words a stochastic process $\{X_t\}$ has the strong Markov property if the information about the behavior of $\{X_t\}$ prior to the stopping time α is irrelevant in predicting its behavior after that time α once X_α is observed.

2.3 Discrete time martingales

Martingales are probably the most important type of stochastic processes used for modeling. They occur naturally in almost any information processing problem involving sequential acquisition of data: for example, the sequence of estimates of a random variable based on increasing observations, and the sequence of likelihood ratios in a sequential hypothesis test are martingales.

The stochastic process X is a *submartingale (supermartingale)* with respect to the filtration $\{\mathcal{F}_n\}$ if it is

1. $\mathcal{F}_n$-adapted,
2. $E[|X_n|] < \infty$ for all n, and
3. $E[X_n \mid \mathcal{F}_{n'}] \ge X_{n'}$ a.s. $(E[X_n \mid \mathcal{F}_{n'}] \le X_{n'}$ a.s.) for all $n' \le n$.

The stochastic process X is a *martingale* if it is a submartingale and a supermartingale.

If we recall the definition of conditional expectation we see that the requirement $E[X_{n+1} \mid \mathcal{F}_n] = X_n$ a.s. implies the following:

$$\int_F E[X_{n+1} \mid \mathcal{F}_n] dP = \int_F X_{n+1} dP, \quad F \in \mathcal{F}_n,$$

and

$$\int_F X_n dP = \int_F X_{n+1} dP, \quad F \in \mathcal{F}_n. \tag{2.3.1}$$

Since $\mathcal{F}_n \subset \mathcal{F}_{n+1} \subset \cdots \subset \mathcal{F}_{n+k}$, it easily seen that

$$\int_F X_n dP = \int_F X_{n+1} dP \cdots = \int_F X_{n+k} dP, \quad F \in \mathcal{F}_n. \tag{2.3.2}$$

and hence with probability 1 $E[X_{n+k} \mid \mathcal{F}_n] = X_n$. Setting $F = \Omega$ and $n = 1, 2, \ldots$ in (2.3.2) gives

$$E[X_1] = E[X_2] = \cdots = E[X_n].$$

A classical example of a martingale X is a player's fortune in successive plays of a fair game. If X_0 is the initial fortune, then "fair" means that, on average, the fortune at some future time n, after more plays, should be neither more nor less than X_0. If the game is favorable to the player, then his fortune should increase on average and X_n is a submartingale. If the game is unfavorable to the player, X_n is a supermartingale.

The following important inequality is used to prove a fundamental result on constructing a uniformly integrable family of random variables by conditioning a fixed (integrable) random variable on a family of sub-σ-fields.

Lemma 2.3.1 *(Jensen's Inequality). Suppose $X \in L^1$. If $\phi : \mathbb{R} \to \mathbb{R}$ is convex and $\phi(X) \in L^1$, then*

$$E[\phi(X) \mid \mathcal{G}] \geq \phi(E[X \mid \mathcal{G}]). \tag{2.3.3}$$

Proof (see, for example, [11]) Any convex function $\phi : \mathbb{R} \to \mathbb{R}$ is the supremum of a family of affine functions, so there exists a sequence (ϕ_n) of real functions with $\phi_n(x) = a_n x + b_n$ for each n, such that $\phi = \sup_n \phi_n$. Therefore $\phi(X) \geq a_n X + b_n$ holds a.s. for each (and hence all) n. So by the positivity of $E[. \mid \mathcal{G}]$, $E[\phi(X) \mid \mathcal{G}] \geq \sup_n (a_n E[X \mid \mathcal{G}] + b_n) = \phi(E[X \mid \mathcal{G}])$ a.s. ∎

Lemma 2.3.2 *Let $X \in L^p$, $p \geq 1$. The family*

$$\mathcal{L} = \{E[X \mid \mathcal{G}] : \mathcal{G} \text{ is a sub-}\sigma\text{-field of } \mathcal{F}\},$$

is uniformly integrable.

Proof Since $\phi(x) = |x|^p$ is convex, Jensen's Inequality 2.3.1 implies that

$$|E[X \mid \mathcal{G}]|^p \leq E[|X|^p \mid \mathcal{G}].$$

Hence

$$E[|E[X \mid \mathcal{G}]|^p] \leq E[E[|X|^p \mid \mathcal{G}]] = E[|X|^p],$$

that is, $E[|E[X \mid \mathcal{G}]|^p] < \infty$ for all $\mathcal{G}$. Thus the family $\mathcal{L}$ is L^p-bounded, hence uniformly integrable by Example 1.3.35. ∎

Specializing Lemma 2.3.2 to filtrations, we obtain an important type of martingale.

Example 2.3.3 Let $\{\mathcal{F}_n\}$ be a filtration, suppose $\mathcal{F}_\infty = \bigvee \mathcal{F}_n$ and $Y \in L^1(\Omega, \mathcal{F}_\infty)$. Define $X_n = E[Y \mid \mathcal{F}_n]$, $n \geq 1$.

Then $\{X_n, \mathcal{F}_n\}$, $n \geq 1$ is a martingale. To check this consider

$$E[X_{n+1} \mid \mathcal{F}_n] = E[E[Y \mid \mathcal{F}_{n+1}] \mid \mathcal{F}_n] = E[Y \mid \mathcal{F}_n] = X_n,$$

using property (1.4.1) of conditional expectations. Conversely, if the stochastic process $\{X_n\}$ is a martingale with respect to the filtration $\{\mathcal{F}_n\}$ and there exists an integrable random variable X such that

$$X_n = E[X \mid \mathcal{F}_n] \quad (P\text{-a.s.}) \quad n \geq 1,$$

then the martingale $\{X_n, \mathcal{F}_n\}$ is called *regular*. Regularity of a martingale $\{X_n, \mathcal{F}_n\}$ is in fact equivalent to the uniform integrability of the process $\{X_n\}$ by Lemma 2.3.2. In turn, this is equivalent to the convergence in L^1 of $\{X_n\}$ to X. (See [11].)   □

Example 2.3.4 Let $(\Omega, \mathcal{F}, P)$ be a probability space equipped with a filtration $\{\mathcal{F}_n\}$. Let $\overline{P}$ be another probability measure on $(\Omega, \mathcal{F})$ absolutely continuous with respect to P when both are restricted to $\mathcal{F}_n$ (i.e. $P(F) = 0$ then $\overline{P}(F) = 0$ for all $F \in \mathcal{F}_n$). Then from the Radon–Nikodym Theorem 1.3.25 there is an $\mathcal{F}_n$-measurable derivative Λ_n such that:

$$\overline{P}(F) = \int_F \Lambda_n(\omega)\mathrm{d}P, \quad F \in \mathcal{F}_n. \tag{2.3.4}$$

Similarly, there is an $\mathcal{F}_{n+1}$-measurable density Λ_{n+1}. Now, $\mathcal{F}_n \subset \mathcal{F}_{n+1}$ so that $F \in \mathcal{F}_{n+1}$ and (2.3.4) remains true if Λ_n is replaced with Λ_{n+1}:

$$\overline{P}(F) = \int_F \Lambda_n(\omega)\mathrm{d}P = \int_F \Lambda_{n+1}(\omega)\mathrm{d}P, \quad F \in \mathcal{F}_n.$$

which implies that $\{\Lambda_n\}$ is an $\{\mathcal{F}_n\}$ martingale.   □

Definition 2.3.5 *Let $\{X_n, \mathcal{F}_n\}$ be a submartingale. The number $C_n[a, b]$ of up-crossings of the interval $[a, b]$ by the sequence $X_1, \ldots, X_n$ is defined to be the largest positive integer k such that we can find*
$$0 \leq s_1 < t_1 < s_2 < t_2 < \cdots < s_k < t_k \leq n \text{ with } X_{s_i} < a, \, X_{t_i} > b, \text{ for } 1 \leq i \leq k.$$

The following theorem is a useful tool in proving convergence results for submartingales.

Theorem 2.3.6 *(Doob). If $\{X_n, \mathcal{F}_n\}$ is a submartingale then for all $n \geq 1$,*

$$E[C_n[a, b]] \leq \frac{E[X_n - a]^+}{b - a},$$

where and $[X_n - a]^+ = \max\{(X_n - a), 0\}$.

Proof   See [36] page 474.   ■

Theorem 2.3.7 *If $\{X_n, \mathcal{F}_n\}$ is a nonnegative martingale then $X_n \to X$ a.s., where X is an integrable random variable.*

Proof   Suppose that the event $\{\omega : \liminf X_n(\omega) < \limsup X_n(\omega)\} = \bigcup_{p<q}\{\omega : \liminf X_n(\omega) < p < q < \limsup X_n(\omega)\}$ has a positive probability. Then there exists values a and b such that

$$P(\{\omega : \liminf X_n(\omega) < a < b < \limsup X_n(\omega)\}) > 0. \tag{2.3.5}$$

This means that $\{X_n\}$ oscillates about or up-crosses the interval $[a, b]$ infinitely many times. However, using Theorem 2.3.6 and the fact that $\sup E[X_n] = E[X_1] < \infty$ we have:

$$\lim_n E[C_n[a, b]] \leq \lim_n \frac{E[X_n - a]^+}{b - a} \leq \frac{E[X_1] + |a|}{b - a} < \infty,$$

which contradicts (2.3.5), that is, $P(\{\omega : \liminf X_n(\omega) < \limsup X_n(\omega)\}) = 0$. Hence $\lim_n X_n = X$ a.s. To finish the proof we must show that $E[|X|] < \infty$. This follows from Fatou's Lemma 1.3.16. $\blacksquare$

Theorem 2.3.8 *Let $(\Omega, \mathcal{F}, P)$ be a probability space equipped with a filtration $\{\mathcal{F}_n\}$. Write $\mathcal{F}_\infty = \bigvee \mathcal{F}_n \subset \mathcal{F}$. Let $\overline{P}$ be another probability measure on $(\Omega, \mathcal{F})$ which is absolutely continuous with respect to P when both are restricted to $\mathcal{F}_n$ for each n (i.e. $P(F) = 0$ then $\overline{P}(F) = 0$ for all $F \in \mathcal{F}_n$). Suppose Λ_n are the corresponding Radon–Nikodym derivatives. Then Λ_n converges to an integrable random variable Λ with probability 1. Moreover, if $\overline{P}$ is absolutely continuous with respect to P on $\mathcal{F}_\infty$ then Λ is the corresponding Radon–Nikodym derivative.*

Proof The first statement of the theorem follows from Theorem 2.3.7; the second statement follows from Theorem 3, page 478 of Shiryayev [36]. See also Example 2.3.4. $\blacksquare$

Returning to Example (1.3.29):

Example 2.3.9 Suppose $(\Omega, \mathcal{F}, P)$ is a probability space on which is defined a sequence of random variables $Y_1, Y_2, \ldots$ and $\mathcal{F}_n = \sigma\{Y_1, Y_2, \ldots, Y_n\}$. Let $\overline{P}$ be another probability measure on $\mathcal{F}$. Suppose that under P and $\overline{P}$ the random vector $\{Y_1, Y_2, \ldots, Y_n\}$ has densities $f_n(.)$ and $\overline{f}_n$ respectively with respect to n-dimensional Lebesgue measure. Then by Theorem 2.3.8 the Radon–Nikodym derivatives

$$\frac{d\overline{P}}{dP}\Big|_{\mathcal{F}_n} \stackrel{\triangle}{=} \Lambda_n = \frac{\overline{f}_n(Y_1, Y_2, \ldots, Y_n)}{f_n(Y_1, Y_2, \ldots, Y_n)}$$

converge to an integrable and $\mathcal{F}_\infty$-measurable random variable Λ. $\square$

Example 2.3.10 If $\{X_n\}$ is an integrable, real valued process with independent increments having mean 0 then it is a martingale with respect to the filtration it generates. If, in addition, X_n^2 is integrable then $X_n^2 - E(X_n^2)$ is a martingale with respect to the same filtration. The proof is left as an exercise. $\square$

Theorem 2.3.11 *If $\{X_n, \mathcal{F}_n\}$ is a martingale and α is a stopping time with respect to the filtration $\mathcal{F}_n$, then $\{X_{\min(n,\alpha)}, \mathcal{F}_n\}$ is a martingale.*

Proof First we have to show that $X_{\min(n,\alpha)}$ is integrable. But $X_{\min(n,\alpha)} = \sum_{k=0}^{n-1} X_k + X_n I_{\{\alpha \geq n\}}$ and by assumption the variables $X_0, \ldots, X_n$ are integrable. Hence $X_{\min(n,\alpha)}$ is integrable. Moreover, $X_{\min(n,\alpha)}$ is $\mathcal{F}_n$-measurable. It remains to show that $E[X_{\min(n+1,\alpha)} \mid \mathcal{F}_n] = X_{\min(n,\alpha)}$. This follows from

$$E[X_{\min(n+1,\alpha)} - X_{\min(n,\alpha)} \mid \mathcal{F}_n] = E[I_{\alpha > n}(X_{n+1} - X_n) \mid \mathcal{F}_n]$$
$$= I_{\{\alpha > n\}} E[(X_{n+1} - X_n) \mid \mathcal{F}_n] = 0,$$

since $\{\alpha > n\} \in \mathcal{F}_n$.   ■

We also have that stopping at an optional time preserves the martingale property.

Theorem 2.3.12 (*Doob Optional Sampling Theorem*). *Suppose $\{X_n, \mathcal{F}_n\}$ is a martingale. Let $\alpha \leq \beta$ (a.s.) be stopping times such that X_α and X_β are integrable. Also suppose that*

$$\liminf \int_{\{\alpha \geq n\}} |X_n| dP \to 0, \tag{2.3.6}$$

and

$$\liminf \int_{\{\beta \geq n\}} |X_n| dP \to 0. \tag{2.3.7}$$

Then

$$E[X_\beta \mid \mathcal{F}_\alpha] = X_\alpha. \tag{2.3.8}$$

In particular $E[X_\beta] = E[X_\alpha]$.

Proof Using the definition of conditional expectation, we have to show that for every $A \in \mathcal{F}_\alpha$,

$$\int_A I_{\{\alpha \leq \beta\}} E[X_\beta \mid \mathcal{F}_\alpha] dP = \int_A I_{\{\alpha \leq \beta\}} X_\beta dP = \int_A I_{\{\alpha \leq \beta\}} X_\alpha dP.$$

However, $\{\alpha \leq \beta\} = \bigcup_{n \geq 0} \{\alpha = n\} \cap \{\beta \geq n\}$. Hence it suffices to show that, for all $n \geq 0$:

$$\int_A I_{\{\alpha = n\} \cap \{\beta \geq n\}} X_\beta dP = \int_A I_{\{\alpha = n\} \cap \{\beta \geq n\}} X_\alpha dP$$

$$= \int_A I_{\{\alpha = n\} \cap \{\beta \geq n\}} X_n dP. \tag{2.3.9}$$

Now, $\{\omega : \beta(\omega) \geq n\} = \{\omega : \beta(\omega) = n\} \bigcup \{\omega : \beta(\omega) \geq n + 1\}$ and in view of (2.3.1), the last integral in (2.3.9) is equal to

$$\int_{A \cap \{\alpha = n\} \cap \{\beta = n\}} X_n dP + \int_{A \cap \{\alpha = n\} \cap \{\beta \geq n+1\}} X_{n+1} dP$$

$$= \int_{A \cap \{\alpha = n\} \cap \{\beta = n\}} X_\beta dP + \int_{A \cap \{\alpha = n\} \cap \{\beta \geq n+1\}} X_{n+1} dP. \tag{2.3.10}$$

Also, $\{\omega : \beta(\omega) \geq n\} = \{\omega : n \leq \beta(\omega) \leq n+1\} \bigcup \{\omega : \beta(\omega) \geq n+2\}$ and using (2.3.1) again, (2.3.10) equals

$$\int_{A \cap \{\alpha=n\} \cap \{n \leq \beta \leq n+1\}} X_\beta dP + \int_{A \cap \{\alpha=n\} \cap \{\beta \geq n+2\}} X_{n+2} dP.$$

Repeating this step k times,

$$\int_A I_{\{\alpha=n\} \cap \{\beta \geq n\}} X_n dP = \int_{A \cap \{\alpha=n\} \cap \{n \leq \beta \leq n+k\}} X_\beta dP$$

$$+ \int_{A \cap \{\alpha=n\} \cap \{\beta \geq n+k+1\}} X_{n+k+1} dP,$$

that is

$$\int_{A \cap \{\alpha=n\} \cap \{n \leq \beta \leq n+k\}} X_\beta dP = \int_{A \cap \{\alpha=n\} \cap \{\beta \geq n\}} X_n dP$$

$$- \int_{A \cap \{\alpha=n\} \cap \{\beta \geq n+k+1\}} X_{n+k+1} dP.$$

Now,

$$X_{n+k+1} = X_{n+k+1}^+ - X_{n+k+1}^-$$

$$= 2X_{n+k+1}^+ - (X_{n+k+1}^+ + X_{n+k+1}^-) = 2X_{n+k+1}^+ - |X_{n+k+1}|$$

so that

$$\int_{A \cap \{\alpha=n\} \cap \{n \leq \beta \leq n+k\}} X_\beta dP = \int_{A \cap \{\alpha=n\} \cap \{\beta \geq n\}} X_n dP$$

$$- 2 \int_{A \cap \{\alpha=n\} \cap \{\beta \geq n+k+1\}} X_{n+k+1}^+ dP$$

$$+ \int_{A \cap \{\alpha=n\} \cap \{\beta \geq n+k+1\}} |X_{n+k+1}| dP. \qquad (2.3.11)$$

Taking the limit when $k \to \infty$ of both sides of (2.3.11) and using (2.3.7), we obtain

$$\int_{A \cap \{\alpha=n\} \cap \{n \leq \beta\}} X_\beta dP = \int_{A \cap \{\alpha=n\} \cap \{n \leq \beta\}} X_n dP,$$

which establishes (2.3.9) and finishes the proof. ∎

Definition 2.3.13 *The stochastic process $\{X_n, \mathcal{F}_n\}$ is a local martingale if there is a sequence of stopping times $\{\alpha_k\}$ increasing to ∞ with probability 1 and such that $\{X_{n \wedge \alpha_k}, \mathcal{F}_n\}$ is a martingale.*

Remark 2.3.14 The interesting fact about local martingales is that they can be obtained rather naturally through a martingale transform (stochastic integral in the continuous time case) which is defined as follows. Suppose $\{Y_n, \mathcal{F}_n\}$ is a martingale and $\{A_n, \mathcal{F}_n\}$ is a predictable process. Then the sequence

$$X_n = A_0 Y_0 + \sum_{k=1}^{n} A_k (Y_k - Y_{k-1})$$

is called a *martingale transform* and is a local martingale.

Proof   To show that $\{X_n, \mathcal{F}_n\}$ is a local martingale we have to find a sequence of stopping times $\{\alpha_k\}$, $k \geq 1$, increasing to infinity (P-a.s.) and such that the "stopped" process $\{X_{\min(n,\alpha_k)}, \mathcal{F}_n\}$ is a martingale. Let $\alpha_k = \inf\{n \geq 0 : |A_{n+1}| > k\}$. Since A is predictable the α_k are stopping times and clearly $\alpha_k \uparrow \infty$ (P-a.s.). Since Y is a martingale and $|A_{\min(n,\alpha_k)} I_{\{\alpha_k > n\}}| \leq k$ then, for all $n \geq 1$,

$$E[|X_{\min(n,\alpha_k)} I_{\{\alpha_k > n\}}| < \infty.$$

Moreover, from Theorem 2.3.11,

$$E[(X_{\min(n+1,\alpha_k)} - X_{\min(n,\alpha_k)}) I_{\{\alpha_k > n\}} \mid \mathcal{F}_n]$$
$$= I_{\{\alpha_k > n\}} A_{\min(n+1,\alpha_k)} E[Y_{\min(n+1,\alpha_k)} - Y_{\min(n,\alpha_k)} \mid \mathcal{F}_n] = 0.$$

This finishes the proof.   ∎

◻

Example 2.3.15   Suppose that you are playing a game using the following "strategy". At each time n your stake is A_n. Write X_n for the state of your total gain through the n-th game with $X_0 = 0$ for simplicity.

Write $\mathcal{F}_n = \sigma\{X_k : 0 \leq k \leq n\}$. We suppose for each n, A_n is $\mathcal{F}_{n-1}$ measurable, that is $A = \{A_n\}$ is predictable with respect to the filtration $\mathcal{F}_n$. This means that $A_n = A_n(X_0, X_1, \ldots, X_{n-1})$ is a function of $X_0, X_1, \ldots, X_{n-1}$.

If we assume that you win (or lose) at time n if a Bernouilli random variable b_n is equal to 1 (or -1), then

$$X_n = \sum_{k=1}^{n} A_k b_k = \sum_{k=1}^{n} A_k \Delta C_k.$$

Here $\Delta C_k = C_k - C_{k-1}$ and $C_k = \sum_{i=1}^{k} b_i$. If C is a martingale with respect to the filtration $\mathcal{F}_n$ (in this case we say that the game is "fair"), then the same thing holds for X because

$$E[X_n \mid \mathcal{F}_{n-1}] = X_{n-1} + A_n E[C_n - C_{n-1} \mid \mathcal{F}_{n-1}]$$
$$= X_{n-1} + A_n(E[C_n \mid \mathcal{F}_{n-1}] - C_{n-1})$$
$$= X_{n-1} + A_n(C_{n-1} - C_{n-1}) = X_{n-1}.$$

◻

2.4 Doob decomposition

A submartingale is a process which "on average" is nondecreasing. Unlike a martingale, which has a constant mean over time, a submartingale has a trend or an increasing predictable part perturbated by a martingale component which is not predictable. This is made more precise by the following theorem due to J. L. Doob.

Theorem 2.4.1 (*Doob Decomposition*). *Any submartingale* $\{X_n\}$ *can be written* (*P-a.s. uniquely*) *as*

$$X_n = Y_n + Z_n, \quad a.s. \tag{2.4.1}$$

where $\{Y_n\}$ *is a martingale and* $\{Z_n\}$ *is a predictable, increasing process, i.e.* $E(Z_n) < \infty$, $Z_1 = 0$ *and* $Z_n \leq Z_{n+1}$ *a.s.* $\forall n$.

Proof Write $\Delta_n = X_n - X_{n-1}$, $y_i = \Delta_i - E[\Delta_i \mid \mathcal{F}_{i-1}]$ and $z_i = E[\Delta_i \mid \mathcal{F}_{i-1}]$, $z_0 = 0$. Then:

$$X_n = \Delta_1 - E[\Delta_1 \mid \mathcal{F}_0] + \Delta_2 - E[\Delta_2 \mid \mathcal{F}_1]$$

$$+ \cdots + \Delta_n - E[\Delta_n \mid \mathcal{F}_{n-1}] + \sum_{i=1}^{n} E[\Delta_i \mid \mathcal{F}_{i-1}]$$

$$= \sum_{i=1}^{n} y_i + \sum_{i=1}^{n} z_i$$

$$\stackrel{\triangle}{=} Y_n + Z_n,$$

To prove uniqueness suppose that there is another decomposition $X_n = Y'_n + Z'_n = \sum_{i=1}^{n} y'_i + \sum_{i=1}^{n} z'_i$. Let $y_n + z_n = x_n = y'_n + z'_n$ and take conditional expectation with respect to $\mathcal{F}_{n-1}$ to get $z_n = z'_n$, because y'_n is a martingale increment and z'_n is predictable. This implies $y'_n = y_n$ and the uniqueness of the decomposition. ∎

Remarks 2.4.2

1. In Theorem 2.4.1 if $\{X_n\}$ is just an $\mathcal{F}_n$-adapted and integrable process the decomposition remains valid but we lose the "increasing" property of the process $\{Z_n\}$.
2. The process $X - Z$ is a martingale; as a result Z is called the *compensator* of the submartingale X.
3. A processes which is the sum of a predictable process and a martingale is called a *semimartingale*.
4. Uniqueness of the decomposition is ensured by the predictability of the process $\{Z_n\}$.

☐

Definition 2.4.3 *A discrete-time stochastic process* $\{X_n\}$, *with finite-state space* $S = \{s_1, s_2, \ldots, s_N\}$, *defined on a probability space* $(\Omega, \mathcal{F}, P)$ *is a Markov chain if*

$$P(X_{n+1} = s_{i_{n+1}} \mid X_0 = s_{i_0}, \ldots, X_n = s_{i_n}) = P(X_{n+1} = s_{i_{n+1}} \mid X_n = s_{i_n}),$$

for all $n \geq 0$ *and all states* $s_{i_0}, \ldots, s_{i_n}, s_{i_{n+1}} \in S$. *This is termed the Markov property.*
 $\{X_n\}$ *is a homogeneous Markov chain if*

$$P(X_{n+1} = s_j \mid X_n = s_i) \stackrel{\triangle}{=} \pi_{ji}$$

is independent of n.

The matrix $\Pi = \{\pi_{ji}\}$ *is called the probability transition matrix of the homogeneous Markov chain and it satisfies the property* $\sum_{j=1}^{N} \pi_{ji} = 1$.

Note that our transition matrix Π is the transpose of the traditional transition matrix defined elsewhere. The convenience of this choice will be apparent later.

The following properties of a homogeneous Markov chain are easy to check.

1. Let $\pi^0 = (\pi_1^0, \pi_2^0, \ldots, \pi_N^0)'$ be the distribution of X_0. Then
$$P(X_0 = s_{i_0}, X_1 = s_{i_1}, \ldots, X_n = s_{i_n}) = \pi_{i_0}^0 \pi_{i_0 i_1} \ldots \pi_{i_{n-1} i_n}.$$
2. Let $\pi^n = (\pi_1^n, \pi_2^n, \ldots, \pi_N^n)'$ be the distribution of X_n. Then

$$\pi^n = \Pi^n \pi^0 = \Pi \pi^{n-1}.$$

Example 2.4.4 Let $\{\eta_n\}$ be a discrete-time Markov chain as in Definition 2.4.3. Consider the filtration $\{\mathcal{F}_n\} = \sigma\{\eta_0, \eta_1, \ldots, \eta_n\}$.

Write $X_n = (I_{(\eta_n = s_1)}, I_{(\eta_n = s_2)}, \ldots, I_{(\eta_n = s_N)})$.

Then X_n is a discrete-time Markov chain with state space the set of unit vectors $e_1 = (1, 0, \ldots, 0)', \ldots, e_N = (0, \ldots, 1)'$ of $\mathbb{R}^N$. However, the probability transitions matrix of X is Π. We can write:

$$E[X_n \mid \mathcal{F}_{n-1}] = E[X_n \mid X_{n-1}] = \Pi X_{n-1}, \tag{2.4.2}$$

from which we conclude that ΠX_{n-1} is the predictable part of X_n, given the history of X up to time $n - 1$ and the nonpredictable part of X_n must be $M_n \overset{\triangle}{=} X_n - \Pi X_{n-1}$. In fact it can be easily shown that $M_n \in \mathbb{R}^N$ is a mean 0, $\mathcal{F}_n$-vector martingale and we have the semimartingale (or Doob decomposition) representation of the Markov chain $\{X_n\}$,

$$X_n = \Pi X_{n-1} + M_n. \tag{2.4.3}$$

$\square$

Definition 2.4.5 *Given two (column) vectors X and Y the tensor or Kronecker product $X \otimes Y$ is the (column) vector obtained by stacking the rows of the matrix XY', where $'$ is the transpose, with entries obtained by multiplying the i-th entry of X by the j-th entry of Y.*

Example 2.4.6 Let $\{X_n\}$ be an order-2 Markov chain (see (2.4.4) below) with state space the standard basis of $\mathbb{R}^2$ $\{e_1, e_2\}$ on a filtered probability space $(\Omega, \mathcal{F}, \mathcal{F}_n, P)$, $\mathcal{F}_n = \sigma\{X_0, X_1, \ldots, X_n\}$ such that

$$P(X_n = e_k \mid \mathcal{F}_{n-1}) = P(X_n = e_k \mid X_{n-2}, X_{n-1}), \tag{2.4.4}$$

and probability transitions matrix

$$\Pi = \{\pi_{k,ji}\}, \quad \sum_k \pi_{k,ji} = 1, \quad i, j, k = 1, 2$$

or

$$\Pi = \begin{bmatrix} \pi_{1,11} & \pi_{1,12} & \pi_{1,21} & \pi_{1,22} \\ \pi_{2,11} & \pi_{2,12} & \pi_{2,21} & \pi_{2,22} \end{bmatrix}.$$

Lemma 2.4.7 *A semimartingale representation (or Doob decomposition) of the order-2 Markov chain X is:*

$$X_n = \Pi(X_{n-2} \otimes X_{n-1}) + M_n, \tag{2.4.5}$$

that is $M_n \triangleq X_n - \Pi X_{n-2} \otimes X_{n-1}$ *is an* $\mathcal{F}_n$-*martingale.* $(X_{n-2} \otimes X_{n-1})$ *is the tensor, or Kronecker, product of the vectors* X_{n-1}, X_{n-2}. *This can be identified with one of the standard unit vectors* $\{e_1, e_2, e_3, e_4\}$ *of* $\mathbb{R}^4$, *that is*

$$e_1 \otimes e_1 = (1, 0, 0, 0)', \quad e_1 \otimes e_2 = (0, 1, 0, 0)',$$
$$e_2 \otimes e_1 = (0, 0, 1, 0)', \quad e_2 \otimes e_2 = (0, 0, 0, 1)'.$$

Proof

$$E[X_n \mid \mathcal{F}_{n-1}] = E[X_n \mid X_{n-2}, X_{n-1}]$$
$$= \sum_{ij} E[X_n \mid X_{n-2} = e_i, X_{n-1} = e_j]I_{\{X_{n-2}=e_i, X_{n-1}=e_j\}}$$
$$= \sum_k \sum_{ij} e_k \pi_{k,ji} I_{\{X_{n-2}=e_i, X_{n-1}=e_j\}}$$
$$= \sum_{ij} (\pi_{1,ji}, \pi_{2,ji}) I_{\{X_{n-2}=e_i, X_{n-1}=e_j\}}$$
$$= \Pi \sum_{ij} e_i \otimes e_j I_{\{X_{n-2}=e_i, X_{n-1}=e_j\}} = \Pi X_{n-2} \otimes X_{n-1}.$$

■
□

2.5 Continuous time martingales

The stochastic process X is a *submartingale (supermartingale)* with respect to the filtration $\{\mathcal{F}_t\}$ if

1. it is $\mathcal{F}_t$-adapted, $E[|X_t|] < \infty$ for all t and
2. $E[X_t \mid \mathcal{F}_{t'}] \geq X_{t'}$ $(E[X_t \mid \mathcal{F}_{t'}] \leq X_{t'})$ for all $t' \leq t$.

The stochastic process X is a *martingale* if it is a submartingale and a supermartingale.
  Since for a martingale $E[X_t \mid \mathcal{F}_s] = X_s$, it follows that

$$E[E[X_t \mid \mathcal{F}_s]] = E[X_s],$$

and $E[X_t] = E[X_s]$ for all $s \geq 0$, so that

$$E[X_t] = E[X_0] \text{ for all } t \geq 0.$$

Example 2.5.1 If X is an integrable random variable on a filtered probability space then $X_t \stackrel{\triangle}{=} E[X \mid \mathcal{F}_t]$ is a martingale, since for $s \leq t$,

$$E[X_t \mid \mathcal{F}_s] = E[E[X \mid \mathcal{F}_t] \mid \mathcal{F}_s] = E[X \mid \mathcal{F}_s] = X_s. \qquad \square$$

An important application of Example 2.5.1 is

Example 2.5.2 Let $(\Omega, \mathcal{F}, P, \overline{P})$ be a probability space with a filtration $\{\mathcal{F}_t, t \geq 0\}$ and two probability measures such that $\overline{P} \ll P$. Then the Radon–Nikodym Theorem asserts the existence of a nonnegative random variable Λ such that for all $F \in \mathcal{F}$,

$$\overline{P}(F) = \int_F \Lambda(\omega) dP(\omega).$$

Then $\Lambda_t \stackrel{\triangle}{=} E[\Lambda \mid \mathcal{F}_t]$ is a nonnegative martingale with mean $E[\Lambda_t] = \int_\Omega \Lambda_t(\omega) dP(\omega) = \int_\Omega \Lambda(\omega) dP(\omega) = 1.$ $\qquad \square$

Example 2.5.3 Let $\{X_t\}$ be a stochastic process adapted to the filtration $\{\mathcal{F}_t\}$ with independent increments, that is, for $s \leq t$, $X_t - X_s$ is independent of the σ-field $\mathcal{F}_s$. Then the process $\{X_t - E[X_t]\}$ is an $\mathcal{F}_t$-martingale since

$$\begin{aligned}
E[X_t - E[X_t] \mid \mathcal{F}_s] &= E[X_t - E[X_t] - (X_s - E[X_s]) + (X_s - E[X_s]) \mid \mathcal{F}_s] \\
&= X_s - E[X_s] + E(X_t - X_s) - E(X_t - X_s) = X_s - E[X_s].
\end{aligned}$$

$\qquad \square$

The following martingale convergence result is proved in, for instance, [6] page 16.

Theorem 2.5.4 *(Martingale Convergence Theorem). Let* $\{X_t, \mathcal{F}_t\}$, $t \geq 0$, *be a martingale with right-continuous sample paths. If* $\sup_t E[|X_t|] < \infty$ *then there is a random variable* $X_\infty \in L^1$ *such that* $\lim_{t \to \infty} X_t = X_\infty$ *a.s. Furthermore, if* $\{X_t, \mathcal{F}_t\}$, $t \geq 0$ *is uniformly integrable then* $(X_t \stackrel{L^1}{\to} X_\infty)$ *and* $E[|X_t|]$ *increases to* $E[|X_\infty|]$ *as* $t \to \infty$.

Theorem 2.5.5 *(Stopped Martingales are Martingales). Let* $\{X_t, \mathcal{F}_t\}$ *be a martingale with right-continuous sample paths and* α *a stopping time. The stopped process* $\{X_{t \wedge \alpha}, t \geq 0\}$ *is also a martingale.*

Proof   See [34] page 189.   ∎

Theorem 2.5.6 *(Optional Stopping). Let* $\{X_t, \mathcal{F}_t, t \geq 0\}$ *be a right-continuous martingale with a last element* X_∞, *and let* $\alpha \leq \beta$ *be two stopping times. Then*

$$E[X_\beta \mid \mathcal{F}_\alpha] = X_\alpha \quad a.s.$$

In particular, we have $E[X_\beta] = X_0$.

Proof   See [21] page 19.   ∎

Now we give a characterization of a uniformly integrable martingale. We need this result to prove Theorem 3.5.3

Theorem 2.5.7 *Suppose* $\{X_t\}$, $0 \le t \le \infty$, *is an adapted right-continuous process such that for every stopping time* α, $E[|X_\alpha|] < \infty$ *and* $E[X_\alpha] = 0$. *Then* $\{X_t\}$ *is a uniformly integrable martingale.*

Proof Consider any time $t \in [0, \infty]$ and $F \in \mathcal{F}_t$. Let

$$\alpha(\omega) = t I_{\{\omega \in F\}} + \infty I_{\{\omega \notin F\}}.$$

Then α is a stopping time and by assumption

$$0 = E[X_\alpha] = E[X_t I_{\{\omega \in F\}}] + E[X_\infty I_{\{\omega \notin F\}}] = E[X_\infty] = E[X_\infty I_{\{\omega \in F\}}] + E[X_\infty I_{\{\omega \notin F\}}].$$

Hence $E[X_t I_{\{\omega \in F\}}] = E[X_\infty I_{\{\omega \in F\}}]$ for all $F \in \mathcal{F}_t$, so $X_t = E[X_\infty \mid \mathcal{F}_t]$ a.s. ∎

Recall that the definition of a martingale involves the integrability of X_t, for all t which in fact is a sufficient condition for the existence of $E[X_t \mid \mathcal{F}_s]$, $s \le t$. However, $E[X_t \mid \mathcal{F}_s]$, $s \le t$ may exist even though $E[|X_t|] = \infty$, in which case $\{X_t, \mathcal{F}_t\}$ is called a *local martingale*. First recall the concept of local properties of deterministic functions.

The (deterministic) function $X_t = e^t/(t-1)$ is locally bounded, i.e. it is bounded on compact sets not containing 1 (closed bounded intervals in $\mathbb{R} - \{1\}$). In fact we can define, for each $n \in \mathbb{N}$:

$$Y_t^n = X_t I_{[|X_t| \le n]} + n I_{[X_t > n]} - n I_{[X_t < -n]}.$$

Clearly Y_t^n is bounded everywhere and equals X_t on closed bounded intervals. However, for Y_t^n to converge to X_t we must allow n to increase to infinity. The same idea is used when $X_t(\omega)$ is a random function or a stochastic process. However, the *localizing sequence* is then a sequence of random variables, in fact stopping times. For example, consider

$$\alpha_n(\omega) = \inf\{t : |X_t(\omega)| > n\},$$

which is the first time the sample path $X_t(\omega)$ leaves the interval $[-n, +n]$. Then define

$$Y_t^n(\omega) = X_{t \wedge \alpha_n(\omega)}(\omega),$$

so that for different ωs there are, for each n, different times t when $X_t(\omega)$ leaves the bounded set $[-n, n]$. As in the deterministic case, the sequence of stopping times $\alpha_n(\omega)$ must increase to infinity for almost all ω. Here $x \wedge y$ stands for the smaller of x and y.

Definition 2.5.8 *The stochastic process* $X = \{X_t\}$, $t \ge 0$, *is said to be square integrable if* $\sup_t E[X_t^2] < \infty$.

Definition 2.5.9 *The stochastic process* $\{X_t, \mathcal{F}_t\}$ *is a local martingale if there is a sequence of stopping times* $\{\alpha_n\}$ *increasing to* ∞ *with probability 1 and such that for each n,* $\{X_{t \wedge \alpha_n}, \mathcal{F}_t\}$ *is a martingale.*

Definition 2.5.10 *The stochastic process* $\{X_t, \mathcal{F}_t\}$ *is a locally square integrable martingale (i.e. locally in L^2) if there is a sequence of stopping times* $\{\alpha_n\}$ *increasing to* ∞ *with probability 1 and such that for each n,* $\{X_{t \wedge \alpha_n}, \mathcal{F}_t\}$ *is a square integrable martingale.*

The following two theorems, whose proofs can be found in [11], are needed in the proof of Theorem 3.5.6.

Theorem 2.5.11 *Let $\{X_t, \mathcal{F}_t\}$ be a local martingale which is zero at time $t = 0$. Then there exists a sequence of stopping times $\{\alpha_n\}$ increasing to ∞ with probability 1 and such that for each n, $\{X_{t \wedge \alpha_n}, \mathcal{F}_t\}$ is a uniformly integrable martingale and $E[X_{t \wedge \alpha_n} \mid \mathcal{F}_t]$ is bounded on the stochastic interval $\{(t, \omega) \in [0, \infty[\times \Omega : 0 \leq t < \alpha_n(\omega))$ (denoted $[\![0, \alpha_n[\![)$.*

Theorem 2.5.12 *Let $\{X_t, \mathcal{F}_t\}$ be a local martingale. Then there exists a sequence of stopping times $\{\alpha_n\}$ increasing to ∞ with probability 1 such that for each n, $X_{\{\alpha_n \wedge t\}} = U_{\{\tau_n \wedge t\}} + V_{\{\alpha_n \wedge t\}}$, where $U_0 = 0$, $U_{\{\alpha_n \wedge t\}}$ is square integrable and $V_{\{\alpha_n \wedge t\}}$ is a martingale of integrable variation which is zero at $t = 0$.*

2.6 Doob–Meyer decomposition

The following definitions are needed in the sequel.

Definition 2.6.1 *Let f be a real valued function on an interval $[a, b]$. The* variation *of f on the interval $[a, b]$ is given by*

$$\lim_{n \to \infty} \sum_{k=1}^{n} |f(t_k^n) - f(t_{k-1}^n)| \stackrel{\triangle}{=} \int_a^b |df|,$$

where $a = t_0^n < t_1^n < \cdots < t_n^n = b$ denotes a sequence of partitions of the interval $[a, b]$ such that $\delta_n = \max(t_k^n - t_{k-1}^n) \to 0$ as $n \to \infty$. If

$$\int_a^b |df| < \infty,$$

then we say that f has finite variation *on the interval $[a, b]$. If*

$$\int_a^b |df| = \infty,$$

then we say that f has infinite variation *on the interval $[a, b]$.*

Definition 2.6.2 *A stochastic process X is of* integrable variation *if*

$$E\left[\int_0^\infty |dX_s|\right] < \infty.$$

Example 2.6.3 A typical example of a continuous function of infinite variation is the following:

$$f(x) = \begin{cases} 0 & \text{for } x = 0, \\ x \sin\left(\dfrac{\pi}{2x}\right) & \text{for } 0 < x \leq 1. \end{cases}$$

Consider the sequence of partitions of the interval $[0, 1]$:

$$\pi_1 = \{0, 1\},$$
$$\pi_2 = \{0, \tfrac{1}{2}, 1\},$$
$$\pi_3 = \{0, \tfrac{1}{3}, \tfrac{1}{2}, 1\},$$
$$\pi_4 = \{0, \tfrac{1}{4}, \tfrac{1}{3}, \tfrac{1}{2}, 1\},$$

$$\cdots$$

$$\pi_n = \left\{0, \frac{1}{n-1}, \frac{1}{n-2}, \cdots, \frac{1}{n-(n-2)}, 1\right\}.$$

Then it can be verified that

$$\int_0^1 |df| = \lim_{n \to \infty} \sum_{k=1}^{n} |f(t_k^n) - f(t_{k-1}^n)| = \infty.$$

□

Another example of a function of infinite variation in any interval containing 0 is

$$f(x) = \frac{(-1)^{[1/x]}}{1 + x},$$

where $[1/x]$ stands for the integral part of $1/x$.

Definition 2.6.4 *An adapted process* $\{X_t, \mathcal{F}_t\}$ *is called a semimartingale if it can be written in the form*

$$X_t = X_0 + M_t + V_t.$$

Here $\{M_t\}$, $t \geq 0$ *is a local martingale with* $M_0 = 0$; $\{V_t\}$ *is an adapted process with paths of finite variation (see Definition 2.6.1), and* $V_0 = 0$. $\{V_t\}$ *is not necessarily predictable.*

Roughly speaking, $\{V_t\}$ is a slowly changing component (trend) and $\{M_t\}$ is a quickly changing component.

Definition 2.6.5 *An adapted process* $\{X_t, \mathcal{F}_t\}$ *is called a* special semimartingale *if it can be written in the form*

$$X_t = X_0 + M_t + V_t.$$

Here $\{M_t\}$, $t \geq 0$, *is a local martingale with* $M_0 = 0$; $\{V_t\}$ *is a predictable process with paths of finite variation, and* $V_0 = 0$.

Theorem 2.6.6 X *is a (special) semimartingale if and only if the stopped process* $X_{t \wedge \tau_n}$ *is a (special) semimartingale, where* $\{\tau_n\}$ *is a sequence of stopping times such that* $\lim_n \tau_n = \infty$

Proof (Elliott [11]). Clearly, if X is a (special) semimartingale then the stopped process $X_{t \wedge \tau_n}$ is a (special) semimartingale for each n.

If S and T are stopping times and $X_{t \wedge S}$ and $X_{t \wedge T}$ are (special) semimartingales then the same is true of $X_{t \wedge (S \vee T)} = X_{t \wedge S} + X_{t \wedge T} - X_{t \wedge (S \wedge T)}$. Therefore, we can assume that $\{\tau_n\}$ is an increasing sequence of stopping times with the stated properties.

If $X_{t \wedge \tau_n}$ is a special semimartingale for each n it has a unique decomposition

$$X_{t \wedge \tau_n} = X_0 + M_t^n + A_t^n.$$

However,

$$(X_{t \wedge \tau_{n+1}})_{t \wedge \tau_n} = X_{t \wedge \tau_n},$$

so $(M^{n+1})_{t \wedge \tau_n} = M^n$ and $(A^{n+1})_{t \wedge \tau_n} = A^n$.

The processes $\{M^n\}$ and $\{A^n\}$ can, therefore, be "pasted" together to give a local martingale M and a predictable process A of locally finite variation, so the process X in this case is a special semimartingale.

In the general case we know that $X_{t \wedge \tau_n}$ is a semimartingale for each n. However, X is certainly a right-continuous process with left limits, so the process $V_t = \sum_{0 < s \leq t} \Delta X_s I_{\{|\Delta X_s| \geq 1\}}$ is of finite variation, as is $Y = X - V - X_0$.

For each n, $Y_{t \wedge \tau_n} = X_{t \wedge \tau_n} - V_{t \wedge \tau_n} - X_0$ is a semimartingale whose jumps are all bounded by 1. Therefore, by Corollary 12.40(c) page 150 in [11], $Y_{t \wedge \tau_n}$ is a special semimartingale. By the first part of this proof Y is then a special semimartingale and $X_t = X_0 + Y_t + V_t$ is a semimartingale. ∎

Definition 2.6.7 *A right-continuous stochastic process $\{X_t\}$ on the stochastic basis $(\Omega, \mathcal{F}, \mathcal{F}_t, P)$ is said to be of class D if the family $\{X_\tau\}$, for all τ which is an a.s. finite stopping time, is uniformly integrable. It is of class DL if it is of class D on each interval $[0, a], a < \infty$.*

Definition 2.6.8 *A right-continuous uniformly integrable supermartingale $\{X_t\}$ is said to be of class D if the set of random variables $\{X_\tau\}$, for τ any stopping time, is uniformly integrable.*

Note that *any* uniformly integrable martingale is of class D. This follows from Doob's Optional Stopping Theorem 2.5.6 because $X_\tau = E[X_\infty \mid \mathcal{F}_\tau]$ a.s.

The proof of the following important theorem can be found, for instance, in [11].

Theorem 2.6.9 *(Doob–Meyer Decomposition). Any class D supermartingale $(X_t, \mathcal{F}_t)$ can be written (P-a.s. uniquely) as*

$$M_t = X_t + A_t, \tag{2.6.1}$$

where $\{M_t, \mathcal{F}_t\}$ is a uniformly integrable martingale and $\{A_t, \mathcal{F}_t\}$ is a predictable, increasing process.

Remarks 2.6.10

1. If we replace class D by class DL in the theorem $\{M_t, \mathcal{F}_t\}$ is no longer a uniformly integrable martingale. (See Theorem 4.10 in [21].)

2. The Doob–Meyer decomposition of a process is the *special semimartingale* representation of that process because of the predictability of the process $\{A_t, \mathcal{F}_t\}$.
3. Recall that in the Doob decomposition for discrete time submartingales the increasing predictable process is given by

$$Z_n = \sum E[X_i - X_{i-1} \mid \mathcal{F}_{i-1}].$$

or

$$\Delta Z_n = E[\Delta X_n \mid \mathcal{F}_{n-1}].$$

By analogy, A_t is obtained if we replace summation with integration in the following manner:

$$A_t = \lim_{h \to 0} \int_0^t E\left[\frac{X_{s+h} - X_s}{h} \mid \mathcal{F}_s\right] ds,$$

or

$$dZ_t = E[dX_t \mid \mathcal{F}_t].$$

4. An interesting consequence of the Doob–Meyer Theorem is that any continuous martingale has unbounded variation. To see this suppose that $(M_t, \mathcal{F}_t)$ is a continuous martingale with bounded variation so that it can be written as a difference of two continuous increasing functions X_t and Z_t,

$$M_t = X_t - Z_t \quad \text{or} \quad X_t = M_t + Z_t,$$

which is a Doob–Meyer decomposition of the submartingale X_t. But $X_t = 0 + X_t$ is another Doob–Meyer decomposition of X_t. By uniqueness $M_t = 0$.   □

Example 2.6.11 Suppose $\tau_1 \leq \tau_2 \leq \ldots$ is a sequence of stopping times such that $\lim_n \tau_n = +\infty$ a.s. Then the *counting* process

$$N_t = \sum_{n \geq 1} I_{\{\tau_n \leq t\}}$$

is an $\mathcal{F}_t$-submartingale and it admits a Doob–Meyer decomposition

$$N_t = Y_t + Z_t.$$

Here, the predictable, increasing process $\{Z_t, \mathcal{F}_t\}$ is called the compensator of N_t.   □

Example 2.6.12 Let X be the single jump process introduced in Example 2.1.4. For $t \geq 0$ define the process

$$\mu(t, A) = I_{\{T \leq t\}} I_{\{Z \in A\}}. \tag{2.6.2}$$

Note that the process $\mu(t, A)$ has sample paths which are identically zero until the jump time T. They then have a unit jump at T and $Z \in A$. We now show that the predictable compensator of μ is given by

$$\mu_p(t, A) = -\int_{]0,T\wedge t]} \frac{dF_s^A}{F_{s-}}, \qquad (2.6.3)$$

where $F_s^A = P[T > s, Z \in A]$ and $F_s = P[T > t, Z \in E]$.

Write $\mathcal{F}_t$ for the completed σ-field generated by $\{X_s\}$, $s \le t$, so that $\mathcal{F}_t$ is generated by $\mathcal{B}([0, t]) \times \mathcal{E}$. Note that $]t, \infty] \times E$ is an atom of $\mathcal{F}_t$.

We have the following result which will be used later (see Lemma 3.8.9).

Lemma 2.6.13 *Suppose τ is an $\mathcal{F}_t$ stopping time with $P(\tau < T) > 0$. Then there is a $t_0 \in [0, \infty[$ such that $\tau \wedge T = t_0 \wedge T$ a.s.*

Proof   Suppose τ takes two values $t_1 < t_2$ on $\{\omega \in \Omega = [0, \infty] \times E : \tau(\omega) \le T(\omega)\}$ with positive probability. Then for $t_1 < t < t_2$,

$$\{\omega \in \Omega = [0, \infty] \times E : \tau(\omega) \le t\} \cap]t, \infty] \times E \ne]t, \infty] \times E,$$

so $\{\tau \le t\} \notin \mathcal{F}_t$. Therefore for some $t_0 \in [0, \infty[$, $\{\tau \le T\} \subset \{t_0 \le T\}$. A similar argument gives the reverse inclusion and the result follows.   ∎

Theorem 2.6.14

$$q(t, A) = \mu(t, A) - \mu_p(t, A)$$

is an $\mathcal{F}_t$-martingale.

Proof   ([11]) For $t > s$,

$$E[q(t, A) - q(s, A) \mid \mathcal{F}_s] = E[\mu(t, A) - \mu_p(t, A) - (\mu(s, A) - \mu_p(s, A)) \mid \mathcal{F}_s]$$
$$= E[\mu(t, A) - \mu(s, A) - (\mu_p(t, A) - \mu_p(s, A)) \mid \mathcal{F}_s]$$

So we must show that

$$E[\mu(t, A) - \mu(s, A) \mid \mathcal{F}_s] = E[\mu_p(t, A) - \mu_p(s, A) \mid \mathcal{F}_s]. \qquad (2.6.4)$$

First note that, in view of (2.6.2), if $T \le s$ both sides of (2.6.4) are zero. Now recall that $]s, \infty] \times E$ is an atom of $\mathcal{F}_s$

$$E[\mu(t, A) - \mu(s, A) \mid \mathcal{F}_s] = E[I_{\{Z \in A\}} I_{\{s < T \le t\}} \mid \mathcal{F}_s]$$
$$= P(T > s, Z \in A \mid T > s, Z \in E) I_{\{T > s, Z \in E\}}$$
$$\quad - P(T > t, Z \in A \mid T > s, Z \in E) I_{\{T > s, Z \in E\}}$$
$$= \frac{F_s^A - F_t^A}{F_s} I_{\{T > s, Z \in E\}}.$$

On the other hand $\mu_p(t, A)$ is a function of T only, and $F(t) = P(T > t)$. Therefore, using (2.6.3),

$$E[\mu_p(t, A) - \mu_p(s, A) \mid \mathcal{F}_s]$$

$$= -E\left[\left. \int_{]0,T\wedge t]} \frac{d F_u^A}{F_{u-}} - \int_{]0,T\wedge s]} \frac{d F_u^A}{F_{u-}} \mid T > s, Z \in E \right] I_{\{T>s, Z\in E\}}$$

$$= -\frac{I_{\{T>s, Z\in E\}}}{P(T > s, Z \in E)} E\left[\left(\int_{]0,T\wedge t]} \frac{d F_u^A}{F_{u-}} - \int_{]0,s]} \frac{d F_u^A}{F_{u-}} \right)(I_{T>t} + I_{s<T\leq t})\right]$$

$$= -\frac{I_{\{T>s, Z\in E\}}}{P(T > s, Z \in E)} E\left[\left(\int_{]0,t]} \frac{d F_u^A}{F_{u-}} - \int_{]0,s]} \frac{d F_u^A}{F_{u-}} \right) I\{T > t, Z \in E\}\right.$$

$$+ \left. \left(\int_{]0,T]} \frac{d F_u^A}{F_{u-}} - \int_{]0,s]} \frac{d F_u^A}{F_{u-}} \right) I_{s<T\leq t} \right]$$

$$= \frac{I\{T > s, Z \in E\}}{F_s}\left[-F_t \int_{]s,t]} \frac{d F_u^A}{F_{u-}} - \int_{]s,t]} \int_{]s,r]} \frac{d F_u^A}{F_{u-}} (-d F_r) \right].$$

Interchanging the order of integration, the double integral is

$$\int_{]s,t]} \int_{]s,r]} \frac{d F_u^A}{F_{u-}} (d F_r) = \int_{]s,t]} \frac{1}{F_{u-}} \int_{]u,t]} d F_r d F_u^A$$

$$= \int_{]s,t]} \frac{1}{F_{u-}} (F_t - F_{u-}) d F_u^A$$

$$= F_t \int_{]s,t]} \frac{d F_u^A}{F_{u-}} - (F_t^A - F_s^A).$$

Therefore (2.6.4) holds and the result follows.   ∎

□

A continuous-time, discrete-state stochastic process of great importance in stochastic modeling is the following.

Definition 2.6.15 *A continuous-time stochastic process $\{X_t\}$, $t \geq 0$, with finite-state space $S = \{s_1, s_2, \ldots, s_N\}$, defined on a probability space $(\Omega, \mathcal{F}, P)$ is a Markov chain if for all $t, u \geq 0$ and $0 \leq r \leq u$,*

$$P(X_{(t+u)} = s_j \mid X_u = s_i, X_r = s_k) = P(X_{(t+u)} = s_j \mid X_u = s_i),$$

for all states $s_i, s_j, s_k \in S$.
    $\{X_t\}$, $t \geq 0$, is a homogeneous Markov chain if

$$P(X_{(t+u)} = s_j \mid X_u = s_i) \stackrel{\triangle}{=} p_{ji}(t)$$

is independent of u.
    The family $P_t = \{p_{ji}(t)\}$ is called the transition semigroup of the homogeneous Markov chain and it satisfies the property $\sum_{j=1}^{N} p_{ji}(t) = 1$.

The following properties are similar to the discrete-time case.

$P_{(t+u)} = P_t P_u$ and $P_0 = I$, where I is the identity matrix.

Let $p_0 = (p_0^1, p_0^2, \ldots, p_0^N)'$ be the distribution of X_0 and $p_t = (p_t^1, p_t^2, \ldots, p_t^N)'$ be the distribution of X_t. Then $p_t = P_t p_0$.

Theorem 2.6.16 *Let* $\{P_t\}$, $t \geq 0$ *be a continuous transition semigroup. Then there exist*

$$q_i \overset{\triangle}{=} \lim_{h \downarrow 0} \frac{1 - p_{ii}(h)}{h} \in [0, \infty],$$

and

$$q_{ji} \overset{\triangle}{=} \lim_{h \downarrow 0} \frac{p_{ji}(h)}{h} \in [0, \infty).$$

Proof   See, for instance, [5] page 334.   ∎

The matrix $A = \{q_{ij}\}$ is called the *infinitesimal generator* of the continuous-time homogeneous Markov chain.

Note that since $\sum_{j=1}^{N} p_{ji}(h) = 1$ it follows immediately that

$$q_i = -\sum_{j \neq i, j=1}^{N} q_{ji}.$$

The differential system

$$\frac{d}{dt} P_t = \lim_{h \downarrow 0} \frac{P_{(t+h)} - P_t}{h} = P_t \lim_{h \downarrow 0} \frac{P_h - I}{h} = P_t A$$

is called *Kolmogorov's forward differential system*.

Similarly, the system $\dfrac{d}{dt} P_t = A P_t$ is called *Kolmogorov's backward differential system*. In this finite-state case, a solution for both systems, with initial condition $P_0 = I$, is e^{tA}.

Example 2.6.17 (Semimartingale representation of a continuous-time Markov chain) Let $\{Z_t\}$ $t \geq 0$ be a continuous-time Markov chain, with state space $\{s_1 \ldots, s_N\}$, defined on a probability space $(\Omega, \mathcal{F}, P)$. S will denote the (column) vector $(s_1, \ldots, s_N)'$.

Suppose $1 \leq i \leq N$, and for $j \neq i$,

$$\pi_i(x) = \prod_{j=1}^{N}(x - s_j),$$

and $\phi_i(x) = \dfrac{\pi_i(x)}{\pi_i(s_i)}$; then $\phi_i(s_j) = \delta_{ij}$ and $\phi = (\phi_1, \ldots, \phi_N)$ is a bijection of the set $\{s_1 \ldots, s_N\}$ with the set $S = \{e_1, \ldots, e_N\}$. Here, for $0 \leq i \leq N$, $e_i = (0, \ldots, 1, \ldots, 0)'$ is the i-th unit (column) vector in $\mathbb{R}^N$. Consequently, without loss of generality, we shall consider a Markov chain on S. If $X_t \in S$ denotes the state of this Markov chain at time $t \geq 0$, then the corresponding value of Z_t is $\langle X_t, S \rangle$, where $\langle ., . \rangle$ denotes the inner product in $\mathbb{R}^N$.

Write $p_t^i = P(X_t = e_i)$, $0 \leq i \leq N$. We shall suppose that for some family of matrices A_t, $p_t = (p_t^1, \ldots, p_t^N)'$ satisfies the forward Kolmogorov equation

$$\frac{\mathrm{d}p_t}{\mathrm{d}t} = A_t p_t,$$

with p_0 known and $A_t = (a_{ij}(t))$, $t \geq 0$.

The fundamental transition matrix associated with A will be denoted by $\Phi(t, s)$, so with I the $N \times N$ identity matrix,

$$\frac{\mathrm{d}\Phi(t, s)}{\mathrm{d}t} = A_t \Phi(t, s), \qquad \Phi(s, s) = I \qquad (2.6.5)$$

$$\frac{\mathrm{d}\Phi(t, s)}{\mathrm{d}s} = -\Phi(t, s)A_s, \qquad \Phi(t, t) = I.$$

(If A_t is constant $\Phi(t, s) = \exp(t - s)A$.)

Consider the process in state $x \in S$ at time s and write $X_{s,t}(x)$ for its state at the later time $t \geq s$. Then

$$E_{s,x}[X_t \mid \mathcal{F}_s] = E_{s,x}[X_t \mid X_s] = E_{s,x}[X_{s,t}(x)] = \Phi(t, s)x.$$

Write $\mathcal{F}_t^s$ for the right-continuous, complete filtration generated by $\sigma\{X_r : s \leq r \leq t\}$, and $\mathcal{F}_t = \mathcal{F}_t^0$. We have the following representation result.

Lemma 2.6.18

$$M_t \overset{\triangle}{=} X_t - X_0 - \int_0^t A_r X_r \mathrm{d}r$$

is an $\{\mathcal{F}_t\}$ martingale.

Proof Suppose $0 \leq s \leq t$. Then

$$E[M_t - M_s \mid \mathcal{F}_s] = E\left[X_t - X_s - \int_s^t A_r X_r \mathrm{d}r \mid \mathcal{F}_s\right]$$

$$= E\left[X_t - X_s - \int_s^t A_r X_r \mathrm{d}r \mid X_s\right]$$

$$= E_{s,X_s}[X_t] - X_s - \int_s^t A_r E_{s,X_s}[X_r]\mathrm{d}r$$

$$= \Phi(t, s)X_s - X_s - \int_s^t A_r \Phi(r, s)X_s \mathrm{d}r = 0$$

by (2.6.5). Therefore, the (special) semimartingale representation of the Markov chain X is

$$X_t = X_0 + \int_0^t A_r X_r \mathrm{d}r + M_t.$$

■
□

2.7 Brownian motion

Let X be a real valued random variable with $E[X^2] < \infty$, $E[X] = \mu$ and $E[X - \mu]^2 = \sigma^2 \neq 0$. Recall that X is Gaussian if its probability density function is given by the function

$$f(x) = \frac{1}{\sqrt{2\pi\sigma^2}} \exp\left(-\frac{(x - \mu)^2}{2\sigma^2}\right), \quad x \in \mathbb{R}.$$

If $X = (X_1, \ldots, X_n)'$ is a vector valued random variable with positive definite covariance matrix $C = \{\text{Cov}(X_i, X_j)\}$, $i, j = 1, \ldots, n$, $E[X] = \mu = (\mu_1, \ldots, \mu_n)'$, then $X = (X_1, \ldots, X_n)'$ is Gaussian if its density function is

$$f(x_1, \ldots, x_n) = \frac{1}{(2\pi)^{n/2}(\det C)^{1/2}} \exp\left(-\frac{(x - \mu)'C^{-1}(x - \mu)}{2}\right),$$

$(x_1, \ldots, x_n) \in \mathbb{R}^n$.

Notice that the first two moments completely characterize a Gaussian random variable and uncorrelatedness implies independence between Gaussian random variables.

A continuous-time, continuous-state space stochastic process $\{B_t\}$ is said to be a *standard one-dimensional Brownian motion* process if $X_0 = 0$ a.s., it has stationary independent increments and for every $t > 0$, B_t is normally distributed with mean 0 and variance t.

These features make the $\{B_t\}$, perhaps the most well-known and extensively studied continuous-time stochastic process. The joint distribution of any finite number of the random variables $B_{t_1}, B_{t_2}, \ldots, B_{t_n}, t_1 \leq t_2 \leq \cdots \leq t_n$ of the process is normal with density

$$f(x_1, x_2, \ldots, x_n) = \frac{1}{\sqrt{2\pi t_1}} \exp\left(-\frac{x_1^2}{2t_1}\right) \prod_{i=1}^{n-1} \frac{1}{\sqrt{2\pi(t_{i+1} - t_i)}}$$

$$\times \exp\left(-\frac{(x_{i+1} - x_i)^2}{2(t_{i+1} - t_i)}\right).$$

The form of the density function $f(x_1, x_2, \ldots, x_n)$ shows that indeed the random variables $B_{t_1}, B_{t_2} - B_{t_1}, \ldots, B_{t_n} - B_{t_{n-1}}$ are independent.

By the independent increment property,

$$P(B_t \leq x \mid B_{t_0} = x_o) = P(B_t - B_{t_0} \leq x - x_0)$$

$$= \frac{1}{\sqrt{2\pi(t - t_0)}} \int_{-\infty}^{x - x_0} \exp\left(-\frac{u^2}{2(t - t_0)}\right) du.$$

If $B_t = (B_t^1, \ldots, B_t^n)'$ is a vector valued Brownian motion process and $x, y \in \mathbb{R}^n$, then

$$f_B(t, x, y) = \frac{1}{(2\pi t)^{n/2}} \exp\left(-\frac{|y - x|^2}{2t}\right)$$

$$= \prod_{i=1}^{n} \frac{1}{\sqrt{2\pi t}} \exp\left(-\frac{(y_i - x_i)^2}{2t}\right),$$

so that the n components of B_t are themselves independent one-dimensional Brownian motion processes.

Some properties of the Brownian motion process

The proofs of the following properties are left as exercises.

If $\{B_t\}$ is a Brownian motion process then:

1. the process $\{-B_t\}$ is a Brownian motion,
2. for any $a \geq 0$, the process $\{B_{t+a} - B_t\}$ is a Brownian motion and the same result holds if a is replaced with a finite valued stopping time $a(\omega)$,
3. for any $a \neq 0$, the process $\{a B_{t/a^2}\}$ is a Brownian motion,
4. the process $\{B_{1/t}\}$, for $t > 0$, is a Brownian motion,
5. Almost all the paths of (one-dimensional) Brownian motion visit any real number infinitely often.

Theorem 2.7.1 *Let $\{B_t\}$ be a standard Brownian motion process and $\mathcal{F}_t = \sigma\{B_s : s \leq t\}$. Then*

1. *$\{B_t\}$ is an $\mathcal{F}_t$-martingale,*
2. *$\{B_t^2 - t\}$ is an $\mathcal{F}_t$-martingale, and*
3. *for any real number σ, $\{\exp(\sigma B_t - \dfrac{\sigma^2}{2}t)\}$ is an $\mathcal{F}_t$-martingale.*

Proof

1. Let $s \leq t$. $E[B_t - B_s \mid \mathcal{F}_s] = E[B_t - B_s] = 0$ because $\{B_t\}$ has independent increments and $E[B_t] = E[B_s] = 0$ by hypothesis.
2. $E[(B_t - B_s)^2 \mid \mathcal{F}_s] = E[B_t^2 - B_s^2] = t - s$. Therefore $E[B_t^2 - t^2 \mid \mathcal{F}_s] = B_s^2 - s$.
3. If Z is a standard normal random variable, with density $\dfrac{1}{\sqrt{2\pi}}e^{-x^2/2}$, and $\lambda \in \mathbb{R}$ then

$$E[e^{\lambda Z}] = \frac{1}{\sqrt{2\pi}}\int_{-\infty}^{\infty} e^{\lambda x}e^{-x^2/2}dx = e^{\lambda^2/2}.$$ Using the independence of increments and stationarity we have, for $s < t$,

$$E[e^{\sigma B_t - \frac{\sigma^2}{2}t} \mid \mathcal{F}_s] = e^{\sigma B_s - \frac{\sigma^2}{2}t} E[e^{\sigma(B_t - B_s)} \mid \mathcal{F}_s]$$
$$= e^{\sigma B_s - \frac{\sigma^2}{2}t} E[e^{\sigma(B_t - B_s)}]$$
$$= e^{\sigma B_s - \frac{\sigma^2}{2}t} E[e^{\sigma B_{t-s}}].$$

Now σB_{t-s} is $N(0, \sigma^2(t - s))$; that is, if Z is $N(0, 1)$ as previously, σB_{t-s} has the same law as $\sigma\sqrt{t - s}Z$ and

$$E[e^{\sigma B_{t-s}}] = E[e^{\sigma\sqrt{t-s}Z}] = e^{\sigma^2(t-s)/2}.$$

Therefore $E[e^{\sigma B_t - \frac{\sigma^2}{2}t} \mid \mathcal{F}_s] = e^{\sigma B_s - \frac{\sigma^2}{2}s}$ and the result follows. ∎

It turns out that Theorem 2.7.1 (2) characterizes a Brownian motion (see Theorem 3.7.3).

Theorem 2.7.2 *(The Strong Markov Property for Brownian Motion) Let $\{B_t\}$ be a Brownian motion process on a filtered probability space $(\Omega, \mathcal{F}, \mathcal{F}_t)$, and let τ be a finite valued stopping time with respect to the filtration $\{\mathcal{F}_t\}$. Then the process $B_{\{\tau+t\}} - B_\tau$, $t \geq 0$, is a Brownian motion independent of $\mathcal{F}_\tau$*

Proof   See [34] page 22.   ∎

Theorem 2.7.3 *(Existence of Brownian Motion) There exists a probability space on which it is possible to define a process $\{B_t\}$, $0 \leq t \leq 1$, which has all the properties of a Brownian motion process.*

Proof   See [34] page 10.   ∎

2.8 Brownian motion process with drift

An important stochastic process in applications is the one-dimensional *Brownian motion with drift*

$$X_t = \mu t + \sigma B_t,$$

where μ is a constant, called the *drift parameter*, and B_t is a standard Brownian motion. Then it is easily seen that X_t has independent increments and that $X_{t+h} - X_t$ is normally distributed with mean μh and variance σh. By the independent increment property we have

$$P(X_t \leq x \mid X_{t_0} = x_o) = P(X_t - X_{t_0} \leq x - x_0)$$

$$= \frac{1}{\sqrt{2\pi(t - t_0)}\sigma} \int_{-\infty}^{x-x_0} \exp\left(-\frac{(u - \mu(t - t_0))^2}{2(t - t_0)\sigma^2}\right) du.$$

2.9 Brownian paths

The sample paths of a Brownian motion process are highly irregular. In fact they model the motion of a microscopic particle suspended in a fluid and subjected to the impacts of the fluid molecules. This phenomenon was first reported by the Scottish botanist Robert Brown in 1828. The path followed by a Brownian particle is very irregular. The sample paths of a Brownian motion process are nowhere differentiable with probability 1. To see this consider the quantity

$$Z_h \triangleq \frac{B_{t+h} - B_t}{h},$$

which is normally distributed with variance $1/h \to \infty$ as $h \to 0$. Hence for every bounded Borel set B,

$$P(Z_h \in B) \to 0 \quad (h \to 0),$$

that is, Z_h does not converge with positive probability to a finite random variable. Using Kolmogorov's Continuity Theorem, which we now state, one can show that almost all sample paths of a Brownian motion process are continuous.

Theorem 2.9.1 (*Kolmogorov–Čentsov Continuity Theorem*). *Suppose that the stochastic process $\{X_t\}$ satisfies the following conditions. For all $T > 0$ there exists constants $\alpha > 0$, $\beta > 0$, $D > 0$ such that:*

$$E[|X_t - X_s|^\alpha] \leq D|t - s|^{1+\beta}; \quad 0 \leq s, t \leq T, \tag{2.9.1}$$

then almost every sample path is uniformly continuous on the interval $[0, T]$.

For the proof see [15] page 57.
   Recall that for a Brownian motion,

$$P(B_t - B_s \leq x) = \frac{1}{\sqrt{2\pi|t - s|}} \int_{-\infty}^{x} \exp\left(-\frac{u^2}{2|t - s|}\right) du.$$

Hence

$$E|B_t - B_s|^4 = \frac{1}{\sqrt{2\pi|t - s|}} \int_{-\infty}^{+\infty} u^4 \exp\left(-\frac{u^2}{2|t - s|}\right) du = 3(t - s)^2,$$

which verifies the Kolmogorov condition with $\alpha = 4$, $D = 3$, $\beta = 1$ and establishes the almost sure continuity of the Brownian motion process. We now show that each portion of almost every sample path of the Brownian motion process B_t has infinite length, i.e. almost all sample paths are of unbounded variation, so that terms in a Taylor series expansion which would ordinarily be of second order get promoted to first order. This is one of the most remarkable properties of a Brownian motion process.

Lemma 2.9.2 *Let B_t be a Brownian motion process and let $a = t_0^n < t_1^n < \cdots < t_n^n = b$ denote a sequence of partitions of the interval $[a, b]$ such that $\delta_n = \max(t_k^n - t_{k-1}^n) \overset{\triangle}{=} \max \Delta t_k^n \to 0$ as $n \to \infty$.*
   Write $(B_{t_k^n} - B_{t_{k-1}^n})^2 = \Delta^2 B_{t_k^n}$ and

$$S_n(B) = \sum_{k=1}^{n} \Delta^2 B_{t_k^n}.$$

Then:

1. $E[S_n(B) - (b - a)]^2 \to 0$ $(\delta_n \to 0)$.
2. If $\delta_n \to 0$ so fast that

$$\sum_{n=1}^{\infty} \delta_n < \infty \tag{2.9.2}$$

then $S_n(B) \to b - a$ (a.s.)

Proof

1.
$$E[S_n(B) - (b-a)]^2 = E[\sum_{k=1}^{n}(\Delta^2 B_{t_k^n} - \Delta t_k^n)]^2$$

$$= \sum_{k=1}^{n} E[\Delta^2 B_{t_k^n}]^2 - 2\Delta t_k^n E[\Delta^2 B_{t_k^n}] + (\Delta t_k^n)^2]$$

$$= \sum_{k=1}^{n} (3(\Delta t_k^n)^2 - 2(\Delta t_k^n)^2 + (\Delta t_k^n)^2)$$

$$= \sum_{k=1}^{n} 2(\Delta t_k^n)^2 \le 2\delta_n \sum_{k=1}^{n} 2(\Delta t_k^n) = 2\delta_n(b-a),$$

which goes to zero as $\delta_n \to 0$ and $E[S_n - (b-a)]^2 \to 0$.

2. By Chebyshev's inequality (1.3.33)

$$P(|S_n(B) - (b-a)| \ge \epsilon) \le \frac{\text{Var}(S_n(B) - (b-a))}{\epsilon^2} \le \frac{2\delta_n(b-a)}{\epsilon^2}. \qquad (2.9.3)$$

In view of (2.9.2) we can sum up both sides of (2.9.3) and use the Borel–Cantelli Lemma (1.2.7) to get

$$P(\limsup\{|S_n(B) - (b-a)| \ge \epsilon\}) = 0,$$

and the event $\{\omega : |S_n(B(\omega)) - (b-a)| \ge \epsilon\}$ occurs only a finite number of times with probability 1 as n increases to infinity. Therefore we have almost sure convergence.

∎

The above argument shows that $B_t(\omega)$ is, a.s., of infinite variation on $[a, b]$. To see this note that

$$b - a \le \limsup \max |B_{t_k^n}(\omega) - B_{t_{k-1}^n}(\omega)| \sum_{k=1}^{n} |B_{t_k^n} - B_{t_{k-1}^n}|.$$

From the sample-path continuity of Brownian motion, $\max |B_{t_k^n}(\omega) - B_{t_{k-1}^n}(\omega)|$ can be made arbitrarily small for almost all ω which implies that $\sum_{k=1}^{n} |B_{t_k^n} - B_{t_{k-1}^n}| \to \infty$ for almost all ω as $n \to \infty$.

There is a simple construction for Brownian motion. Take a sequence $X_1, X_2, \ldots,$ of i.i.d. $N(0, 1)$ random variables and an orthonormal basis $\{\phi_n\}$ for $L^2[0, 1]$. That is

$$\langle \phi_n, \phi_n \rangle_{L^2} = \int_0^t \phi_n^2(s)ds = 1,$$

and

$$\langle \phi_m, \phi_n \rangle_{L^2} = \int_0^t \phi_m(s)\phi_n(s)ds = 0,$$

if $m \neq n$. For $t \in [0, 1]$ define

$$B_t^n = \sum_{k=1}^{n} X_k \int_0^t \phi_k(s) ds.$$

Using the Parseval equality it is seen that

$$E[B_t^n - B_t^m]^2 \to 0 \quad (n, m \to \infty).$$

The completeness of $L^2[0, 1]$ implies the existence of a limit process B_t with the same covariance function as a Brownian motion. It can also be shown that B_t^n converges uniformly in $t \in [0, 1]$ to B_t with probability 1 (a.s.), that is, $\{B_t\}$ has continuous sample paths a.s.

2.10 Poisson process

A continuous-time, discrete-state space, stochastic process which keeps the count of the occurrences of some specific event (or events) $\{N_t\}_{t \geq 0}$ is called a *counting process*.

The Poisson process is a counting process which, like the Brownian motion, has independent increments, but its sample paths are not continuous. They are increasing step functions with each step having height 1 and a random waiting time between two consecutive jumps. The times between successive jumps are independent and exponentially distributed with parameter $\lambda > 0$. The joint probability distribution of any finite number of values $N_{t_1}, N_{t_2}, \ldots, N_{t_n}$ of the process is

$$P[N_{t_1} = k_1, \ldots, N_{t_n} = k_n]$$

$$= \frac{(\lambda t_1)^{k_1} \exp(-\lambda t_1)}{k_1!} \prod_{i=1}^{n-1} \frac{[\lambda(t_{i+1} - t_i)]^{k_{i+1} - k_i}}{(k_{i+1} - k_i)!} \exp(-\lambda(t_{i+1} - t_i)),$$

provided that $t_1 \leq t_2 \leq \cdots \leq t_n$ and $k_1 \leq k_2 \leq \cdots \leq k_n$. The Poisson process is a.s. continuous at any point, as shown by

$$P(\omega : \lim_{\epsilon \to 0} |N_{t+\epsilon}(\omega) - N_t(\omega)| = 0) = \lim_{\epsilon \to 0} e^{-\lambda \epsilon} = 1.$$

However, the probability of continuity at *all* points in any interval is less than 1 so that it is not (a.s.) sample path continuous.

Like any process with independent increments, the Poisson process is Markovian (see 2.2.4). However, the independent increment assumption is stronger than the Markov property.

2.11 Problems

1. Show that the Borel σ-field $\mathcal{B}(\mathbb{R}^\infty)$ coincides with the smallest σ-field containing the open sets in $\mathbb{R}^\infty$ in the metric $\rho_\infty(x^1, x^2) = \sum_k 2^{-k} \frac{|x_k^1 - x_k^2|}{1 + |x_k^1 - x_k^2|}$.

2. Suppose that at time 0 you have \$$a$ and your component has \$$b$. At times $1, 2, \ldots$ you bet a dollar and the game ends when somebody has \$0. Let S_n be a random walk

on the integers $\{\ldots, -2, -1, 0, +1, +2, \ldots\}$ with $P(X = -1) = q$, $P(X = +1) = p$. Let $\alpha = \inf\{n \geq 1 : S_n = -a, +b\}$, i.e. the first time you or your component is ruined, then $\{S_{n \wedge \alpha}\}_{n=0}^{\infty}$ is the running total of your profit. Show that if $p = q = 1/2$, $\{S_{n \wedge \alpha}\}$ is a bounded martingale with mean 0 and that the probability of your ruin is $b/(a + b)$. Show that if the game is not fair ($p \neq q$) then S_n is not a martingale but $Y_n \stackrel{\Delta}{=} \left(\dfrac{q}{p}\right)^{S_n}$ is a martingale. Find the probability of your ruin and check that if $a = b = 500$, $p = .499$ and $q = .501$ then $P(\text{ruin}) = .8806$ and it is almost 1 if $p = 1/3$.

3. Show that if $\{X_n\}$ is an integrable, real valued process, with independent increments and mean 0, then it is a martingale with respect to the filtration it generates; and if in addition X_n^2 is integrable, $X_n^2 - E(X_n^2)$ is a martingale with respect to the same filtration.

4. Let $\{X_n\}$ be a sequence of i.i.d. random variables with $E[X_n] = 0$ and $E[X_n^2] = 1$. Show that $S_n^2 - n$ is an $\mathcal{F}_n = \sigma\{X_1, \ldots, X_n\}$-martingale, where $S_n = \sum_{i=1}^{n} X_i$.

5. Let $\{y_n\}$ be a sequence of independent random variables with $E[y_n] = 1$. Show that the sequence $X_n = \prod_{k=0}^{n} y_n$ is a martingale with respect to the filtration $\mathcal{F}_n = \sigma\{y_0, \ldots, y_n\}$.

6. Let $\{X_n\}$ and $\{Y_n\}$ be two sequences of i.i.d. random variables with $E[X_n] = E[Y_n] = 0$, $E[X_n^2] < \infty$, $E[Y_n^2] < \infty$ and $\text{Cov}(X_n, Y_n) \neq 0$. Show that

$$\left\{S_n^X S_n^Y - \sum_{i=1}^{n} \text{Cov}(X_i, Y_i)\right\}$$

is an $\mathcal{F}_n = \sigma\{X_1, \ldots, X_n, Y_1, \ldots, Y_n\}$-martingale, where $S_n^X = \sum_{i=1}^{n} X_i$, $S_n^Y = \sum_{i=1}^{n} Y_i$.

7. Show that two square integrable martingales X and Y are orthogonal if and only if $X_0 Y_0 = 0$ and the process $\{X_n Y_n\}$ is a martingale.

8. Show that the square integrable martingales X and Y are orthogonal if and only if for every $0 \leq m \leq n$,

$$E[X_n Y_n \mid \mathcal{F}_m] = E[X_n \mid \mathcal{F}_m] E[Y_n \mid \mathcal{F}_m].$$

9. Let $\{B_t\}$ be a standard Brownian motion process ($B_0 = 0$, a.s., $\sigma^2 = 1$). Show that the conditional density of $\{B_t\}$ for $t_1 < t < t_2$,

$$P(B_t \in dx \mid B_{t_1} = x_1, B_{t_2} = x_2),$$

is a normal density with mean and variance

$$\mu = x_1 + \frac{x_2 - x_1}{t_2 - t_1}(t - t_1), \qquad \sigma^2 = \frac{(t_2 - t)(t - t_1)}{t_2 - t_1}.$$

10. Let $\{B_t\}$ be a standard Brownian motion process. Show that the density of $\alpha = \inf\{t, B_t = b\}$, the first time the process B_t hits level $b \in \mathbb{R}$ (see Example 2.2.5), is given by

$$f_\alpha(t) = \frac{|b|}{\sqrt{2\pi t^3}} e^{-b/2t}; \qquad t > 0.$$

11. Let $\{B_t\}$ be a Brownian motion process with drift μ and diffusion coefficient σ^2. Let

$$x_t = e^{B_t}, \quad t \geq 0.$$

Show that

$$E[x_t \mid x_0 = x] = xe^{t(\mu + \frac{1}{2}\sigma^2)},$$

and

$$\text{var}[x_t \mid x_0 = x] = x^2 e^{2t(\mu + \frac{1}{2}\sigma^2)}\left[e^{t\sigma^2} - 1\right].$$

12. Let N_t be a standard Poisson process and $Z_1, Z_2 \ldots$ a sequence of i.i.d. random variables such that $P(Z_i = 1) = P(Z_i = -1) = 1/2$. Show that the process

$$X_t = \sum_{i=1}^{N_t} Z_i$$

is a martingale with respect to the filtration $\{\mathcal{F}_t\} = \sigma\{X_s, s \leq t\}$.

13. Show that the process $\{B_t^2 - t, \mathcal{F}_t^B\}$ is a martingale, where B is the standard Brownian motion process and $\{\mathcal{F}_t^B\}$ its natural filtration.

14. Show that the process $\{(N_t - \lambda t)^2 - \lambda t\}$ is a martingale, where N_t is a Poisson process with parameter λ.

15. Show that the process

$$I_t = \int_0^t f(\omega, s) dM_s$$

is a martingale. Here $f(.)$ is an adapted, bounded, continuous sample paths process and $M_t = N_t - \lambda t$ is the Poisson martingale.

16. Referring to Example 2.4.4, define the processes

$$N_n^{sr} = \sum_{k=1}^{n} I_{(\eta_{k-1}=s, \eta_k=r)} = \sum_{k=1}^{n} \langle X_{k-1}, e_s \rangle \langle X_k, e_r \rangle, \tag{2.11.1}$$

and

$$O_n^r = \sum_{k=1}^{n} I_{(\eta_k=r)} = \sum_{k=1}^{n} \langle X_k, e_r \rangle. \tag{2.11.2}$$

Show that 2.11.1 and 2.11.2 are increasing processes and give their Doob decompositions.

17. Let $\{X_k, \mathcal{F}_k\}$, for $0 \leq k \leq n$ be a martingale and α a stopping time. Show that $E[X_\alpha] = E[X_0]$.

18. Let α be a stopping time with respect to the filtration $\{X_n, \mathcal{F}_n\}$. Show that

$$\mathcal{F}_\alpha = \{A \in \mathcal{F}_\infty : A \cap \{\omega : \alpha(\omega) \leq n\} \in \mathcal{F}_n \quad \forall n \geq 0\}$$

is a σ-field and that α is $\mathcal{F}_\alpha$-measurable.

19. Let $\{X_n\}$ be a stochastic process adapted to the filtration $\{\mathcal{F}_n\}$ and B a Borel set. Show that

$$\alpha_B = \inf\{n \geq 0, \quad X_n \in B\}$$

is a stopping time with respect $\mathcal{F}_n$.

20. Show that if α_1, α_2, are two stopping times such that $\alpha_1 \leq \alpha_2$ (a.s.) then $\mathcal{F}_{\alpha_1} \subset \mathcal{F}_{\alpha_2}$.

21. Show that if α is a stopping time and a is a positive constant, then $\alpha + a$ is a stopping time.

22. Show that if $\{\alpha_n\}$ is a sequence of stopping times and the filtration $\{\mathcal{F}_t\}$ is right-continuous then $\inf \alpha_n$, $\liminf \alpha_n$ and $\limsup \alpha_n$ are stopping times.

3

Stochastic calculus

3.1 Introduction

It is known that if a function f is continuous and a function g is right continuous with left limits, of bounded variation (see Definition 2.6.1), then the Riemann–Stieltjes integral of f with respect to g on $[0, t]$ is well-defined and equals

$$\int_0^t f(s) dg(s) = \lim_{\delta_n \to 0} \sum_{k=1}^n f(\tau_k^n)(g(t_k^n) - g(t_{k-1}^n)),$$

where $0 = t_0^n < t_1^n < \cdots < t_n^n = t$ denotes a sequence of partitions of the interval $[0, t]$ such that $\delta_n = \max(t_k^n - t_{k-1}^n) \to 0$ as $n \to \infty$ and $t_{k-1}^n \leq \tau_k^n \leq t_k^n$.

The Lebesgue–Stieltjes integral with respect to g can be defined by constructing a measure μ_g on the Borel field $\mathcal{B}([0, \infty))$, starting with the definition

$$\mu_g((a, b]) = g(b) - g(a),$$

and then starting with the integral of simple functions f with respect to μ_g, as in Chapter 1.

For right continuous left limited stochastic processes with bounded variation sample paths, path-by-path integration is defined for each sample path by fixing ω and performing Lebesgue–Stieltjes integration with respect to the variable t. If a continuous (local) martingale X has bounded variation, its quadratic variation is zero (see Remark 2.6.10(4)).

However, continuous (local) martingales have unbounded variation, so that the Stieltjes definition cannot be used in stochastic integration to define path-by-path integrals. We assume the dependence of f as ω is constant in time.

3.2 Quadratic variations

Discrete-time processes

Definition 3.2.1 *The stochastic process $X = \{X_n\}$, $n \geq 0$, is said to be square integrable if $\sup_n E[X_n^2] < \infty$.*

Definition 3.2.2 *Let $\{X_n\}$ be a discrete time, square integrable stochastic process on a filtered probability space $(\Omega, \mathcal{F}, \{\mathcal{F}_n\}, P)$.*

1. *The nonnegative, increasing process defined by*

$$[X, X]_n = X_0^2 + \sum_{k=1}^{n}(X_k - X_{k-1})^2$$

is called the optional quadratic variation *of $\{X_n\}$.*
The predictable quadratic variation *of $\{X_n\}$ relative to the filtration $\{\mathcal{F}_n\}$ and probability measure P is defined by*

$$\langle X, X \rangle_n = E(X_0^2) + \sum_{k=1}^{n} E[(X_k - X_{k-1})^2 \mid \mathcal{F}_{k-1}].$$

2. *Given two square integrable processes $\{X_n\}$ and $\{Y_n\}$ the* optional covariation *process is defined by*

$$[X, Y]_n = X_0 Y_0 + \sum_{i=1}^{n}(X_i - X_{i-1})(Y_i - Y_{i-1}),$$

and the predictable covariation *process is defined by*

$$\langle X, Y \rangle_n = E[X_0 Y_0] + \sum_{i=1}^{n} E[(X_i - X_{i-1})(Y_i - Y_{i-1}) \mid \mathcal{F}_{i-1}].$$

Example 3.2.3 Let $X_1, X_2, \ldots$ be a sequence of i.i.d. normal random variables with mean 0 and variance 1, and consider the process $Z_0 = 0$ and $Z_n = \sum_{k=1}^{n} X_k$. Then it is left as an exercise to show that

$$[Z, Z]_n = \sum_{k=1}^{n} X_k^2,$$

$$\langle Z, Z \rangle_n = n,$$

$$E([Z, Z]_n) = E\left(\sum_{k=1}^{n} X_k^2\right) = n.$$

Here $\langle Z, Z \rangle_n$ is not random and is equal to the variance of Z_n. $\square$

Example 3.2.4 Let $\Omega = \{\omega_i, \ 1 \le i \le 8\}$ and the time index be $n = 1, 2, 3$. Suppose we are given a probability measure $P(\omega_i) = 1/8, i = 1, \ldots, 8$, a filtration

$$\mathcal{F}_0 = \{\Omega, \emptyset\},$$
$$\mathcal{F}_1 = \sigma\{\{\omega_1, \omega_2, \omega_3, \omega_4\}, \{\omega_5, \omega_6, \omega_7, \omega_8\}\},$$
$$\mathcal{F}_2 = \sigma\{\{\omega_1, \omega_2\}, \{\omega_3, \omega_4\}, \{\omega_5, \omega_6\}, \{\omega_7, \omega_8\}\},$$
$$\mathcal{F}_3 = \sigma\{\omega_1, \omega_2, \omega_3, \omega_4, \omega_5, \omega_6, \omega_7, \omega_8\},$$

and a stochastic process X given by:

$$X = \begin{array}{|c|c|c|c|}
\hline
X_0(\omega_1) & X_0(\omega_2) & \cdots & X_0(\omega_8) \\
\hline
X_1(\omega_1) & X_1(\omega_2) & \cdots & X_1(\omega_8) \\
\hline
X_2(\omega_1) & X_2(\omega_2) & \cdots & X_2(\omega_8) \\
\hline
X_3(\omega_1) & X_3(\omega_2) & \cdots & X_3(\omega_8) \\
\hline
\end{array}$$

which is adapted to the filtration $\{\mathcal{F}_i, i = 0, 1, 2, 3\}$, that is

$$X = \begin{array}{|c|c|c|c|c|c|c|c|}
\hline
x_0 & x_0 & x_0 & x_0 & x_0 & x_0 & x_0 & x_0 \\
\hline
x_{1,1} & x_{1,1} & x_{1,1} & x_{1,1} & x_{1,2} & x_{1,2} & x_{1,2} & x_{1,2} \\
\hline
x_{2,1} & x_{2,1} & x_{2,2} & x_{2,2} & x_{2,3} & x_{2,3} & x_{2,4} & x_{2,4} \\
\hline
x_{3,1} & x_{3,2} & x_{3,3} & x_{3,4} & x_{3,5} & x_{3,6} & x_{3,7} & x_{3,8} \\
\hline
\end{array}.$$

In this simple example the stochastic process

$$\langle X, X \rangle_n = E(X_0^2) + \sum_{k=1}^{n} E[(X_k - X_{k-1})^2 \mid \mathcal{F}_{k-1}]$$

can be explicitly calculated:

$$\langle X, X \rangle_0 = E(X_0^2) = x_0^2,$$

$$\langle X, X \rangle_1 = E(X_0^2) + E[(X_1 - X_0)^2 \mid \mathcal{F}_0] = x_0^2 + E[(X_1 - X_0)^2]$$

$$= x_0^2 + \frac{4}{8}(x_{1,1} - x_0)^2 + \frac{4}{8}(x_{1,2} - x_0)^2.$$

Note that $\langle X, X \rangle_0$ and $\langle X, X \rangle_1$ are both $\mathcal{F}_0$-measurable, that is, they are constants.

$$\langle X, X \rangle_2(\omega) = E(X_0^2) + E[(X_1 - X_0)^2] + E[(X_2 - X_1)^2 \mid \mathcal{F}_1](\omega)$$

$$= \langle X, X \rangle_1 + E[(X_2 - X_1)^2 \mid \{\omega_1, \omega_2, \omega_3, \omega_4\}] I_{\{\omega_1, \omega_2, \omega_3, \omega_4\}}$$

$$+ E[(X_2 - X_1)^2 \mid \{\omega_5, \omega_6, \omega_7, \omega_8\}] I_{\{\omega_5, \omega_6, \omega_7, \omega_8\}}$$

$$= \langle X, X \rangle_1 + \frac{(x_{2,1} - x_{1,1})^2 2/8 + (x_{2,2} - x_{1,1})^2 2/8}{P\{\omega_1, \omega_2, \omega_3, \omega_4\} = 4/8} I_{\{\omega_1, \omega_2, \omega_3, \omega_4\}}$$

$$+ \frac{(x_{2,3} - x_{1,2})^2 2/8 + (x_{2,4} - x_{1,2})^2 2/8}{P\{\omega_5, \omega_6, \omega_7, \omega_8\} = 4/8} I_{\{\omega_5, \omega_6, \omega_7, \omega_8\}}$$

$$= x_0^2 + \frac{4}{8}(x_{1,1} - x_0)^2 + \frac{4}{8}(x_{1,2} - x_0)^2$$

$$+ \frac{(x_{2,1} - x_{1,1})^2 + (x_{2,2} - x_{1,1})^2}{2} I_{\{\omega_1, \omega_2, \omega_3, \omega_4\}}$$

$$+ \frac{(x_{2,3} - x_{1,2})^2 + (x_{2,4} - x_{1,2})^2}{2} I_{\{\omega_5, \omega_6, \omega_7, \omega_8\}}.$$

Note that $\langle X, X \rangle_2$ is $\mathcal{F}_1$-measurable.

$$\langle X, X \rangle_3(\omega) = E(X_0^2) + E[(X_1 - X_0)^2] + E[(X_2 - X_1)^2 \mid \mathcal{F}_1](\omega)$$
$$+ E[(X_3 - X_2)^2 \mid \mathcal{F}_2](\omega)$$
$$= \langle X, X \rangle_2 + E[(X_3 - X_2)^2 \mid \mathcal{F}_2](\omega)$$
$$= \langle X, X \rangle_2 + \frac{(x_{3,1} - x_{2,1})^2 + (x_{3,2} - x_{2,1})^2}{2} I_{\{\omega_1, \omega_2\}}$$
$$+ \frac{(x_{3,3} - x_{2,2})^2 + (x_{3,4} - x_{2,2})^2}{2} I_{\{\omega_3, \omega_4\}}$$
$$+ \frac{(x_{3,5} - x_{2,3})^2 + (x_{3,6} - x_{2,3})^2}{2} I_{\{\omega_5, \omega_6\}}$$
$$+ \frac{(x_{3,7} - x_{2,4})^2 + (x_{3,8} - x_{2,4})^2}{2} I_{\{\omega_7, \omega_8\}}.$$

Note that $\langle X, X \rangle_3$ is $\mathcal{F}_2$-measurable. $\square$

Theorem 3.2.5 *If $\{X_n\}$ is a square integrable martingale then X^2 is a submartingale and $X^2 - \langle X, X \rangle$ is a martingale, i.e. $\langle X, X \rangle$ is the unique predictable, increasing process in the Doob decomposition of X^2.*

Proof From Jensen's inequality 2.3.3,

$$E[X_n^2 \mid \mathcal{F}_{n-1}] \geq (E[X_n \mid \mathcal{F}_{n-1}])^2 = X_{n-1}^2.$$

Hence X^2 is a submartingale.

The rest of the proof is left as an exercise.   ∎

Theorem 3.2.6 *If X and Y are (square integrable) martingales, then $XY - [X, Y]$ and $XY - \langle X, Y \rangle$ are martingales.*

Proof

$$E(X_n Y_n - [X, Y]_n \mid \mathcal{F}_{n-1})$$
$$= -[X, Y]_{n-1} + E(X_n Y_n - (X_n - X_{n-1})(Y_n - Y_{n-1}) \mid \mathcal{F}_{n-1})$$
$$= -[X, Y]_{n-1} - X_{n-1} Y_{n-1} + E(X_n Y_n - X_n Y_n + X_n Y_{n-1}$$
$$+ X_{n-1} Y_n \mid \mathcal{F}_{n-1})$$
$$= -[X, Y]_{n-1} - X_{n-1} Y_{n-1} + 2 X_{n-1} Y_{n-1}$$
$$= X_{n-1} Y_{n-1} - [X, Y]_{n-1}.$$

The proof for $XY - \langle X, Y \rangle$ is similar.   ∎

Two martingales X and Y are *orthogonal* if and only if $\langle X, Y \rangle_n = 0$ for, all n.

Example 3.2.7 Returning to Example 2.3.15, we call the stochastic process $X = \sum_{k=1}^{n} A_k b_k = \sum_{k=1}^{n} A_k \Delta C_k$ a *stochastic integral* with predictable integrand A and integrator the martingale C.

Note that the predictability of the integrand is a rather natural requirement. ☐

In discrete time the stochastic integral is usually called the *martingale transform* and it is usually written

$$(A \bullet C)_n = \sum_{k=1}^{n} A_k \Delta C_k.$$

Stochastic integrals can be defined for more general integrands and integrators.

Theorem 3.2.8 *For any discrete time process $X = \{X_n\}$ we have:*

$$\sum_{k=1}^{n} X_{k-1} \Delta X_k = \frac{1}{2}(X_n^2 - [X, X]_n).$$

Proof

$$2\sum_{k=1}^{n} X_{k-1} \Delta X_k + [X, X]_n = \sum_{k=1}^{n} [2X_{k-1}(X_k - X_{k-1}) + (X_k - X_{k-1})^2] = X_n^2.$$

■

In order to recover the analog of the familiar form of the integral $\int X_s \mathrm{d}X_s = \frac{1}{2}(X_t^2 - X_0^2)$ we should replace the integrand X_{n-1} by a non-predictable one, $(X_{n-1} + X_n)/2$. This is a discrete-time *Stratonovitch integral* and:

$$\sum_{k=1}^{n} \frac{X_{k-1} + X_k}{2} \Delta X_k = \frac{1}{2}(X_n^2 - X_0^2).$$

However, we then lose the martingale property of the stochastic integral.

The following result, which is proved using the identity

$$[X, Y]_n = \frac{1}{2}([X + Y, X + Y]_n - [X, X]_n - [Y, Y]_n),$$

is the *integration (or summation) by parts formula.*

Theorem 3.2.9

$$X_n Y_n = \sum_{k=1}^{n} X_{k-1} \Delta Y_k + \sum_{k=1}^{n} Y_{k-1} \Delta X_k + [X, Y]_n.$$

We now state the rather trivial discrete-time version of the so-called Itô formula of stochastic calculus.

Theorem 3.2.10 *For a real valued differentiable function f and a stochastic process X we have*

$$f(X_n) = f(X_0) + \sum_{k=1}^{n} f'(X_{k-1}) \Delta X_k + \sum_{k=1}^{n} [f(X_k) - f(X_{k-1}) - f'(X_{k-1}) \Delta X_k].$$

Continuous-time processes

We begin by recalling few definitions and results regarding deterministic functions.

Definition 3.2.11 *The quadratic variation $S_n(f)$ of a function f on an interval $[a, b]$ is*

$$S_n(f) = \sum_{k=1}^{n}(f(t_k^n) - f(t_{k-1}^n))^2,$$

where $a = t_0^n < t_1^n < \cdots < t_n^n = b$ denotes a sequence of partitions of the interval $[a, b]$ such that $\delta_n = \max(t_k^n - t_{k-1}^n) \to 0$ as $n \to \infty$.

Lemma 3.2.12 *If f is a continuous real valued function of bounded variation (see Definition 2.6.1) then its quadratic variation on any interval $[a, b]$ is 0, that is*

$$\lim_{n\to\infty} S_n(f) = \lim_{n\to\infty} \sum_{k=1}^{n}(f(t_k^n) - f(t_{k-1}^n))^2 = 0,$$

where $a = t_0^n < t_1^n < \cdots < t_n^n = b$ denotes a sequence of partitions of the interval $[a, b]$ such that $\delta_n = \max(t_k^n - t_{k-1}^n) \to 0$ as $n \to \infty$.

Proof   Since f is continuous and of bounded variation there exists $M > 0$ such that for $\varepsilon > 0$ we can choose a partition so fine that
$$\max_k(|f(t_k^n) - f(t_{k-1}^n)|) < \frac{\varepsilon}{nM}.$$

$$S_n(f) < M \sum_{k=1}^{n} |f(t_k^n) - f(t_{k-1}^n)| < Mn\frac{\varepsilon}{nM} = \varepsilon,$$

and the result follows.   ∎

Let $\{X_t, \mathcal{F}_t\}$ be a square integrable martingale. Then $\{X_t^2, \mathcal{F}_t\}$ is a nonnegative submartingale, hence of class DL and from the Doob–Meyer decomposition there exists a unique predictable increasing process $\{\langle X, X\rangle_t, \mathcal{F}_t\}$ such that

$$X_t^2 = M_t + \langle X, X\rangle_t,$$

where $\{M_t, \mathcal{F}_t\}$ is a right-continuous martingale and $\langle X, X\rangle_0 = X_0^2$.

Lemma 3.2.13 *Suppose $X = \{X_t, \mathcal{F}_t\}$ is a square integrable martingale. Then:*

1. *$X = X^c + X^d$, where X^c is the continuous martingale part of X and X^d is the purely discontinuous martingale part of X. This decomposition is unique.*
2. *$E[\sum_s \Delta X_s^2] \leq E[X_\infty^2]$, where $X_\infty = \lim_{t\to\infty} X_t$.*
3. *For any t, $\sum_{s \leq t} \Delta X_s^2 < \infty$ a.s.*

Proof   See [11] page 97.   ∎

The following result is analogous to Lemma 3.2.13.

Lemma 3.2.14 *Suppose* $X = \{X_t, \mathcal{F}_t\}$ *is a local martingale. Then:*

1. $X = X^c + X^d$, *where* X^c *is the continuous local martingale part of* X *and* X^d *is the purely discontinuous local martingale part of* X. *This decomposition is unique.*
2. *For any* t,

$$\sum_{s \leq t} \Delta X_s^2 < \infty \quad a.s.$$

Proof See [11] page 119. ∎

Definition 3.2.15 *Let* $X = \{X_t, \mathcal{F}_t\}$ *be a square integrable martingale.*

1. $\langle X, X \rangle$ *is called the* predictable quadratic variation *of* X.
2. *The optional increasing process*

$$[X, X]_t = \langle X^c, X^c \rangle_t + \sum_{s \leq t} (\Delta X_s)^2$$

is called the optional quadratic variation *of* X. $X = X^c + X^d$ *is the unique decomposition given by Lemma 3.2.13.*

Example 3.2.16 If $\{N_t\}$ is a Poisson process with parameter λ, $\Delta N_s = 0$ or 1 for all $s \geq 0$ and $\langle N^c, N^c \rangle_t = 0$. Therefore

$$[N, N]_t = \sum_{0 \leq s \leq t} (\Delta N_s)^2 = N_t.$$

Since $\{N_t - \lambda t\}$ is a martingale that is 0 at 0, we have

$$\langle N, N \rangle_t = \lambda t.$$

□

Theorem 3.2.17 *If* $X = \{X_t, \mathcal{F}_t\}$ *is a continuous local martingale, there exists a unique increasing process* $\langle X, X \rangle$, *vanishing at zero, such that* $X^2 - \langle X, X \rangle$ *is a continuous local martingale.*

Proof See [32] page 124. ∎

Definition 3.2.18 *Suppose* $X = X_0 + M + V$ *is a semimartingale (see Definition 2.6.4). Then the* optional quadratic variation *of* X *is the process*

$$[X, X]_t = \langle X^c, X^c \rangle_t + \sum_{s \leq t} (\Delta X_s)^2.$$

By definition V *has finite variation in* $[0, t]$,

$$\sum_{s \leq t} (\Delta V_s)^2 \leq K \sum_{s \leq t} |\Delta V_s| < \infty,$$

for some K. *Also, from Lemma 3.2.14* $\sum_{s \leq t} (\Delta M_s)^2 < \infty$. *Therefore* $\sum_{s \leq t} (\Delta X_s)^2$ *is a.s. finite because* $(\Delta X_s)^2 \leq (\Delta M_s)^2 + (\Delta V_s)^2$.

Lemma 3.2.19 *Almost every sample path of $[X, X]$ is right-continuous with left limits and of finite variation on each compact subset of $\mathbb{R}$. Further, $[X, X]_t < \infty$ a.s. for each $t \in [0, \infty)$.*

Proof See [11].   ∎

Definition 3.2.20 *Suppose $\{X_t, \mathcal{F}_t\}$ and $\{Y_t, \mathcal{F}_t\}$ are two square integrable martingales. Then*

$$\langle X, Y \rangle = \frac{1}{2}(\langle X + Y, X + Y \rangle - \langle X, X \rangle - \langle Y, Y \rangle).$$

$\langle X, Y \rangle$ *is the unique predictable process of integrable variation (see Definition 2.6.2) such that*

$$XY - \langle X, Y \rangle$$

is a martingale and $X_0 Y_0 = \langle X, Y \rangle_0$.

   Two square integrable martingales X and Y are called orthogonal martingales if $\langle X, Y \rangle_t = 0$, a.s., holds for every $t \geq 0$.

Remark 3.2.21 From the definition, the orthogonality of two square integrable martingales X and Y implies that XY is a martingale. Conversely, from the identity

$$E[(X_t - X_s)(Y_t - Y_s) \mid \mathcal{F}_s] = E[X_t Y_t - X_t Y_s - X_s Y_t + X_s Y_s \mid \mathcal{F}_s]$$
$$= E[X_t Y_t - X_s Y_s \mid \mathcal{F}_s] = E[\langle X, Y \rangle_t - \langle X, Y \rangle_s \mid \mathcal{F}_s],$$

if XY is a martingale the two square integrable martingales X and Y are orthogonal.   □

Definition 3.2.22 *Suppose $\{X_t, \mathcal{F}_t\}$ and $\{Y_t, \mathcal{F}_t\}$ are two square integrable martingales. Define*

$$[X, Y] = \frac{1}{2}([X + Y, X + Y] - [X, X] - [Y, Y]).$$

Then $[X, Y]$ is of integrable variation (see Definition 2.6.2),

$$XY - [X, Y]$$

is a martingale and $X_0 Y_0 = [X, Y]_0 = \Delta X_0 \Delta Y_0$.

Remark 3.2.23 From the definition

$$[X, Y]_t = \langle X^c, Y^c \rangle_t + \sum_{s \leq t} \Delta X_s \Delta Y_s.$$

                                                                                    □

Definition 3.2.24 *Suppose $X = \{X_t, \mathcal{F}_t\}$ is a local martingale and let $X = X^c + X^d$ be its unique decomposition into a continuous local martingale and a totally discontinuous local martingale. Then the* optional quadratic variation *of X is the increasing process*

$$[X, X]_t = \langle X^c, X^c \rangle_t + \sum_{s \leq t} (\Delta X_s)^2.$$

If X, Y are local martingales,

$$[X, Y] = \frac{1}{2}([X + Y, X + Y] - [X, X] - [Y, Y])$$

$$= \langle X^c, Y^c \rangle_t + \sum_{s \leq t} \Delta X_s \Delta Y_s.$$

We end this section with the following useful inequalities. Write

$$\mathcal{H}^2 = \{\text{uniformly integrable (see Definition 1.3.34) martingales } \{M_t\}$$

$$\text{such that } \sup_t |M_t| \in L^2\} \tag{3.2.1}$$

Theorem 3.2.25 *Suppose $X, Y \in \mathcal{H}^2$ and f, g are measurable processes. (See Definition 2.1.9.) If $1 < p < \infty$ and $1/p + 1/q = 1$ then*

$$E\left[\int_0^\infty |f_s||g_s||d\langle X, Y \rangle_s|\right] \leq \left\|\left(\int_0^\infty f_s^2 d\langle X, X \rangle_s\right)^{1/2}\right\|_p \left\|\left(\int_0^\infty g_s^2 d\langle Y, Y \rangle_s\right)^{1/2}\right\|_q,$$

and

$$E\left[\int_0^\infty |f_s||g_s||d[X, Y]_s|\right] \leq \left\|\left(\int_0^\infty f_s^2 d[X, X]_s\right)^{1/2}\right\|_p \left\|\left(\int_0^\infty g_s^2 d[Y, Y]_s\right)^{1/2}\right\|_q.$$

Proof See [11] page 102. ∎

Theorem 3.2.26 *(Time-Change for Martingales). If M is an $\mathcal{F}_t$-continuous local martingale vanishing at 0 and such that $\langle M, M \rangle_\infty = \infty$ and if we set*

$$T_t = \inf\{s : \langle M, M \rangle_s > t\},$$

then $B_t = M_{T_t}$ is an $\mathcal{F}_{T_t}$-Brownian motion and $M_t = B_{\langle M, M \rangle_t}$.

Proof See [32] page 181. ∎

3.3 Simple examples of stochastic integrals

Example 3.3.1 Suppose $\{X_t\}$, $t \geq 0$, is a stochastic process representing the random price of some asset.

Consider a partition $0 = t_0^n < t_1^n < \cdots < t_n^n = t$ of the interval $[0, t]$.

Suppose ξ_{t_i}, $i = 0, 1, \ldots, n - 1$, is the amount of the asset which is bought at time t_i for the price X_{t_i}. This amount ξ_{t_i} is held until time t_{i+1} when it is sold for price $X_{t_{i+1}}$. The amount gained (or lost) is, therefore, $\xi_{t_i}(X_{t_{i+1}} - X_{t_i})$. Then $\xi_{t_{i+1}}$ is bought at time t_{i+1}. Clearly ξ_{t_i} should be predictable with respect to the filtration $\{\mathcal{F}_t^X\}$ generated by X. Then

$$\sum_{i=0}^{n-1} \xi_{t_i}(X_{t_{i+1}} - X_{t_i}) \stackrel{\Delta}{=} \int_0^t \xi_s dX_s$$

is the total increase (or loss) in the trader's wealth from holding these amounts of the asset. □

Example 3.3.2 Since the sample paths of a Poisson process N_t are increasing and of finite variation we can write

$$\int_0^t X_s(\omega)\mathrm{d}N_s(\omega) = \sum_{k=1}^{\infty} X_{\alpha_k(\omega)}(\omega)I_{(\alpha_k \leq t)}(\omega),$$

where α_k is the time of the k-th jump. Recall that the number of jumps in any finite interval $[0, t]$ is finite with probability 1. Hence the infinite series has only finitely many nonzero terms for almost all ω. $\square$

Example 3.3.3 Stochastic integration with respect to the family of martingales $q(t, A)$ related to the single jump process (See Examples 2.1.4 and 2.6.12) is simply ordinary (Stieltjes) integration with respect to the measures μ and μ_p applied to suitable integrands. Recall that μ picks out the jump time T and the location Z of the stochastic process X, that is $\mu(\mathrm{d}s, \mathrm{d}z)$ is non-zero only when $T \in \mathrm{d}s$ and $Z \in \mathrm{d}z$. Therefore, we may write for any suitable real valued function g defined on $\Omega = [0, \infty] \times E$:

$$\int_{\Omega} g(s, z)q(\mathrm{d}s, \mathrm{d}z) = \int_{\Omega} g(s, z)\mu(\mathrm{d}s, \mathrm{d}z) - \int_{\Omega} g(s, z)\mu_p(\mathrm{d}s, \mathrm{d}z),$$

where

$$\int_{\Omega} g(s, z)\mu(\mathrm{d}s, \mathrm{d}z) \stackrel{\triangle}{=} g(T, Z),$$

since the random measure μ picks out the jump time T and the location Z only. We say that $g \in L^1(\mu)$ if

$$\|g\|_{L^1(\mu)} \stackrel{\triangle}{=} E\left[\int_E |g|\mathrm{d}\mu\right] = E[g(T, Z)] < \infty.$$

We say that $g \in L^1_{\mathrm{loc}}(\mu)$ if $gI_{\{t < \tau_n\}} \in L^1(\mu)$ for some sequence of stopping times $\tau_n \uparrow \infty$ a.s.

Using (2.1.1) and (2.1.2) we have

$$\mu_p(t, A) = -\int_{]0, T \wedge t]} \frac{\mathrm{d}F_s^A}{F_{s-}} = \int_{]0, T \wedge t]} \lambda(A, s)\mathrm{d}\Lambda(s). \tag{3.3.1}$$

Hence

$$\int_{\Omega} g(s, z)\mu_p(\mathrm{d}s, \mathrm{d}z) = \int_{]0, T]} \int_E g(s, z)\lambda(\mathrm{d}z, s)\mathrm{d}\Lambda(s).$$

We also have

$$\int_{\Omega} g(s, z)\mu_p(\mathrm{d}s, \mathrm{d}z) = \int_{]0, T]} \int_E g(s, z)\frac{P(\mathrm{d}s, \mathrm{d}z)}{F_{s-}}.$$

Define

$$M_t^g \stackrel{\triangle}{=} \int_E I_{\{s \leq t\}}g(s, z)q(\mathrm{d}s, \mathrm{d}z)$$

$$= \int_E I_{\{s \leq t\}}g(s, z)\mu(\mathrm{d}s, \mathrm{d}z) - \int_E I_{\{s \leq t\}}g(s, z)\mu_p(\mathrm{d}s, \mathrm{d}z),$$

or, from the definition,

$$M_t^g = g(T, Z)I_{\{T \le t\}} + \int_{]0,T \wedge t]} \int_E g(s, z)\frac{P(ds, dz)}{F_{s-}}.$$

Theorem 3.3.4 M_t^g is an $\mathcal{F}_t$-martingale for $g \in L^1(\mu)$.

Proof For $t > s$,

$$E[M_t^g - M_s^g \mid \mathcal{F}_s] = E\left[g(T, Z)(I_{\{T \le t\}} - I_{\{T \le s\}})\right.$$
$$\left. + \int_{]0,T \wedge t]} \int_E g(u, z)\frac{P(du, dz)}{F_{u-}} - \int_{]0,T \wedge s]} \int_E g(u, z)\frac{P(du, dz)}{F_{u-}} \mid \mathcal{F}_s\right].$$

So we must show that

$$E[g(T, Z)(I_{\{T \le t\}} - I_{\{T \le s\}}) \mid \mathcal{F}_s]$$
$$= -E\left[\int_{]0,T \wedge t]} \int_E g(u, z)\frac{P(du, dz)}{F_{u-}}\right.$$
$$\left. - \int_{]0,T \wedge s]} \int_E g(u, z)\frac{P(du, dz)}{F_{u-}} \mid \mathcal{F}_s\right]. \tag{3.3.2}$$

First note that if $T \le s$ both sides of (3.3.2) are zero. Now

$$E[g(T, Z)(I_{\{T \le t\}} - I_{\{T \le s\}}) \mid \mathcal{F}_s] = E[g(T, Z)I_{\{s < T \le t\}} \mid T > s]I_{\{T > s\}}$$
$$= \frac{I_{\{T > s\}}}{F_s} \int_{]s,t]} \int_E g(u, z)P(du, dz),$$

and

$$-E\left[\int_{]0,T \wedge t]} \int_E g(u, z)\frac{P(du, dz)}{F_{u-}} - \int_{]0,T \wedge s]} \int_E g(u, z)\frac{P(du, dz)}{F_{u-}} \mid \mathcal{F}_s\right]$$
$$= -E\left[\int_{]0,T \wedge t]} \int_E g(u, z)\frac{P(du, dz)}{F_{u-}}\right.$$
$$\left. - \int_{]0,s]} \int_E g(u, z)\frac{P(du, dz)}{F_{u-}} \mid T > s\right]I_{\{T > s\}}$$
$$= -\frac{I_{\{T > s\}}}{P(T > s)}E\left[\left[\int_{]0,T \wedge t]} \int_E g(u, z)\frac{P(du, dz)}{F_{u-}}\right.\right.$$
$$\left.\left. - \int_{]0,s]} \int_E g(u, z)\frac{P(du, dz)}{F_{u-}}\right](I_{\{T > t\}} + I_{\{s < T \le t\}})\right]$$
$$= \frac{I_{\{T > s\}}}{F_s}\left[-F_t \int_{]s,t]} \int_E g(u, z)\frac{P(du, dz)}{F_{u-}}\right.$$
$$\left. + \int_{]s,t]} \int_{]s,r]} \int_E g(u, z)\frac{P(du, dz)}{F_{u-}}dF_r\right].$$

Interchanging the order of integration, the triple integral is

$$= \int_{]s,t]} \int_{]s,r]} \int_E g(u,z) \frac{P(du,dz)}{F_{u-}} dF_r$$

$$= \int_{]s,t]} \frac{1}{F_{u-}} \int_{]u,t]} dF_r \int_E g(u,z) P(du,dz)$$

$$= \int_{]s,t]} \int_E \frac{1}{F_{u-}} (F_t - F_{u-}) g(u,z) P(du,dz)$$

$$= F_t \int_{]s,t]} \int_E g(u,z) \frac{P(du,dz)}{F_{u-}} - \int_{]s,t]} \int_E g(u,z) P(du,dz).$$

Therefore (3.3.2) holds and the result follows.   ∎

□

3.4 Stochastic integration with respect to a Brownian motion

Let $B = \{B_t, t \geq 0\}$ be a Brownian motion process and let $0 = t_0^n < t_1^n < \cdots < t_n^n = t$ denote a sequence of partitions of the interval $[0, t]$ such that $\delta_n = \max(t_k^n - t_{k-1}^n) \to 0$ as $n \to \infty$. Write formally

$$I_t = \int_0^t B_s dB_s.$$

If the usual integration-by-parts formula were true for stochastic integrals $I_t = B_t^2 - \int_0^t B_s dB_s$, so $I_t = \frac{1}{2} B_t^2$. (This assumes the existence, in some sense, of the limit, as $\delta_n = \max(t_k^n - t_{k-1}^n) \to 0 \, (n \to \infty)$, of the Riemann–Stieltjes sums $S_n = \sum_{k=1}^n B_{\tau_k^n}(B_{t_k^n} - B_{t_{k-1}^n})$, where $t_{k-1}^n \leq \tau_k^n \leq t_k^n$.)

Now S_n can be written as

$$S_n = \frac{1}{2} B_t^2 + S_n', \tag{3.4.1}$$

where

$$S_n' = -\frac{1}{2} \sum_{k=1}^n (B_{t_k^n} - B_{t_{k-1}^n})^2 + \sum_{k=1}^n (B_{\tau_k^n} - B_{t_{k-1}^n})^2$$

$$+ \sum_{k=1}^n (B_{t_k^n} - B_{\tau_k^n})(B_{\tau_k^n} - B_{t_{k-1}^n}).$$

To see this write

$$B_{\tau_k^n}(B_{t_k^n} - B_{t_{k-1}^n}) = (B_{\tau_k^n} - B_{t_{k-1}^n} + B_{t_{k-1}^n})(B_{t_k^n} - B_{\tau_k^n} + B_{\tau_k^n} - B_{t_{k-1}^n})$$

$$= (B_{t_k^n} - B_{\tau_k^n})(B_{\tau_k^n} - B_{t_{k-1}^n}) + (B_{\tau_k^n} - B_{t_{k-1}^n})^2$$

$$+ B_{t_{k-1}^n}(B_{t_k^n} - B_{t_{k-1}^n}). \tag{3.4.2}$$

The last term in (3.4.2) is written

$$
\begin{aligned}
B_{t_{k-1}^n}(B_{t_k^n} - B_{t_{k-1}^n}) &= (B_{t_{k-1}^n} - B_{t_k^n} + B_{t_k^n})(B_{t_k^n} - B_{t_{k-1}^n}) \\
&= -(B_{t_k^n} - B_{t_{k-1}^n})^2 + B_{t_k^n}(B_{t_k^n} - B_{t_{k-1}^n}) \\
&= -\frac{1}{2}[2(B_{t_k^n} - B_{t_{k-1}^n})^2 - 2B_{t_k^n}(B_{t_k^n} - B_{t_{k-1}^n})] \\
&= -\frac{1}{2}(B_{t_k^n} - B_{t_{k-1}^n})^2 - \frac{1}{2}[(B_{t_k^n} - B_{t_{k-1}^n}) - B_{t_k^n}]^2 + \frac{1}{2}B_{t_k^n}^2 \\
&= -\frac{1}{2}(B_{t_k^n} - B_{t_{k-1}^n})^2 - \frac{1}{2}B_{t_{k-1}^n}^2 + \frac{1}{2}B_{t_k^n}^2.
\end{aligned}
$$

Using this form of S_n one can show that if $\tau_k^n = (1 - \alpha)t_k + \alpha t_{k-1}^n, 0 \le \alpha \le 1$, then

$$
\lim_{\delta_n \to 0} S_n \overset{L^2}{=} \frac{B_t^2}{2} + (\alpha - \tfrac{1}{2})t = I_t(\alpha),
$$

where S_n is given by (3.4.1).

It is interesting to notice that the stochastic integral

$$
I_t(\alpha) = \frac{B_t^2}{2} + (\alpha - \tfrac{1}{2})t
$$

is an $\mathcal{F}_t$-martingale if and only if $\alpha = 0$. When $\alpha = 0$ the integrand $B_{t_{k-1}^n}$ is $\mathcal{F}_{t_{k-1}^n}$-measurable and so does not anticipate future events in $\mathcal{F}_{t_k^n}$. Then, because B has independent increments, $B_{t_{k-1}^n}$ is independent of the integrator $B_{t_k^n} - B_{t_{k-1}^n}$ which gives $E[S_n] = 0$.

K. Itô [17] has given a definition of the integral $\int_0^t f(s, \omega)\mathrm{d}B_s(\omega)$ for the class of predictable, locally square integrable stochastic processes $\{f(t, \omega)\}$.

The next important step was given by H. Kunita and S. Watanabe in 1967 [24]. They extended the definition of Itô by replacing the Brownian motion process by an arbitrary square integrable martingale $\{X_t\}$ employing the quadratic variation processes $\langle X, X \rangle_t$.

The stochastic (Itô) integral with respect to a Brownian motion integrator will be defined for two classes of integrands. The larger class of integrands gives an integral which is a local martingale. The more restricted class of integrands gives an integral which is a martingale.

Suppose $(\Omega, \mathcal{F}, P)$ is a probability space and $B = \{B_t, t \ge 0\}$ is a standard Brownian motion. Write $\mathcal{F}_t^0 = \sigma\{B_u : u \le t\}$ and $\{\mathcal{F}_t, t \ge 0\}$ for the right continuous, complete filtration generated by B.

Let H be the set of all adapted, measurable processes $\{f(\omega, t), \mathcal{F}_t\}$ such that with probability 1,

$$
\int_0^t f^2(\omega, s)\mathrm{d}s < \infty, \quad \forall t \ge 0,
$$

and let $\{H^2, ||.||_{H^2}\}$ be the normed space of all adapted, measurable processes $\{f(\omega, t), \mathcal{F}_t\}$ such that

$$E\left[\int_0^t f^2(\omega, s)ds\right] < \infty, \quad \forall t \geq 0,$$

where $||f||_{H^2} = E\left[\int_0^t f^2(\omega, s)ds\right]^{1/2}$, for $f \in H^2$.

It is clear that $H^2 \subset H$, since for a nonnegative random variable X, if

$$P(X = \infty) \neq 0 \quad \text{then} \quad E[X] = \infty,$$

in other words, if

$$E[X] < \infty \quad \text{then} \quad P(X = \infty) = 0.$$

In our case the nonnegative random variable is $\int_0^t f^2(\omega, s)ds$.

As in the definition of the (deterministic) Stieltjes integral, a natural way to define the stochastic integral is to start with *simple functions*, that is, piecewise constant functions.

Definition 3.4.1 *A (bounded and predictable) function $f(\omega, t)$ is simple on the interval $[0, t]$ if $f(0, \omega)$ is constant and for $s \in (0, t]$,*

$$f(s, \omega) = \sum_{k=0}^{n-1} f_k(\omega) I_{(t_k, t_{k+1}]}(s),$$

where $0 = t_0, \ldots, t_n = t$ is a partition of the interval $[0, t]$ independent of ω, each $f_k(\omega)$ is $\mathcal{F}_{t_k}$ measurable and $E[f_k^2] < \infty$.

For any simple function $f(\omega, t) \in H$ (or $f(\omega, t) \in H^2$) the Itô stochastic integral is defined as

$$I(f) = \int_0^t f(\omega, s)dB_s(\omega) \overset{\triangle}{=} \sum_k f(t_k, \omega)(B_{t_{k+1}}(\omega) - B_{t_k}(\omega))$$

$$= \sum_k f_k(\omega)(B_{t_{k+1}}(\omega) - B_{t_k}(\omega)).$$

Note that each f_k is $\mathcal{F}_{t_k}$-measurable and hence independent of the integrator $(B_{t_{k+1}} - B_{t_k})$ because of the independent increment property of the Brownian motion $B = \{B_t, t \geq 0\}$.

In order to define the integral for functions in $\{H^2, ||.||_{H^2}\}$ we need a few preliminary results.

Lemma 3.4.2 *([16]). Let $(\Omega, \mathcal{F}, \mathcal{F}_t, P)$ be a filtered probability space. Let $\mathcal{L}$ be a linear space of real and bounded measurable stochastic processes such that*

1. *$\mathcal{L}$ contains all bounded, left-continuous adapted processes,*
2. *if $\{X^n\}$ is a monotone increasing sequence of processes in $\mathcal{L}$ such that $X = \sup_n X^n$ is bounded, then $X \in \mathcal{L}$.*

Then $\mathcal{L}$ contains all bounded predictable processes.

Proof See [16] page 21.   ∎

Lemma 3.4.3 *Let $\mathcal{S}^2$ be the set of all simple processes in H^2. Then*

1. *$\mathcal{S}^2$ is dense in H^2.*
2. *For $f \in \mathcal{S}^2$, $\|I(f)\|_{L^2} = \|f\|_{H^2}$.*
3. *For $f \in \mathcal{S}^2$, $E[I(f)] = 0$.*

Proof

1. Let $f \in H^2$ and for $K > 0$ set $F^K = fI_{[-K,K]}$. Then $f^K \in H^2$ and $\|f - f^K\|_{H^2} \to 0$ as $K \to \infty$. Therefore suppose that $f \in H^2$ is bounded. Let $\mathcal{L} = \{f \in H^2 : f$ is bounded and there exists $f_n \in \mathcal{S}^2$ such that $\|f - f_n\|_{H^2} \to 0, n \to \infty\}$. $\mathcal{L}$ is linear and is closed under monotone increasing sequences. If f is left-continuous bounded and adapted one can set $f_n(0, \omega) = f(0, \omega)$, and for $t > 0$,

$$f_n(t, \omega) = f(k/2^n, \omega)I_{(k/2^n, (k+1)/2^n]}(t), \quad k = 0, 1, \ldots.$$

Then $f_n \in \mathcal{S}^2$ and by bounded convergence $\|f - f_n\|_{H^2} \to 0, n \to \infty$. Now, in view of Lemma 3.4.2, $\mathcal{L}$ contains all bounded, predictable processes and $\mathcal{L}$ contains all bounded processes in H^2. (See [16] Remark 1.1, page 45.)

2. $\|I(f)\|_{L^2} = E[I(f)]^2 = E[(\sum_k f(t_k, \omega)(B_{t_{k+1}}(\omega) - B_{t_k}(\omega)))^2] \stackrel{\triangle}{=} E[(\sum_k A_k)^2]$

$$= \sum_k E[(A_k)^2] + 2\sum_{i<j} E[A_i A_j] = \sum_k E[E[(A_k)^2 \mid \mathcal{F}_k]]$$

$$+ 2\sum_{i<j} E[E[A_i A_j \mid \mathcal{F}_j]] = \sum_k E[E[f^2(t_k, \omega)(B_{t_{k+1}} - B_{t_k})^2 \mid \mathcal{F}_k]]$$

$$= E[\sum_k f^2(t_k, \omega)(t_{k+1} - t_k)] = E[\int_0^t f^2(\omega, s)ds] = \|f\|_{H^2}^2.$$

The proof of the last part of the lemma is left as an exercise. ∎

Theorem 3.4.4 *Suppose that $f(\omega, t) \in H^2$. Then there exists an (a.s. unique) L^2-random variable $I(f)$ such that $I(f_n) \stackrel{L^2}{\to} I(f)$ independently of the choice of the sequence of simple functions $f_n(\omega, t) \in H^2$, that is*

$$\int_0^t f_n(\omega, s)dB_s(\omega) \stackrel{L^2}{\to} \int_0^t f(\omega, s)dB_s(\omega). \tag{3.4.3}$$

The left hand side of (3.4.3) is also called the Itô integral of f.

Proof In view of Lemma 3.4.3 we have that for $f(\omega, t) \in H^2$ there exists a sequence of simple functions $f_n \in \mathcal{S}^2$ such that

$$\lim f_n \stackrel{H^2}{\to} f,$$

and we see that

$$\|I(f_n) - I(f_m)\|_{L^2} = \|I(f_n - f_m)\|_{L^2} = \|f_n - f_m\|_{H^2} \to 0.$$

However, L^2 is complete, so that the Cauchy sequence $I(f_n)$ has a limit $I(f) \in L^2$.

Suppose that $\{f'_n\}$ is a second sequence converging to f but $I(f'_n)$ converges to another integral $I'(f)$. Then

$$||f_n - f||_{H^2} + ||f - f'_n||_{H^2} \geq ||f_n - f'_n||_{H^2} = ||I(f_n) - I(f'_n)||_{L^2}.$$

However, $||f_n - f||_{H^2} + ||f - f'_n||_{H^2} \to 0$ by assumption and therefore $||I(f_n) - I(f'_n)||_{L^2} \to 0$, which establishes the uniqueness of the limit $I(f)$. ∎

Remark 3.4.5 Since $I(f_n) \overset{L^2}{\to} I(f)$, then $\lim E[I(f_n)] = E[I(f)]$ and in view of Lemma 3.4.3 we have:

1. For $f \in H^2$, $E[I(f)] = 0$.
2. For $f \in H^2$, $||I(f)||_{L^2} = ||f||_{H^2}$. □

3.5 Stochastic integration with respect to general martingales

Recall that $\mathcal{H}^2$ is given by (3.2.1). Write

$$\begin{aligned}
\mathcal{S} &= \{\text{bounded simple predictable} \\
&\qquad \text{processes (Definition 3.4.1)}\}, & (3.5.1) \\
\mathcal{H}_0^2 &= \{\{M_t\} \in \mathcal{H}^2 : M_0 = 0 \text{ a.s.}\}, & (3.5.2) \\
\mathcal{H}^{2,c} &= \{\{M_t\} \in \mathcal{H}^2 \text{ and } \{M_t\} \text{ is continuous}\}, & (3.5.3) \\
\mathcal{H}_0^{2,c} &= \{\{M_t\} \in \mathcal{H}^{2,c} : M_0 = 0 \text{ a.s.}\}. & (3.5.4)
\end{aligned}$$

Suppose $X \in \mathcal{H}^2$. Then the integral

$$\int_0^t f(s, \omega) dX_s \overset{\triangle}{=} f_0 X_0 + \sum_0^n f_k(X_{t_{k+1} \wedge t} - X_{t_k \wedge t}) \text{ exists for } f \in \mathcal{S}.$$

Lemma 3.5.1 *([11]).* $\displaystyle\int_0^t f(s) dX_s \in \mathcal{H}^2$ *and*

$$E\left[\left(\int_0^\infty f(s) dX_s\right)^2\right] = E\left[\int_0^\infty f^2(s, \omega) d\langle X, X \rangle_s\right]$$
$$= E\left[\int_0^\infty f^2(s, \omega) d[X, X]_s\right].$$

Proof By definition

$$\int_0^t f(s, \omega) dX_s = f_0 X_0 + \sum_0^n f_k(X_{t_{k+1} \wedge t} - X_{t_k \wedge t}).$$

By the optional stopping theorem, for $s \leq t$:

$$E\left[\int_0^t f(z, \omega) dX_z \mid \mathcal{F}_s\right] = \int_0^s f(z, \omega) dX_z.$$

For $k < \ell$, so that $k + 1 \le \ell$,

$$E[f_k f_\ell (X_{t_{k+1} \wedge t} - X_{t_k \wedge t})(X_{t_{\ell+1} \wedge t} - X_{t_\ell \wedge t})]$$
$$= E[E[f_k f_\ell (X_{t_{k+1} \wedge t} - X_{t_k \wedge t})(X_{t_{\ell+1} \wedge t} - X_{t_\ell \wedge t}) \mid \mathcal{F}_{t_\ell}]]$$
$$= 0.$$

Therefore

$$E\left[\left(\int_0^t f(s)\mathrm{d}X_s\right)^2\right] = E[\sum_0^n f_k^2 (X_{t_{k+1} \wedge t}^2 - X_{t_k \wedge t}^2)$$

$$= E[\sum_0^n f_k^2 (\langle X, X \rangle_{t_{k+1} \wedge t} - \langle X, X \rangle_{t_k \wedge t})$$

$$= E\left[\int_0^t f^2(s, \omega)\mathrm{d}\langle X, X \rangle_s\right]$$

$$\le E\left[\int_0^\infty f^2(s, \omega)\mathrm{d}\langle X, X \rangle_s\right] < \infty,$$

because f is bounded and $X \in \mathcal{H}^2$. The integrals on the right are Stieltjes integrals. Therefore, by Lebesgue's Theorem, letting $t \to \infty$:

$$E\left[\left(\int_0^\infty f(s)\mathrm{d}X_s\right)^2\right] = E\left[\int_0^\infty f^2(s, \omega)\mathrm{d}\langle X, X \rangle_s\right].$$

Finally note that $\langle X, X \rangle - [X, X]$ is a martingale of integrable variation and the result follows.   ∎

Theorem 3.5.2 *Write $L^2(\langle X, X \rangle)$ for the space of predictable processes $\{f(\omega, t)\}$ such that*

$$\|f\|^2_{\langle X, X \rangle} = E\left[\int_0^\infty f^2(\omega, s)\mathrm{d}\langle X, X \rangle_s\right] < \infty.$$

Then the map $f \to \int_0^t f\mathrm{d}X$ of S into $\mathcal{H}^2$ extends in a unique manner to a linear isometry of $L^2(\langle X, X \rangle)$ into $\mathcal{H}^2$.

Proof   Suppose that the space S is endowed with the seminorm $\|.\|_{\langle X, X \rangle}$. Then from Lemma 3.5.1 the map $f \to \int_0^t f\mathrm{d}X$ of S into $\mathcal{H}^2$ is an isometry. However, S is dense in $\mathcal{H}^2$ and this map extends in a unique manner to an isometry of $L^2(\langle X, X \rangle)$ into $\mathcal{H}^2$.   ∎

The following characterization is due to Kunita and Watanabe [24]. (See [11] page 107.)

Theorem 3.5.3 *Suppose $f \in L^2(\langle X, X \rangle)$.*

1. Then for every $Y \in \mathcal{H}^2$,

$$E\left[\int_0^\infty |f(s)||\mathrm{d}\langle X, Y \rangle_s|\right] < \infty,$$

$$E\left[\int_0^\infty |f(s)||\mathrm{d}[X, Y]_s|\right] < \infty.$$

2. *The stochastic integral* $I_t = \int_0^t f(s)\mathrm{d}X_s$ *is characterized as the unique element of* $\mathcal{H}^2$
such that for every $Y \in \mathcal{H}^2$,

$$E[I_\infty Y_\infty] = E\left[\int_0^\infty f(s)\mathrm{d}\langle X, Y\rangle_s\right] = E\left[\int_0^\infty f(s)\mathrm{d}[X, Y]_s\right].$$

3. *Furthermore, for every* $Y \in \mathcal{H}^2$,

$$\langle I, Y\rangle_t = \int_0^t f(s)\mathrm{d}\langle X, Y\rangle_s,$$

$$\langle I, Y\rangle_t = \int_0^t f(s)\mathrm{d}[X, Y]_s.$$

Proof

1. Follows from Theorem 3.2.25.
2. The linear functional on $L^2(\langle X, X\rangle)$ defined by

$$f \rightarrow E\left[I_\infty Y_\infty - \int_0^\infty f(s)\mathrm{d}\langle X, Y\rangle_s\right]$$

is continuous by Theorem 3.2.25 and it is zero on the space of simple processes $\mathcal{S}$ which is dense in $L^2(\langle X, X\rangle)$. Therefore it is zero on $L^2(\langle X, X\rangle)$ by continuity. The second identity follows because $\langle X, Y\rangle - [X, Y]$ is a martingale of integrable variation.
3. Note that

$$j_t \overset{\triangle}{=} I_t Y_t - \int_0^t f(s)\mathrm{d}\langle X, Y\rangle_s \le \sup_t |I_t Y_t| + \int_0^\infty |f(s)||\mathrm{d}\langle X, Y\rangle_s| \in L^1.$$

Applying the identity in part (ii) it is seen that, for any stopping time T, $E[J_T] = 0$. In view of Theorem 2.5.7, j_t is a martingale. However, $\langle I, Y\rangle_t$ is the unique predictable process of integrable variation such that $I_t Y_t - \langle I, Y\rangle_t$ is a martingale. Therefore, $\langle I, Y\rangle_t = \int_0^t f(s)\mathrm{d}\langle X, Y\rangle_s$.
To prove the last identity, decompose X and Y into their continuous and totally discontinuous parts and then use a similar argument. (See [11] page 108.) ∎

Note that the first identity in (2) uniquely characterizes the stochastic integral I. This is because the right hand side is a continuous linear functional Y (given f and X), whilst the left hand side is just the inner product of I and Y in the Hilbert space $\mathcal{H}^2$. Consequently, given f and X there is a unique $I \in \mathcal{H}^2$ which gives this linear functional.

Definition 3.5.4 *A process* $\{f(t, \omega)\}$ *is locally bounded process if* $\{f(0, \omega)\}$ *is a.s. finite, and if there is a sequence of stopping times* $\tau_n \uparrow \infty$ *and constants* K_n *such that*

$$|f(t, \omega)|I_{\{0 < t \le \tau_n\}} \le K_n \quad a.s.$$

Definition 3.5.5 *A martingale* X *is a locally uniformly integrable martingale if there is a sequence of stopping times* $\tau_n \uparrow \infty$ *such that the stopped martingale* $X_{\{\tau_n \wedge t\}}$ *is a uniformly integrable martingale.*

Theorem 3.5.6 *([11] page 121)*

1. *Suppose X is a locally uniformly integrable martingale and $\{f(t, \omega)\}$ is a predictable locally bounded process. There is then a unique local martingale $\{I_t \stackrel{\triangle}{=} \int_0^t f(s) \mathrm{d}X_s\}$ such that for every bounded martingale Y,*

$$[I, Y]_t = \int_0^t f(s) \mathrm{d}[X, Y]_s.$$

   (Here the right hand side is just a Stieltjes integral on each sample path.)
2. *$I_0 = f(0)X_0$, $I_t^c = \int_0^t f(s) \mathrm{d}X_s^c$ and the processes $\Delta(I_t)$ and $f(t, \omega)\Delta X_t$ are indistinguishable.*
3. *If the local martingale X is also of locally integrable variation (see Definition 2.6.2) then I_t can be calculated as the Stieltjes integral along each sample path.*

Proof Assume that $X_0 = 0$ and $f(0) = 0$. There is a sequence of stopping times $\tau_n \uparrow \infty$ such that the stopped martingale $X_{\{\tau_n \wedge t\}}$ is a uniformly integrable and bounded martingale by Theorem 2.5.11. Using Theorem 2.5.12 we can write $X_{\{\tau_n \wedge t\}} = U_{\{\tau_n \wedge t\}} + V_{\{\tau_n \wedge t\}}$, where $U_0 = 0$, $U_{\{\tau_n \wedge t\}}$ is square integrable and $V_{\{\tau_n \wedge t\}}$ is a martingale of integrable variation which is zero at $t = 0$. The stochastic integral

$$\int_0^t f(s) \mathrm{d}X_{\{\tau_n \wedge t\}} = \int_0^t f(s) \mathrm{d}U_{\{\tau_n \wedge t\}} + \int_0^t f(s) \mathrm{d}V_{\{\tau_n \wedge t\}}$$

is defined by Theorem 3.5.2. Furthermore, this integral a uniformly integrable martingale.

If $n < m$ (so that $\tau_n \le \tau_m$ a.s.), because $X_{\{\tau_n \wedge t\}}$ is equal to $X_{\{\tau_m \wedge t\}}$ stopped at τ_n we have $\int_0^t f(s) \mathrm{d}X_{\{\tau_n \wedge t\}}$ is equal to $\int_0^t f(s) \mathrm{d}X_{\{\tau_m \wedge t\}}$ stopped at τ_n.

A process $\{I_t \stackrel{\triangle}{=} \int_0^t f(s) \mathrm{d}X_s\}$ is then defined by putting $\{I_{\{\tau_n \wedge t\}} = \int_0^t f(s) \mathrm{d}X_{\{\tau_n \wedge t\}}\}$ and it is seen that I_t is a local martingale.

The rest of the proof is left as an exercise. ∎

3.6 The Itô formula for semimartingales

Write

$$\mathcal{V} = \{\{V_t\} \text{ which are adapted, right-continuous with left limits}$$
$$\text{(corlol or càdlàg) and for which almost every sample path}$$
$$\text{is of finite variation on each compact subset of } [0, \infty[]\}, \tag{3.6.1}$$

$$\mathcal{V}_0 = \{\{V_t\} \in \mathcal{V} : V_0 = 0 \text{ a.s}\}, \tag{3.6.2}$$

$$\mathcal{A} = \{\{V_t\} \in \mathcal{V} : E\left[\int_0^\infty |\mathrm{d}V_t|\right] < \infty\}, \tag{3.6.3}$$

$$\mathcal{A}_0 = \{\{V_t\} \in \mathcal{A} : V_0 = 0 \text{ a.s.}\}. \tag{3.6.4}$$

Theorem 3.6.1 *Suppose $X = X_0 + M + V$ is a semimartingale and $\{f(t, \omega)\}$ is a predictable locally bounded process. Then the process*

$$I_t = \int_0^t f(s, \omega) \mathrm{d}X_s(\omega) \stackrel{\triangle}{=} f(0)X_0 + \int_0^t f(s, \omega) \mathrm{d}M_s(\omega) + \int_0^t f(s, \omega) \mathrm{d}V_s(\omega)$$

is a semimartingale. I_t is independent of the decomposition X, and the processes I_t^c and
$\int_0^t f(s)dX_s^c$ and $\Delta(I_t)$ and $f(t)\Delta X_t$ are indistinguishable.

Proof Suppose $X = X_0 + \hat{M} + \hat{V}$ is a second decomposition of X. Then $M - \hat{M} = V - \hat{V}$ is a local martingale which is locally of integrable variation. Therefore, by Theorem 3.5.6(3) the stochastic integral $\int_0^t f(s)d(M - \hat{M})_s$ is equal to the Stieltjes integral $\int_0^t f(s)d(V - \hat{V})_s$, and so

$$f(0)X_0 + \int_0^t f(s, \omega)dM_s(\omega) + \int_0^t f(s, \omega)dV_s(\omega) = f(0)X_0 + \int_0^t f(s, \omega)d\hat{M}_s(\omega)$$
$$+ \int_0^t f(s, \omega)d\hat{V}_s(\omega).$$

Because $X^c = M^c$ the processes $\int_0^t f(s)dX_s^c$ and I^c are indistinguishable by Theorem 3.5.6(ii). Similarly

$$f(t)\Delta X_t = f(t)(\Delta M_t + \Delta V_t) = \Delta\left(\int_0^t f(s)dM_s\right) + \Delta\left(\int_0^t f(s)dV_s\right) = \Delta(I_t).$$

∎

The Itô formula is first established for a continuous, bounded, real semimartingale.

Theorem 3.6.2 *Suppose $X = X_0 + M + V$ is a semimartingale such that $|X_0| \leq K$ a.s., $M \in \mathcal{H}_0^{2,c}$ (see (3.5.4)) and bounded by K, $V \in \mathcal{V}_0$ (see (3.6.3)), V is continuous and*

$$\int_0^\infty |dV_s| \leq K \quad \text{a.s.}$$

Let F be a twice continuously differentiable function on $\mathbb{R}$. Then

$$F(X_t) = F(X_0) + \int_0^t F'(X_{s-})dM_s + \int_0^t F'(X_{s-})dV_s + \frac{1}{2}\int_0^t F''(X_s)d\langle M, M \rangle_s.$$

$$(3.6.5)$$

That is, the processes on the left and right hand sides are indistinguishable.

Proof Write

$$I_1 = \int_0^t F'(X_{s-})dM_s, \quad I_2 = \int_0^t F'(X_{s-})dV_s, \quad I_3 = \int_0^t F''(X_s)d\langle M, M \rangle_s.$$

Now $|X| \leq 3K$. If $a, b \in [-3K, +3K]$,

$$F(b) - F(a) = (b - a)F'(a) + \frac{1}{2}(b - a)^2 F''(a) + r(a, b),$$

where, because F'' is uniformly continuous on $[-3K, +3K]$,

$$|r(a, b)| \leq \epsilon(|b - a|)(b - a)^2.$$

Here $\epsilon(s)$ is an increasing function of s such that $\lim_{s \to 0} \epsilon(s) = 0$.

A *stochastic* subdivision of $[0, t]$ is now defined by putting $t_0 = 0$,

$$t_{i+1} = t \wedge (t_i + a) \wedge \inf\{s > t_i : |M_s - M_{t_i}| > a \text{ or } |V_s - V_{t_i}| > a\},$$

where a is any positive real number.

Then as $a \to 0$ the steps of the subdivision, $\sup(t_{i+1} - t_i)$, converge uniformly to 0, and the random variables $\sup |M_{t_{i+1}} - M_{t_i}| \leq a$, $\sup |V_{t_{i+1}} - V_{t_i}| \leq a$, tend uniformly to 0. Therefore the variation of X on the interval $[t_i, t_{i+1}]$ is bounded by $4a$. Now

$$F(X_t) - F(X_0) = \sum_i F'(X_{t_i})(X_{t_{i+1}} - X_{t_i}) + \frac{1}{2} \sum_i F''(X_{t_i})(X_{t_{i+1}} - X_{t_i})^2$$

$$+ \sum_i r(X_{t_i}, X_{t_{i+1}}) \stackrel{\triangle}{=} S_1 + \frac{1}{2} S_2 + R, \quad \text{say}.$$

We shall show that as $a \to 0$, $S_1 \stackrel{P}{\to} I_1 + I_2$, $S_2 \stackrel{P}{\to} I_3$, and $R \stackrel{P}{\to} 0$.
Write

$$S_1 = \sum_i F'(X_{t_i})(M_{t_{i+1}} - M_{t_i}) + \sum_i F'(X_{t_i})(V_{t_{i+1}} - V_{t_i})$$

$$\stackrel{\triangle}{=} U_1 + U_2.$$

Step 1. We show that $U_1 \stackrel{L^2}{\to} I_1$.
  Write

$$I_1 = \sum_i \int_{t_i}^{t_{i+1}} F'(X_s) dM_s.$$

The martingale property implies different terms in the sum are mutually orthogonal, so

$$\|U_1 - I_1\|_2^2 = \sum_i \left\| \int_{t_i}^{t_{i+1}} (F'(X_s) - F'(X_{t_i})) dM_s \right\|_2^2$$

$$= E\left[\sum_i \int_{t_i}^{t_{i+1}} (F'(X_s) - F'(X_{t_i}))^2 d\langle M, M \rangle_s \right]$$

$$\leq E[\{\sup_t \sup_{t_i \leq s \leq t_{i+1}} (F'(X_s) - F'(X_{t_i}))^2\} \langle M, M \rangle_t].$$

By uniform continuity, the supremum tends uniformly to zero. $\langle M, M \rangle_t$ is integrable, so the result follows by the Monotone Convergence Theorem 1.3.15.

Step 2. We show that $U_2 \stackrel{L^1}{\to} I_2$.

$$|U_2 - I_2| \leq \sum_i \int_{t_i}^{t_{i+1}} |(F'(X_s) - F'(X_{t_i}))| |dV_s|$$

$$\leq \{\sup_t \sup_{t_i \leq s \leq t_{i+1}} |F'(X_s) - F'(X_{t_i})| \int_0^t |dV_s|\}.$$

Again by uniform continuity of F' and the Monotone Convergence Theorem 1.3.15, $||U_2 - I_2||_1$ converges to 0.

Step 3. Writing

$$S_2 = \sum_i F''(X_{t_i})(V_{t_{i+1}} - V_{t_i})^2 + 2\sum_i F''(X_{t_i})(V_{t_{i+1}} - V_{t_i})(M_{t_{i+1}} - M_{t_i})$$

$$+ \sum_i F''(X_{t_i})(M_{t_{i+1}} - M_{t_i})^2$$

$$\stackrel{\triangle}{=} V_1 + V_2 + V_3, \text{ respectively.}$$

We first show that V_1 and V_2 converge to 0 both a.s. and in L^1. However, if $C > \sup\{|F'(x)| + |F''(x)| : -3K \le x \le 3K\}$,

$$|V_1| \le C \sup_i |V_{t_{i+1}} - V_{t_i}| \int_0^t |dV_s| \le aCK.$$

Step 4. We show that $V_3 \stackrel{P}{\to} I_3$.

First recall that M is bounded by K, so

$$E[\langle M, M\rangle_\infty - \langle M, M\rangle_t \mid \mathcal{F}_t] = E[M_\infty^2 \mid \mathcal{F}] - M_t^2 \le K^2.$$

Therefore

$$E[\langle M, M\rangle_\infty^2] = 2E\left[\int_0^\infty (\langle M, M\rangle_\infty - \langle M, M\rangle_t)d\langle M, M\rangle_t\right]$$

$$= 2E\left[\int_0^\infty (E[M_\infty^2 \mid \mathcal{F}] - M_t^2)d\langle M, M\rangle_t\right]$$

$$\le 2K^2 E[\langle M, M\rangle_\infty] \le 2K^4.$$

Consequently $\langle M, M\rangle_\infty \in L^2$ and the martingale $M^2 - \langle M, M\rangle$ is actually in $\mathcal{H}_0^2$. Write

$$J_3 = \sum_i F''(X_{t_i})(\langle M, M\rangle_{t_{i+1}} - \langle M, M\rangle_{t_i}).$$

Then the same argument as Step 2 shows that

$$J_3 \stackrel{L^1}{\to} I_3 = \int_0^t F''(X_s)d\langle M, M\rangle_s.$$

Therefore, $J_3 \stackrel{P}{\to} I_3$. We shall show that $||V_3 - J_3||_{L_2}^2 \to 0$.
Because $M^2 - \langle M, M\rangle$ is a martingale,

$$E[(M_{t_{i+1}} - M_{t_i})^2 - \langle M, M\rangle_{t_{i+1}} + \langle M, M\rangle_{t_i} \mid \mathcal{F}_{t_i}] = 0.$$

Therefore, distinct terms in the sum defining $V_3 - J_3$ are orthogonal and

$$||V_3 - J_3||_2^2 = \sum_i E[F''(X_{t_i})^2((M_{t_{i+1}} - M_{t_i})^2 - \langle M, M\rangle_{t_{i+1}} + \langle M, M\rangle_{t_i})^2].$$

However, $F''(X_{t_i})^2 \leq C^2$ and $(\alpha - \beta)^2 \leq 2(\alpha^2 + \beta^2)$, so

$$\|V_3 - J_3\|_2^2 \leq 2C^2 \sum_i E(M_{t_{i+1}} - M_{t_i})^4$$
$$+ 2C^2 \sum_i E(\langle M, M \rangle_{t_{i+1}} - \langle M, M \rangle_{t_i})^2.$$

The second sum here is treated similarly to V_1 in Step 3: because $\langle M, M \rangle$ is uniformly continuous on $[0, t]$, $\sup_i(\langle M, M \rangle_{t_{i+1}} - \langle M, M \rangle_{t_i}) \overset{a.s.}{\to} 0$ as $a \to 0$ and is bounded by $\langle M, M \rangle_t$. Therefore

$$2C^2 \sum_i E(\langle M, M \rangle_{t_{i+1}} - \langle M, M \rangle_{t_i})^2$$
$$\leq 2C^2 E(\sup_i(\langle M, M \rangle_{t_{i+1}} - \langle M, M \rangle_{t_i})\langle M, M \rangle_t).$$

Now $\langle M, M \rangle_t \in L^2$, so the second sum converges to zero by Lebesgue's Dominated Convergence Theorem 1.3.17.

For the first sum,

$$2C^2 \sum_i E(M_{t_{i+1}} - M_{t_i})^4 \leq 2C^2 E(\sup_i(M_{t_{i+1}} - M_{t_i})^2 \sum_i (M_{t_{i+1}} - M_{t_i})^2)$$
$$\leq 2C^2 a^2 E(\sum_i (M_{t_{i+1}} - M_{t_i})^2) = 2C^2 a^2 E[M_t^2].$$

which again converges to zero as $a \to 0$. (Note that it is only here, where we use the fact that $|M_{t_{i+1}} - M_{t_i}| \leq a$, that the random character of the partition $\{t_i\}$ is used.)

We have, thus, shown that $V_3 - J_3 \overset{L^2}{\to} 0$. However, $J_3 \overset{P}{\to} I_3$ so $V_3 \overset{P}{\to} I_3$.

Step 5. Finally, we show that the remainder term R converges to 0 as $a \to 0$.

We have observed that the remainder term r in the Taylor expansion is such that

$$|r(a, b)| \leq \epsilon(|b - a|)(b - a)^2,$$

where ϵ is an increasing function and $\lim_{s \to 0} \epsilon(s) = 0$. Therefore,

$$|R| \leq \sum_i (X_{t_{i+1}} - X_{t_i})^2 \epsilon(|X_{t_{i+1}} - X_{t_i}|)$$
$$\leq 2\epsilon(2a) \sum_i ((V_{t_{i+1}} - V_{t_i})^2 + (M_{t_{i+1}} - M_{t_i})^2).$$

Now

$$E\left[\sum_i (M_{t_{i+1}} - M_{t_i})^2 \right] = E[M_t^2]$$

is independent of the partition, and

$$E\left[\sum_i (V_{t_{i+1}} - V_{t_i})^2 \right] \leq a E\left[\sum_i |V_{t_{i+1}} - V_{t_i}| \right] \leq K a.$$

Because $\lim_{a \to 0} \epsilon(2a) = 0$,

$$\lim_{a \to 0} |E(R)| \leq \lim_{a \to 0} E(|R|) = 0.$$

For a fixed t, therefore,

$$F(X_t) = F(X_0) + \int_0^t F'(X_{s-})dM_s + \int_0^t F'(X_{s-})dV_s$$

$$+ \frac{1}{2}\int_0^t F''(X_s)d\langle M, M\rangle_s,$$

almost surely. Because all the processes are right-continuous with left limits the two sides are indistinguishable (see Definition 2.1.5).   ■

The differentiation rule will next be proved for a function F which is twice continuously differentiable, and which has bounded first and second derivatives, and a semimartingale X of the form

$$X_t = X_0 + M_t + V_t,$$

where $X_0 \in L^1$, $M \in \mathcal{H}_0^2$ and $V \in \mathcal{A}_0$. That is, the following result will be proved after the lemmas and remarks below.

Theorem 3.6.3 *Suppose $X = X_0 + M + V$ is a semimartingale such that $X_0 \in L^1$ a.s., $M \in \mathcal{H}_0^2$, $V \in \mathcal{A}_0$ and F is twice continuously differentiable with bounded first and second derivatives. Then the following two processes, the left and right hand sides, are indistinguishable:*

$$F(X_t) = F(X_0) + \int_0^t F'(X_{s-})dX_s + \frac{1}{2}\int_0^t F''(X_{s-})d\langle X^c, X^c\rangle_s$$

$$+ \sum_{0<s\leq t} (F(X_s) - F(X_{s-}) - F'(X_{s-})\Delta X_s). \qquad (3.6.6)$$

Remarks 3.6.4

1. Note that the series is absolutely convergent because, if C is a bound for $|F''|$, then by Taylor's theorem

$$\sum_{0<s\leq t} |F(X_s) - F(X_{s-}) - F'(X_{s-})\Delta X_s| \leq \frac{C}{2}\sum_{0<s\leq t}(\Delta X_s)^2,$$

and the right hand side is finite as in Definition 3.2.18.
Also, because $\langle X^c, X^c\rangle$ is a continuous process,

$$\int_0^t F''(X_{s-})d\langle X^c, X^c\rangle_s = \int_0^t F''(X_s)d\langle X^c, X^c\rangle_s.$$

2. The first integral on the right of (3.6.6) is a well-defined stochastic integral since the integrand is predictable and locally bounded. Similar remarks apply to the second integral.
3. The series on the right of (3.6.6) is a correction term to balance the number of jumps on both sides of the equation.

4. Another form of the differentiation rule is the following:

$$F(X_t) = F(X_0) + \int_0^t F'(X_{s-})dX_s + \frac{1}{2}\int_0^t F''(X_{s-})d[X, X]_s$$

$$+ \sum_{0<s\leq t} (F(X_s) - F(X_{s-}) - F'(X_{s-})\Delta X_s - \frac{1}{2}F''(X_{s-})\Delta X_s^2).$$

This representation is of interest because whilst the predictable quadratic variation process $\langle X^c, X^c \rangle$ depends on the underlying probability measure, the optional quadratic variation process $[X, X]$ does not.

5. If $\{X_t\}$ is a deterministic function of bounded variation which is right-continuous and has left limits, we require F to be only once continuously differentiable and then:

$$F(X_t) = F(X_0) + \int_0^t F'(X_{s-})dX_s + \sum_{0<s\leq t} (F(X_s) - F(X_{s-}) - F'(X_{s-})\Delta X_s)$$

$$= \int_0^t F'(X_{s-})dX_s^c + \sum_{0<s\leq t} (F(X_s) - F(X_{s-})).$$

$\square$

Lemma 3.6.5 *Suppose the differentiation rule of Theorem 3.6.6 is true for all semimartingales of the form*

$$X_t = X_0 + N_t + B_t,$$

where X_0 belongs to some dense set in L^1, N belongs to some dense set in $\mathcal{H}_0^2$ and B belongs to some dense set in $\mathcal{A}_0$.

Then Theorem 3.6.6 is true for general semimartingales of the stated form.

Proof   See [11] page 133.   ∎

Lemma 3.6.6 *The semimartingale we need consider can be further restricted so that, if $X_t = X_0 + M_t + V_t$, then $M \in \mathcal{H}_0^{2,c}$ is bounded, $V \in \mathcal{A}$ has at most N jumps and $\int_0^\infty |dA_s^c|$, is bounded, and X_0 is bounded.*

Proof   See [11] page 137.   ∎

We now prove Theorem 3.6.6.

Proof of Theorem 3.6.6 From Lemma 3.6.5 and Lemma 3.6.6, the result of Theorem 3.6.6 will follow if it can be proved that for a semimartingale

$$X_t = X_0 + M_t + V_t,$$

where $X_0 \in L^1$ is bounded, $M \in \mathcal{H}_0^2$ and M is bounded, $V \in \mathcal{A}_0$ has at most N jumps and $\int_0^\infty |dV_s| < \infty.$

However, note that the two sides of the differentiation formula have the same jump at time t: because $\langle X^c, X^c \rangle$ is continuous the jump of the right hand side is

$$F'(X_{t-})\Delta X_t + (F(X_t) - F(X_{t-}) - F'(X_{t-})\Delta X_t) = F(X_t) - F(X_{t-}),$$

which is the jump of the left hand side at t. Consider the continuous semimartingale

$$\bar{X}_t = X_0 + M_t + V_t^c.$$

Then from Theorem 3.6.2 the differentiation rule is true for $\bar{X}$. Furthermore, if the jumps of X are indexed in increasing order as $0 < S_1 \le \cdots \le S_N \le \infty$, then $X_t = \bar{X}_t$ on the stochastic interval $\{(t, \omega) \in [0, \infty[\times \Omega : 0 \le t < S_1(\omega)\}$ (denoted $[[0, S_1[[)$. Therefore, $X_{t-} = \bar{X}_{t-}$ on the stochastic interval $\{(t, \omega) \in [0, \infty[\times \Omega : 0 \le t \le S_1(\omega)\}$ (denoted $[[0, S_1]]$) so

$$\int_0^t F'(X_{s-})dM_s = \int_0^t F'(\bar{X}_{s-})dM_s$$

on $[[0, S_1]]$. Also,

$$\int_0^t F'(X_{s-})dV_s = \int_0^t F'(\bar{X}_{s-})dV_s^c$$

on $[[0, S_1]]$.

Because $X^c = \bar{X}^c = M$ and the formula is true for $\bar{X}$ (on $[[0, \infty[[)$, the differentiation formula is true on $[[0, S_1[[$. However, the two sides of the formula have equal jumps at S_1, so the formula is true on $[[0, S_1]]$. The same reasoning establishes the formula on $\{(t, \omega) \in [0, \infty[\times \Omega : S_1(\omega) \le t \le S_2(\omega))$ ($[[S_1, S_2]]$), and so on, up to $[[S_N, \infty[[$. Theorem 3.6.6 is, therefore, proved.   ∎

The differentiation rule will now be given when X is a general semimartingale and F is twice continuously differentiable, not necessarily having bounded derivatives.

Theorem 3.6.7 *Suppose X a semimartingale and F is a twice continuously differentiable function. Then $F(X)$ is a semimartingale, and with equality denoting indistinguishability:*

$$F(X_t) = F(X_0) + \int_0^t F'(X_{s-})dX_s + \frac{1}{2}\int_0^t F''(X_s)d\langle X^c, X^c \rangle_s$$

$$+ \sum_{0 < s \le t} (F(X_s) - F(X_{s-}) - F'(X_{s-})\Delta X_s).$$

Proof   See [11] page 138.   ∎

Remark 3.6.8 Note that $F'(X_s)$ is right-continuous with left limit, and so $F'(X_{s-})$ is predictable. Also, by considering stopping times such as

$$S = \inf\{t : F'(X_s) \ge n\},$$

we see that $F'(X_{s-})$ is locally bounded. Similar remarks apply to $F''(X_{s-})$. Therefore, the two integrals are well-defined.

For any $\omega \in \Omega$ the trajectory $X_s(\omega)$, $0 \le s \le t$, remains in a compact interval $[-C(t, \omega), C(t, \omega)]$. On such a compact interval the second derivative of F is bounded by some constant $K(t, \omega)$. Therefore, if $s \leq t$,

$$|F(X_s) - F(X_{s-}) - F'(X_{s-})\Delta X_s| \leq \frac{1}{2}K(t, \omega)(\Delta X_s)^2(\omega).$$

As in Definition 3.2.18, we know that $\sum_{s \leq t}(\Delta X_s)^2(\omega)$ is finite almost surely. Therefore, for any t the sum occurring on the right hand side of the differentiation rule is a.s. absolutely convergent. $\qquad\square$

Theorem 3.6.9 If $\{X_t\} = \{(X_t^1, \ldots, X_t^n)\}$ is an n-vector semimartingale and F is a twice continuously differentiable function with respect to all arguments, we have

$$F(X_t^1, \ldots, X_t^n) = F(X_0) + \sum_i \int_0^t \frac{\partial F(X_{s-})}{\partial X_s^i} dX_s^i + \frac{1}{2}\sum_{i,j} \int_0^t \frac{\partial^2 F(X_{s-})}{\partial X^i \partial X^j} d\langle X^{ic}, X^{jc}\rangle_s$$

$$+ \sum_{0 < s \leq t} \left[F(X_s) - F(X_{s-}) - \sum_i \frac{\partial F(X_{s-})}{\partial X_s^i}\Delta X_s^i \right]. \qquad (3.6.7)$$

Remark 3.6.10 If $\{X_t\}$ is a deterministic right-continuous with left limits function of bounded variation, then

$$\Lambda_t = 1 + \int_0^t \Lambda_{s-} dX_s$$

has the unique exponential solution $(\Lambda_0 = 1)$

$$\Lambda_t = e^{X_t - X_0} \prod_{s \leq t}(1 + \Delta X_s)e^{\{-\Delta X_s\}}.$$

The next example is a generalization of the exponential formula to special semimartingales. $\qquad\square$

Example 3.6.11 We shall apply Theorem 3.6.9 to show that if X_t is a special semimartingale, then

$$\Lambda_t = 1 + \int_0^t \Lambda_{s-} dX_s \qquad (3.6.8)$$

has the unique stochastic exponential solution $(\Lambda_0 = 1)$

$$\Lambda_t = e^{X_t - \frac{1}{2}\langle X^c, X^c\rangle_t} \prod_{s \leq t}(1 + \Delta X_s)e^{\{-\Delta X_s\}}$$

$$\stackrel{\triangle}{=} e^{Y_{1t}} Y_{2t}, \qquad (3.6.9)$$

where

$$Y_{1t} = X_t - \frac{1}{2}\langle X^c, X^c\rangle_t,$$

$$Y_{2t} = \prod_{s \leq t}(1 + \Delta X_s)e^{-\Delta X_s}.$$

First note that the infinite product Y_{2t} is finite (see Lemma 13.7 of [11]).

Write $\Lambda_t = f(Y_{1t}, Y_{2t})$. Using rule (3.6.7),

$$f(Y_{1t}, Y_{2t}) = \Lambda_0 + \sum_{i=1}^{2} \int_0^t \frac{\partial f(Y_{1s-}, Y_{2s-})}{\partial Y_i} dY_{is} \tag{3.6.10}$$

$$+ \frac{1}{2} \sum_{i,j=1}^{2} \frac{\partial^2 f(Y_{1s-}, Y_{2s-})}{\partial Y_i \partial Y_j} d\langle Y_i^c, Y_j^c \rangle_s + \sum_{0 < s \le t} \left[f(Y_{1s}, Y_{2s}) - f(Y_{1s-}, Y_{2s-}) \right.$$

$$\left. - \sum_{i=1}^{2} \frac{\partial f(Y_{1s-}, Y_{2s-})}{\partial Y_i} \Delta Y_{is} \right]. \tag{3.6.11}$$

Because Y_{2t} is a purely discontinuous process and of bounded variation the second integral of the sum in (3.6.10) is equal to

$$\sum_{0 < s \le t} e^{Y_{1s-}} \Delta Y_{2s}.$$

Now $\langle Y_i^c, Y_j^c \rangle = 0$ except for $i = j = 1$ because the continuous part of Y_{2t} is identically zero and since $Y_1^c = X^c$, (3.6.11) becomes

$$\frac{1}{2} \int_0^t \Lambda_{s-} d\langle X^c, X^c \rangle_s.$$

In the last expression, using (3.6.9), we have

$$f(Y_{1s}, Y_{2s}) - f(Y_{1s-}, Y_{2s-}) = \Lambda_s - \Lambda_{s-}$$

$$= e^{X_s - X_{s-} + X_{s-} - \frac{1}{2}\langle X^c, X^c \rangle_s} \prod_{r \le s-} (1 + \Delta X_r) e^{-\Delta X_r} (1 + \Delta X_s) e^{-\Delta X_s}$$

$$- e^{X_{s-} - \frac{1}{2}\langle X^c, X^c \rangle_s} \prod_{r \le s-} (1 + \Delta X_r) e^{-\Delta X_r}$$

$$= e^{X_{s-} - \frac{1}{2}\langle X^c, X^c \rangle_s} \prod_{r \le s-} (1 + \Delta X_r)(1 + \Delta X_s)$$

$$= \Lambda_{s-} \Delta X_s.$$

That is,

$$\Lambda_s = \Lambda_{s-}(1 + \Delta X_s).$$

Putting these results together gives (3.6.8). For the proof of the uniqueness see Theorem 13.5 of [11]. □

Example 3.6.12 Consider the following "log-Poisson plus log-normal" process, with its jump part driven by a finite sum of independent Poisson processes N_t^i, $i = 1, \ldots, n$ with time varying jump sizes a_t^i and intensities λ_t^i:

$$X_t = X_0 + \sigma \int_0^t X_{s-} dB_s + \sum_{i=1}^{n} \int_0^t X_{s-} a_s^i (dN_s^i - \lambda_s^i ds).$$

Applying the result of Example 3.6.11 we see that X has the form

$$X_t = X_{s-} \exp\left[\sigma(B_t - B_s) - \frac{\sigma}{2}(t - s) - \sum_{i=1}^{n} \int_s^t a_s^i \lambda_s^i \, ds\right] \prod_{s \leq r \leq t} (1 + a_r^i \Delta N_r^i).$$

□

Example 3.6.13 If X_t and Y_t are two semimartingales, the product rule gives

$$X_t Y_t = X_0 Y_0 + \int_0^t X_{s-} dY_s + \int_0^t Y_{s-} dX_s + [X, Y]_t,$$

and

$$X_t^2 = X_0^2 + 2 \int_0^t X_{s-} dX_s + [X, X]_t,$$

or

$$\int_0^t X_{s-} dX_s = \frac{1}{2}(X_t^2 - [X, X]_t).$$

Applying the preceding result to a Poisson process we obtain

$$\int_0^t N_{s-} dN_s = \frac{1}{2}(N_t^2 - [N, N]_t) = \frac{1}{2}(N_t^2 - N_t).$$

□

Theorem 3.6.14 *(Lévy's characterization of the Poisson process) Suppose $\{Q_t\}$ is a purely discontinuous martingale on the filtered probability space $(\Omega, \mathcal{F}, P, \mathcal{F}_t)$, $t \geq 0$, all of whose jumps equal 1. If $Q_t^2 - t$ is a martingale then $\{N_t = Q_t + t\}$ is a Poisson process.*

Proof We can suppose $N_0 = Q_0 = 0$. Because Q_t is purely discontinuous,

$$E[[Q, Q]_t] = E[Q_t^2] < \infty,$$

but because all the jumps equal $+1$,

$$[Q, Q]_t = \sum_{s \leq t} \Delta Q_s^2 = \sum_{s \leq t} \Delta Q_s.$$

Write $N_t = \sum_{s \leq t} \Delta Q_s$. Then N_t is integrable, because $[Q, Q]_t$ is. Furthermore, Q is a compensated sum of jumps, by Theorem 9.24 of [11], so N_t has a predictable compensator λ_t:

$$Q_t = N_t - \lambda_t.$$

However,

$$N_t - t = [Q, Q]_t - t = (Q_t^2 - t) + ([Q, Q]_t - Q_t^2),$$

and is therefore, a martingale. Consequently $\lambda_t = t$. That is,

$$Q_t = N_t - t. \tag{3.6.12}$$

Applying the differentiation rule to the martingale Q_t and the function $f(x) = e^{iux}$, $u \in \mathbb{R}$, from time 0 to the first jump time T_1, we have

$$e^{iu Q_{T_1}} = 1 + iu \int_0^{T_1} e^{iu Q_{v-}} dQ_v + (e^{iu Q_{T_1}} - e^{iu Q_{T_1-}} - iue^{iu Q_{T_1-}} \Delta Q_{T_1})$$

$$= 1 + iu \int_0^{T_1} e^{iu Q_{v-}} dQ_v + (e^{iu Q_{T_1}} - e^{-iu T_1} - iue^{-iu T_1}),$$

since $Q_{T_1-} = N_{T_1-} - T_1 = -T_1$ by (3.6.12) and $\Delta Q_{T_1} = 1$. Also note that $e^{iu Q_{T_1}}$ cancels out. Hence

$$(1 + iu)e^{-iu T_1} = 1 + iu \int_0^{T_1} e^{iu Q_{v-}} dQ_v.$$

Taking the conditional expectation with respect to $\mathcal{F}_0$ we have

$$(1 + iu)E[e^{-iu T_1} \mid \mathcal{F}_0] = 1.$$

Therefore, T_1 is independent of $\mathcal{F}_0$ and is exponentially distributed with parameter 1.

A time translation and a similar argument shows that $T_n - T_{n-1}$ is independent of $\mathcal{F}_{n-1}$, and is exponentially distributed with parameter 1. Therefore N_t is a Poisson process. ■

3.7 The Itô formula for Brownian motion

In this section, the Itô formula obtained above for general semimartingales is specialized to the Brownian motion and the related Itô processes. (See Definition 3.7.7.)

Theorem 3.7.1 *Let f be a twice continuously differentiable function on $\mathbb{R}$ and let $B = \{B_t, t \geq 0\}$ be a Brownian motion. Then, in view of Theorem 3.6.7, $f(B_t)$ is a semimartingale given by the formula*

$$f(B_t) = f(B_0) + \int_0^t f'(B_s) dB_s + \frac{1}{2} \int_0^t f''(B_s) d\langle B, B \rangle_s$$

$$= f(B_0) + \int_0^t f'(B_s) dB_s + \frac{1}{2} \int_0^t f''(B_s) ds.$$

Example 3.7.2

$$f(B_t) = B_t^2 = B_0^2 + \int_0^t 2B_s dB_s + \frac{1}{2} \int_0^t 2ds.$$

Then

$$d(B_t^2) = 2B_t dB_t + dt.$$

□

We prove the converse of Theorem 2.7.1.

Theorem 3.7.3 *Let $\{W_t, \mathcal{F}_t\}$, $t \geq 0$ be a continuous (scalar) local martingale such that $\{W_t^2 - t\}$, $t \geq 0$, is a local martingale. Then $\{W_t, \mathcal{F}_t\}$ is a Brownian motion.*

Proof We must show that for $0 \le s \le t$ the random variable $W_t - W_s$ is independent of $\mathcal{F}_s$ and is normally distributed with mean 0 and covariance $t - s$. In terms of characteristic functions this means we must show that for any real u,

$$E[e^{iu(W_t - W_s)} \mid \mathcal{F}_s] = E[e^{iu(W_t - W_s)}]$$
$$= e^{-u^2(t-s)/2}.$$

Consider the (complex-valued) function $f(x) = e^{iux}$. Applying the Itô rule to the real and imaginary parts of $f(x)$ we have

$$f(W_t) = e^{iuW_t} = f(W_s) + \int_s^t iue^{iuW_r} dW_r - \frac{1}{2} \int_s^t u^2 e^{iuW_r} dr, \qquad (3.7.1)$$

because $d\langle W \rangle_r = dr$ by hypothesis. Furthermore, the real and imaginary parts of $\int_s^t iue^{iuW_r} dW_r$ are in fact square integrable martingales because the integrands are bounded by 1. Consequently, $E[\int_s^t iue^{iuW_r} dW_r \mid \mathcal{F}_s] = 0$. For any $A \in \mathcal{F}_s$ we may multiply (3.7.1) by $I_A e^{-iuW_s}$ and take expectations to deduce

$$E[e^{iu(W_t - W_s)} I_A] = P(A) - \frac{1}{2} \int_0^t E[e^{iu(W_r - W_s)} I_A] dr.$$

Solving this equation, we see

$$E[e^{iu(W_t - W_s)} I_A] = P(A)e^{-u^2(t-s)/2},$$

and the result follows. ∎

If the function f is a function of both time and space the Itô rule has the following form.

Theorem 3.7.4 *Let $f(.,.)$ be continuously differentiable in the first argument and twice continuously differentiable in the second argument and let $\{B_t\}$ be a Brownian motion. Then $f(t, B_t)$ is given by the formula*

$$f(t, B_t) = f(0, B_0) + \int_0^t \frac{\partial f(s, B_s)}{\partial s} ds + \int_0^t \frac{\partial f(s, B_s)}{\partial B} dB_s + \frac{1}{2} \int_0^t \frac{\partial^2 f(s, B_s)}{\partial B^2} ds$$
$$= f(0, B_0) + \int_0^t \frac{\partial f(s, B_s)}{\partial B} dB_s + \int_0^t \left[\frac{\partial f(s, B_s)}{\partial s} + \frac{1}{2} \frac{\partial^2 f(s, B_s)}{\partial B^2} \right] ds.$$

One can write formally the differential expression:

$$df(t, B_t) = \frac{\partial f(t, B_t)}{\partial t} dt + \frac{\partial f(t, B_t)}{\partial B} dB_t + \frac{1}{2} \frac{\partial^2 f(t, B_t)}{\partial B^2} dt.$$

Here the differentials satisfy:

$$(dB_t)^2 = dt, \quad (dB_t)^n = 0, \quad n > 2, \quad dt\, dB_t = 0.$$

Example 3.7.5

$$f(t, B_t) = \exp(B_t - \frac{1}{2}t),$$

$$df(t, B_t) = f(t, B_t)dB_t + (-\frac{1}{2}f(t, B_t) + \frac{1}{2}f(t, B_t))dt$$
$$= f(t, B_t)dB_t. \qquad (3.7.2)$$

Hence the function $\exp(B_t - \frac{1}{2}t)$ is the solution of the exponential equation (3.7.2).
Since

$$E\left[\int_0^t (\exp(B_s - \frac{1}{2}s))^2 ds\right] < \infty,$$

the Itô integral $\int_0^t \exp(B_s - \frac{1}{2}s)dB_s$ is a martingale. However,

$$\int_0^t \exp(B_s - \frac{1}{2}s)dB_s = \exp(B_t - \frac{1}{2}t) - 1,$$

so the process $X_t = \exp(B_t - \frac{1}{2}t)$ is a martingale.   □

Example 3.7.6 Given two adapted, measurable, processes X_t and Y_t such that with probability 1,

$$\int_0^t X_s^2 ds < \infty,$$

and

$$\int_0^t Y_s^2 ds < \infty,$$

we have:

$$\langle \int_0^t X_s dB_s, \int_0^t Y_s dB_s \rangle = \int_0^t X_s Y_s d\langle B_s, B_s \rangle = \int_0^t X_s Y_s ds.$$

In particular,

$$\langle \int_0^t X_s dB_s, \int_0^t X_s dB_s \rangle = \int_0^t X_s^2 d\langle B_s, B_s \rangle = \int_0^t X_s^2 ds.$$

□

Definition 3.7.7 *An Itô process is a (special) semimartingale of the form:*

$$X_t = X_0 + \int_0^t \mu(\omega, s)ds + \int_0^t \sigma(\omega, s)dB_s(\omega).$$

Here B_t is a Brownian motion, and $\{\mu(\omega, t)\}$, $\{\sigma(\omega, t)\}$ are adapted, measurable processes such that with probability 1,

$$\int_0^t \mu(\omega, s)ds < \infty, \quad \forall t \geq 0$$

and

$$\int_0^t \sigma^2(\omega, s)ds < \infty, \quad \forall t \geq 0.$$

Given two Itô processes

$$X_t = X_0 + \int_0^t \alpha(\omega, s)ds + \int_0^t \beta(\omega, s)dB_s(\omega),$$

and

$$Y_t = Y_0 + \int_0^t \mu(\omega, s)ds + \int_0^t \sigma(\omega, s)dB_s(\omega),$$

we have

$$[X, Y]_t = \langle X, Y \rangle_t = X_0 Y_0 + \int_0^t \beta(\omega, s)\sigma(\omega, s)ds.$$

Given an adapted, measurable process Y_t such that with probability 1:

$$\int_0^t |Y_s \mu(s)|ds < \infty,$$

and

$$\int_0^t (Y_s \sigma(s))^2 ds < \infty,$$

and an Itô process

$$X_t = X_0 + \int_0^t \mu(\omega, s)ds + \int_0^t \sigma(\omega, s)dB_s(\omega),$$

we can define the stochastic integral

$$\int_0^t Y_s dX_s = \int_0^t Y_s \mu(\omega, s)ds + \int_0^t Y_s \sigma(\omega, s)dB_s(\omega).$$

Remarks 3.7.8

1. From the definition of an Itô process it follows that the process $\int_0^t Y_s dX_s$ is an Itô process.

2. The process $\int_0^t Y_s dX_s$ is a continuous semimartingale. $\quad\square$

Theorem 3.7.9 *Let f be a twice continuously differentiable function on $\mathbb{R}$ and let*

$$X_t = X_0 + \int_0^t \mu(\omega, s)ds + \int_0^t \sigma(\omega, s)dB_s(\omega) \tag{3.7.3}$$

be an Itô process. Then $f(X_t)$ is given by the formula

$$f(X_t) = f(X_0) + \int_0^t f'(X_s)dX_s + \frac{1}{2}\int_0^t f''(X_s)d\langle X, X \rangle_t. \tag{3.7.4}$$

Proof See any of [6], [11], [16], [30], [34]. $\quad\blacksquare$

Or in differential form,

$$df(X_t) = f'(X_t)dX_t + \frac{1}{2}f''(X_t)d\langle X, X \rangle_t.$$

Here, $d\langle X, X \rangle_t = \langle \sigma(t)dB_t, \sigma(t)dB_t \rangle = \sigma(t)^2 dt.$

If (3.7.3) is substituted in (3.7.4) we have

$$f(X_t) = f(X_0) + \int_0^t f'(X_s)\sigma(\omega, s) dB_s(\omega) + \int_0^t \left(\frac{1}{2} f''(X_s)\sigma(s)^2 + \mu(\omega, s) f'(X_s) \right) ds.$$

Remark 3.7.10 Note that $\int_0^t f'(X_s)\sigma(\omega, s) dB_s(\omega)$ is perhaps only a local martingale even if $E\left[\int_0^t \sigma^2(s) ds\right] < \infty$ because $f'(X_t)\sigma(t)$ satisfies the weaker condition $\int_0^t (f'(X_s)\sigma(s))^2 ds < \infty$. This is guaranteed by the continuity of $f'(X_t)$ in t. Consequently, local martingales arise naturally in the context of Itô stochastic calculus. □

Example 3.7.11 Consider the Itô process

$$X_t = \int_0^t \sigma(\omega, s) dB_s(\omega),$$

and suppose $f(X_t) = X_t^2$. By the Itô formula,

$$df(X_t) = 0dt + 2X_t 0dt + 2X_t\sigma(t) dB_t + \frac{1}{2} 2\sigma(s)^2 dt$$

$$= 2X_t\sigma(t) dB_t + \sigma(s)^2 dt. \qquad \square$$

Example 3.7.12 Solve the following linear stochastic differential equation:

$$dX_t = \mu X_t dt + \sigma X_t dB_t, \quad (\mu, \sigma \in \mathbb{R}). \tag{3.7.5}$$

Assume that X_0 is independent of B_t and $E[X_0]^2 < \infty$. We must find an adapted, measurable process X_t such that

$$E\left[\int_0^t X_s^2 ds\right] < \infty,$$

and (3.7.5) holds.

Let $f(X_t) = \log X_t$ and apply the Itô formula:

$$\log X_t = \log X_0 + \int_0^t \left(\mu X_s \frac{1}{X_s} - \frac{1}{2}\sigma^2 X_s^2 \frac{1}{X_s^2} \right) ds + \int_0^t \sigma X_s \frac{1}{X_s} dB_s$$

$$= \log X_0 + (\mu - \frac{1}{2}\sigma^2)t + \sigma B_t.$$

Therefore,

$$X_t = X_0 \exp\left\{ (\mu - \frac{1}{2}\sigma^2)t + \sigma B_t \right\}.$$

As a Borel function of B_t, X_t is adapted and since it is continuous, it is measurable. Now

$$X_t^2 = X_0^2 \exp\{(2\mu - \sigma^2)t + 2\sigma B_t\},$$

and using the assumptions,

$$E[X_t^2] = E[X_0^2]\exp\{(2\mu - \sigma^2)t\}E[\exp\{2\sigma B_t\}].$$

Recall that B_t is an $N(0, t)$ random variable so that

$$E[\exp\{2\sigma B_t\}] = \exp\{2\sigma^2 t\}.$$

Therefore

$$E[X_t^2] = E[X_0^2] \exp\{(2\mu + \sigma^2)t\}.$$

Consequently,

$$E\left[\int_0^t X_s^2 ds\right] = \int_0^t E[X_s^2] ds$$

$$= E[X_0^2] \int_0^t \exp\{(2\mu + \sigma^2)s\} ds < \infty.$$

$\square$

Setting $\mu = 0$ and $\sigma = 1$ in (3.7.5) gives the equation

$$dX_t = X_t dB_t. \tag{3.7.6}$$

This has the solution

$$X_t = \exp\{B_t - \frac{1}{2}t\}. \tag{3.7.7}$$

The process X given by (3.7.7) is called the *stochastic exponential* of the Brownian motion process B.

For a general Itô process,

$$Z_t = Z_0 + \int_0^t \mu(\omega, s) ds + \int_0^t \sigma(\omega, s) dB_s(\omega),$$

consider the equation

$$dX_t = X_t dZ_t,$$

with X_0 given and $\mathcal{F}_0$ measurable, that is, X_0 is a constant.

Then the unique solution of the equation is the process

$$X_t = X_0 \exp\{\int_0^t \sigma(s) dB_s + \int_0^t (\mu(s) - \frac{1}{2}\sigma^2(s)) ds\}$$

$$= \exp\{X_t - \frac{1}{2}\langle X, X\rangle_t\}.$$

X_t is then called the stochastic exponential of the Itô process Z.

A generalization of Theorem 3.7.9 is:

Theorem 3.7.13 *Suppose* $f(., .)$ *is continuously differentiable in the first argument and twice continuously differentiable in the second argument, and consider the Itô process*

$$X_t = X_0 + \int_0^t \mu(\omega, s) ds + \int_0^t \sigma(\omega, s) dB_s(\omega). \tag{3.7.8}$$

Then $f(t, X_t)$ is given by the formula

$$f(t, X_t) = f(0, X_0) + \int_0^t \frac{\partial f(X_s)}{\partial X_s} dX_s$$

$$+ \int_0^t \frac{\partial f(X_s)}{\partial s} ds + \frac{1}{2} \int_0^t \frac{\partial^2 f(X_s)}{\partial X_s^2} d\langle X, X \rangle_t. \tag{3.7.9}$$

Again in differential form:

$$df(t, X_t) = \frac{\partial f(X_t)}{\partial X_t} dX_t + \frac{\partial f(X_t)}{\partial t} dt + \frac{1}{2} \frac{\partial^2 f(X_t)}{\partial X_t^2} d\langle X, X \rangle_t.$$

Substituting (3.7.8) in (3.7.9) we have:

$$f(t, X_t) = f(0, X_0) + \int_0^t \frac{\partial f(X_s)}{\partial s} ds + \int_0^t \sigma(s) \frac{\partial f(X_s)}{\partial X_s} dB_s$$

$$+ \int_0^t \left\{ \mu(s) \frac{\partial f(X_s)}{\partial X_s} + \frac{1}{2} \sigma(s)^2 \frac{\partial^2 f(X_s)}{\partial X_s^2} \right\} ds.$$

The following theorem gives the multi-dimensional Itô formula. (See [25].)

Theorem 3.7.14 *Let $f(t, x_1, \ldots, x_n)$ be continuously differentiable in the first argument and twice continuously differentiable in the other arguments. Suppose $X^1, \ldots, X^n$ are Itô processes of the form:*

$$dX_t^1 = \mu_1(t)dt + \sigma_{11}(t)dB_t^1 + \sigma_{12}(t)dB_t^2 + \cdots + \sigma_{1m}(t)dB_t^m$$
$$dX_t^2 = \mu_2(t)dt + \sigma_{21}(t)dB_t^1 + \sigma_{22}(t)dB_t^2 + \cdots + \sigma_{2m}(t)dB_t^m$$
$$\cdots$$
$$dX_t^n = \mu_n(t)dt + \sigma_{n1}(t)dB_t^1 + \sigma_{n2}(t)dB_t^2 + \cdots + \sigma_{nm}(t)dB_t^m.$$

We, therefore, require that with probability 1,

$$\int_0^t |\mu_i(s)| ds < \infty, \quad i = 1, \ldots, n,$$

and

$$\int_0^t |\sigma_{kl}(t)|^2 ds < \infty, \quad k = 1, \ldots, n; \quad l = 1, \ldots, m.$$

Suppose $B^1, \ldots, B^m$ are m independent Brownian motions.

$$f(t, X_t^1, \ldots, X_t^n) = f(0, X_0^1, \ldots, X_0^n) + \int_0^t \left\{ \frac{\partial f}{\partial s} + \sum_i \mu_i(s) \frac{\partial f}{\partial X_s^i} \right.$$

$$+ \frac{1}{2} \text{Tr} \left[\sigma'(s) \left(\frac{\partial}{\partial X} \left[\frac{\partial}{\partial X} \right]' f \right) \sigma(s) \right] \right\} ds$$

$$+ \int_0^t \sum_{ij} \frac{\partial f}{\partial X_s^i} \sigma_{ij}(t) dB_s^j. \tag{3.7.10}$$

Here $\sigma(t) = \{\sigma_{ij}(t)\}$ *is an* $n \times m$ *matrix,* $\mathbf{X} = (X^1, \ldots, X^n)$, $\left(\dfrac{\partial}{\partial \mathbf{X}} \left[\dfrac{\partial}{\partial \mathbf{X}} \right]' f \right)$ *is the matrix*
$\left\{ \dfrac{\partial^2 f}{\partial X^i \partial X^j} \right\}$, $i, j = 1, \ldots, n$ *and* $\mathbf{Tr}(A)$ *is the trace of the matrix* A, *i.e. the sum of the diagonal entries of the matrix* A.

We can write (3.7.10) in differential form:

$$df(t, X_t^1, \ldots, X_t^n) = \frac{\partial f}{\partial t}dt + \sum_i \mu_i(t)\frac{\partial f}{\partial X_t^i}dt$$

$$+ \sum_i \sigma_i(t)\frac{\partial f}{\partial X_t^i}dB_t^i + \frac{1}{2}\sum_{i,j}\frac{\partial^2 f}{\partial X^i \partial X^j}d\langle X^i, X^j \rangle.$$

Example 3.7.15 Suppose that

$$dX_t^1 = \mu_1(t)dt + \sigma_1(t)dB_t^1$$
$$dX_t^2 = \mu_2(t)dt + \sigma_2(t)dB_t^2$$

are two Itô processes and that $f(X_t^1, X_t^2) = X_t^1 X_t^2$. From the Itô rule (3.7.10)

$$d(X_t^1 X_t^2) = X_t^2 dX_t^1 + X_t^1 dX_t^2 + \sigma_1(t)\sigma_2(t)dt$$
$$= X_t^2 dX_t^1 + X_t^1 dX_t^2 + d\langle X^1, X^2 \rangle_t,$$

or equivalently

$$X_t^1 X_t^2 = X_0^1 X_0^2 + \int_0^t X_s^2 dX_s^1 + \int_0^t X_s^1 dX_s^2 + \langle X^1, X^2 \rangle_t. \qquad \square$$

3.8 Representation results

Measurable, adapted processes $\{f(\omega, t), \mathcal{F}_t\}$ such that

$$E\left[\int_0^t f^2(\omega, s)ds \right] < \infty, \qquad \forall t \geq 0$$

generate martingales $\{X_t, \mathcal{F}_t\}$ via the formula

$$X_t = X_0 + \int_0^t f(\omega, s)dB_s(\omega). \tag{3.8.1}$$

The following theorem (Davis [8]) gives a converse result in the sense that any (square) martingale $\{X_t, \mathcal{F}_t\}$ can be represented as an Itô integral similar to (3.8.1).

Theorem 3.8.1 *Suppose* $\{B_t\}$, $t \geq 0$, *is a Brownian motion on the filtered probability space* $(\Omega, \mathcal{F}, P, \mathcal{F}_t)$. *Write* $\mathcal{G}_t^0 = \sigma\{B_s : s \leq t\}$ *and* $\{\mathcal{G}_t\}$ *for the completion of* $\{\mathcal{G}_t^0\}$, *so that the filtration* $\{\mathcal{G}_t\}$ *is certainly right-continuous.*

Then every random variable $X \in L^2(\Omega, \mathcal{G}_\infty)$ can be represented as a stochastic integral

$$X = E[X \mid \mathcal{G}_0] + \int_0^\infty f_s \, \mathrm{d}B_s,$$

where $\{f_t\}$ is a $\mathcal{G}_t$-predictable process and $E[\int_0^\infty f_s^2 \mathrm{d}s] < \infty$. Furthermore,

$$E[X \mid \mathcal{G}_t] = E[X \mid \mathcal{F}_t] = E[X \mid \mathcal{G}_0] + \int_0^t f_s \, \mathrm{d}B_s.$$

If $\{B_t\} = \{B_t^1, \ldots, B_t^n\}$, $t \geq 0$, is an n-dimensional Brownian motion, then $f_t = (f_t^1, \ldots, f_t^n)$ is a $\mathcal{G}_t$-predictable process such that $E[\int_0^\infty (f_s^i)^2 \mathrm{d}s] < \infty$, $i = 1, \ldots, n$ and $X \in L^2(\Omega, \mathcal{G}_\infty)$ has representation

$$X_t = E[X \mid \mathcal{G}_0] + \sum_{i=1}^n \int_0^t f_s^i \, \mathrm{d}B_s^i.$$

Proof   See [8].   ∎

Theorem 3.8.2 *Suppose $\{N_t\}$, $t \geq 0$, is a Poisson process on the filtered probability space $(\Omega, \mathcal{F}, P, \mathcal{F}_t)$. Write $\mathcal{G}_t^0 = \sigma\{N_s : s \leq t\}$ and $\{\mathcal{G}_t\}$ for the completion of $\{\mathcal{G}_t^0\}$, so that the filtration $\{\mathcal{G}_t\}$ is certainly right-continuous.*

Then every random variable $X \in L^2(\Omega, \mathcal{G}_\infty)$ can be represented as stochastic integral

$$X = E[X \mid \mathcal{G}_0] + \int_0^\infty f_s \, \mathrm{d}Q_s,$$

where $Q_t = N_t - t$, $\{f_t\}$ is a $\mathcal{G}_t$-predictable process and $E[\int_0^\infty f_s^2 \mathrm{d}s] < \infty$. Furthermore,

$$E[X \mid \mathcal{G}_t] = E[X \mid \mathcal{F}_t] \quad a.s. \text{ for all } t.$$

Proof   See [8], [11].   ∎

Representation results for Markov chains

Consider a finite state Markov process $\{X_t\}$, $t \geq 0$, defined on a probability space (Ω, F, P). We have noted in Example 2.6.17 that, without loss of generality, the state space of X can be identified with the set

$$S = \{e_1, \ldots, e_N\}$$

of standard unit vectors in $\mathbb{R}^N$.

Recall that

$$p_t^i = P(X_t = e_i), \quad 0 \leq i \leq N,$$

and

$$\frac{\mathrm{d}p_t}{\mathrm{d}t} = A_t p_t. \tag{3.8.2}$$

$A_t = (a_{ij}(t))$, $t \geq 0$.

Write $\mathcal{F}_t^s$ for the right-continuous, complete filtration generated by $\sigma\{X_r : s \le r \le t\}$, and $\mathcal{F}_t = \mathcal{F}_t^0$.

We saw in Lemma 2.6.18 that

$$V_t \overset{\triangle}{=} X_t - X_0 - \int_0^t A_r X_r dr \qquad (3.8.3)$$

is an $\{\mathcal{F}_t\}$-martingale.

Lemma 3.8.3

$$X_t = \Phi(t, 0)\left(X_0 + \int_0^t \Phi(r, 0)^{-1} dV_r\right). \qquad (3.8.4)$$

Proof   Differentiating (3.8.4) verifies the result.   ∎

If x, y are (column) vectors in $\mathbb{R}^N$ we shall write $\langle x, y \rangle = x'y$ for their scalar (inner) product. Consider $0 \le i, j \le N$ with $i \ne j$. Then, because the Markov chain is piecewise constant, $dX_s = \Delta X_s$ and

$$\langle X_{s-}, e_i \rangle e_j' dX_s = \langle X_{s-}, e_i \rangle e_j' \Delta X_s$$
$$= \langle X_{s-}, e_i \rangle e_j' (X_s - X_{s-}) = I(X_{s-} = e_i, \ X_s = e_j).$$

Therefore,

$$\int_0^t \langle X_{s-}, e_i \rangle e_j' dX_s = \sum_{0 < s \le t} I(X_{s-} = e_i, \ X_s = e_j) \overset{\triangle}{=} J_t^{ij},$$

which equals the number of times X jumps from e_i to e_j in the interval $[0, t]$. Define the martingale

$$V_t^{ij} \overset{\triangle}{=} \int_0^t \langle X_{s-}, e_i \rangle e_j' dV_s.$$

(Note the integrand is predictable.) Then

$$V_t^{ij} = \int_0^t \langle X_{s-}, e_i \rangle e_j' dX_s - \int_0^t \langle X_{s-}, e_i \rangle e_j' A_s X_{s-} ds$$
$$= J_t^{ij} - \int_0^t \langle X_{s-}, e_i \rangle a_{ji}(s) ds$$
$$= J_t^{ij} - \int_0^t \langle X_s, e_i \rangle a_{ji}(s) ds,$$

because $X_s = X_{s-}$ for each ω, except for countably many s. That is, for $i \ne j$,

$$J_t^{ij} = \int_0^t \langle X_s, e_i \rangle a_{ji}(s) ds + V_t^{ij}.$$

The process $\int_0^t \langle X_s, e_i \rangle a_{ji}(s) ds$ is, therefore, the compensator of the counting process J_t^{ij}.

For a fixed j, $0 \le j \le N$, write $\mathcal{J}_t^j$ for the number of jumps into state e_j up to time t. Then, for $i \neq j$,

$$\mathcal{J}_t^j = \sum_{i=1}^N \mathcal{J}_t^{ij} = \sum_{i=1}^N \int_0^t \langle X_s, e_i \rangle a_{ji}(s) ds + V_t^j,$$

where V_t^j is the martingale $\sum_{i=1}^N V_t^{ij}$. Finally, write $\mathcal{J}_t$ for the total number of jumps (of all kinds) of the process X up to time t. Then

$$\mathcal{J}_t = \sum_{j=1}^N \mathcal{J}_t^j = \sum_{i,j=1}^N \int_0^t \langle X_s, e_i \rangle a_{ji}(s) ds + Q_t,$$

where Q_t is the martingale $\sum_{j=1}^N V_t^j$. However,

$$a_{ii}(s) = -\sum_{j=1}^N a_{ji}(s),$$

so

$$\mathcal{J}_t = -\sum_{i=1}^N \int_0^t \langle X_s, e_i \rangle a_{ii}(s) ds + Q_t. \tag{3.8.5}$$

Before we state the next result we need the following definition.

Definition 3.8.4 *If $M = (M^1, \ldots, M^N)$ is a vector, $\mathbb{R}^N$-valued, square integrable martingale, the quadratic predictable variation process of M is the (unique) predictable matrix valued process $\langle M, M \rangle \in \mathbb{R}^{N \times N}$ such that*

$$MM' - \langle M, M \rangle$$

is a martingale. Here MM' is the (Kronecker) product of the (column) vector M with the (row) vector M', so that MM' can be identified with the matrix valued process with entries $(M^i M^j)$.

Lemma 3.8.5 *The quadratic predictable variation process of the (vector) martingale V (see Definition 3.8.4) is given by the matrix valued process*

$$\langle V, V \rangle_t = \operatorname{diag} \int_0^t A_r X_{r-} dr - \int_0^t (\operatorname{diag} X_{r-}) A_r' dr - \int_0^t A_r (\operatorname{diag} X_{r-}) dr.$$

Proof   Recall $X_t \in S$ is one of the unit vectors e_i. Therefore,

$$X_t X_t' = \operatorname{diag} X_t. \tag{3.8.6}$$

Now by the product rule

$$X_t X_t' = X_0 X_0' + \int_0^t X_{r-} (A_r X_{r-})' dr + \int_0^t X_{r-} dV_r' + \int_0^t (A_r X_{r-}) X_{r-}' dr$$

$$+ \int_0^t dV_r X_{r-}' + \langle V, V \rangle_t + ([V, V]_t - \langle V, V \rangle_t),$$

where $[V, V]_t - \langle V, V \rangle_t$ is an $\{\mathcal{F}_t\}$-martingale. However, a simple calculation using 3.8.6 shows

$$X_{r-}(A_r X_{r-})' = (\mathrm{diag}\,X_{r-})A'_r,$$

and

$$(A_r X_{r-})X'_{r-} = A_r(\mathrm{diag}\,X_{r-})'.$$

Therefore,

$$X_t X'_t = X_0 X'_0 + \int_0^t (\mathrm{diag}\,X_{r-})A'_r \mathrm{d}r$$

$$+ \int_0^t A_r(\mathrm{diag}\,X_{r-})\mathrm{d}r + \langle V, V \rangle_t + \mathrm{martingale}. \tag{3.8.7}$$

Also, from (3.8.6),

$$X_t X'_t = \mathrm{diag}\,X_t = \mathrm{diag}\,X_0 + \mathrm{diag} \int_0^t A_r X_{r-}\mathrm{d}r + \mathrm{diag}\,V_t. \tag{3.8.8}$$

The semimartingale decompositions (3.8.7) and (3.8.8) must be the same, so equating the predictable terms,

$$\langle V, V \rangle_t = \mathrm{diag} \int_0^t A_r X_{r-}\mathrm{d}r - \int_0^t (\mathrm{diag}\,X_{r-})A'_r \mathrm{d}r - \int_0^t A_r(\mathrm{diag}\,X_{r-})\mathrm{d}r.$$

∎

We next note the following representation result:

Remark 3.8.6 A time varying function $f(t, X_t)$ of $X_t \in S$ takes only the values $f(t, e_1), \ldots, f(t, e_N)$ for each t. Writing $f_i(t) = f(t, e_i)$, $1 \le i \le N$, we see f can be represented by the vector

$$f(t) = (f_1(t), \ldots, f_N(t))' \in \mathbb{R}^N,$$

so that $f(t, X_t) = \langle f(t), X_t \rangle$, where $\langle \, , \, \rangle$ denotes the inner product in $\mathbb{R}^N$. □

Therefore, we have the following differentiation rule and representation result:

Lemma 3.8.7 *Suppose the components of $f(t)$ are differentiable in t. Then*

$$f(t, X_t) = f(0, X_0) + \int_0^t \langle f'(r), X_r \rangle \mathrm{d}r + \int_0^t \langle f(r), A_r X_{r-} \rangle \mathrm{d}r$$

$$+ \int_0^t \langle f(r), \mathrm{d}V_r \rangle. \tag{3.8.9}$$

Here, $\int_0^t \langle f(r), \mathrm{d}V_r \rangle$ is an $\mathcal{F}_t$-martingale. Also, using (3.8.4),

$$f(t, X_t) = \langle f(t), \Phi(t, 0)X_0 \rangle + \int_0^t \langle f(t), \Phi(t, r)\mathrm{d}V_r \rangle. \tag{3.8.10}$$

The single jump process

Here we discuss representation results for the single jump process. (See Examples 2.1.4 and 2.6.12.)

Lemma 3.8.8 *Suppose $\{M_t\}$ is a uniformly integrable $\{\mathcal{F}_t\}$-martingale (see Example 2.6.12 for the definition of the filtration $\{\mathcal{F}_t\}$) such that $M_0 = 0$ a.s. Then there is an $\mathcal{F} = \bigvee_t \mathcal{F}_t$-measurable function $h : \Omega \to \mathbb{R}$ such that $h \in L^1(P)$ and*

$$M_t = h(T, Z)I_{\{T \leq t\}} - I_{\{T > t\}} \frac{1}{F_t} \int_{]0,t]} \int_E h(s, z)P(ds, dz) \quad a.s.$$

Proof If $\{M_t\}$ is a uniformly integrable $\{\mathcal{F}_t\}$-martingale, then $\{M_t\} = E[h \mid \mathcal{F}_t]$ for some $\mathcal{F} = \bigvee_t \mathcal{F}_t$ measurable random variable h. From the definition of $\mathcal{F} = \bigvee_t \mathcal{F}_t$, h is of the form $h(T, Z)$. However,

$$E[h(T, Z) \mid \mathcal{F}_t] = h(T, Z)I_{\{T \leq t\}} + I_{\{T > t\}} \frac{1}{F_t} \int_{]t,\infty]} \int_E h(s, z)P(ds, dz).$$

Because

$$0 = M_0 = E[h] = \int_\Omega h(s, z)P(ds, dz)$$

$$= \int_{]0,t]} \int_E h(s, z)P(ds, dz) + \int_{]t,\infty]} \int_E h(s, z)P(ds, dz),$$

the result follows. ∎

Lemma 3.8.9 *Suppose $\{M_t\}$, $t \geq 0$, is a local martingale of $\{\mathcal{F}_t\}$.*

1. *If $c = \infty$, or $c < \infty$ and $F_{c-} = 0$, then $\{M_t\}$ is a martingale on $[0, c[$.*
2. *If $c < \infty$ and $F_{c-} > 0$, then $\{M_t\}$ is a uniformly integrable martingale.*

Proof

1. Let $\{T_k\}$ be an increasing sequence of stopping times such that $\lim T_k = \infty$ a.s. and $\{M_{t \wedge T_k}\}$ is a uniformly integrable martingale. If there is a k such that $T_k \geq T$ a.s. then $M_t = M_{t \wedge T_k}$ a.s. is a uniformly integrable martingale. Otherwise, suppose for each k that $P(T_k < T) > 0$. Then, by Lemma 2.6.13, there is a sequence $\{t_k\}$ such that $T_k \wedge T = t_k \wedge T$ for each k, and because $P(T > t_k) > 0$ we have $t_k \leq c$, otherwise we should have $T_k \geq T$. Because $\lim T_k = \infty$ we see that $\lim_k P(T > t_k) = 0$, so $\lim t_k = c$. Now $\{M_t\}$ is stopped at time T so $M_{t \wedge T_k} = M_{t \wedge t_k}$. Consequently $\{M_t\}$, $t \leq t_k$, is a uniformly integrable martingale, and $\{M_t\}$ is certainly a martingale on $[0, c[$.
2. Suppose now that $c < \infty$ and $F_{c-} > 0$. Then $P(T = c) > 0$. Because $\lim T_k = \infty$ a.s. there is a k such that $P(T = c, T_k > c) > 0$. Consequently, for such a k, $T_k \geq T$ a.s. and the process $\{M_{t \wedge T_k}\} = \{M_t\}$ is a uniformly integrable martingale. ∎

Write $L^1(\mu)$ for the set of measurable functions $g : \Omega \to \mathbb{R}$ such that $E\left[\int_{[0,\infty] \times E} |g| d\mu\right] < \infty$, and $L^1_{\text{loc}}(P)$ for the set of measurable functions $g : \Omega \to \mathbb{R}$ such that $I_{\{s \le t\}} g(s, x) \in L^1(P)$ for all $t < c$.

We have the following martingale representation result (see [11]).

Theorem 3.8.10 $\{M_t\}$ *is a local $\mathcal{F}_t$-martingale with $M_0 = 0$ a.s. if and only if $M_t = M^g_t$ for some $g \in L^1_{\text{loc}}(P)$, where*

$$M^g_t = \int_\Omega I_{\{s \le t\}} g(s, x) q(ds, dx).$$

Proof Suppose $g \in L^1_{\text{loc}}(P)$. Then there is an increasing sequence of stopping times $\{T_k\}$ such that $\lim T_k = \infty$ a.s. and $I_{\{s < T_k\}} g(s, x) \in L^1(P)$ for each k. Calculations similar to Theorem 2.6.14 show that $M^g_{t \wedge T_k}$ is a uniformly integrable $\{\mathcal{F}_t\}$-martingale.

Conversely, suppose $\{M_t\}$ is a local $\{\mathcal{F}_t\}$-martingale with $M_0 = 0$ a.s. We consider two situations:

1. Suppose $c < \infty$ and $F_{c-} > 0$. From Lemma 3.8.9 $\{M_t\}$ is a uniformly integrable martingale, and so is of the form

$$M_t = h(T, Z) I_{\{T \le t\}} - I_{\{T > t\}} \frac{1}{F_t} \int_{]0,t]} \int_E h(s, z) P(ds, dz), \qquad (3.8.11)$$

where $h(T, Z) = M_\infty$. Define

$$g(t, Z) = h(t, Z) - I_{\{t < c\}} \frac{1}{F_t} \int_{]0,t]} \int_E h(s, z) P(ds, dz) \quad \text{if } t < \infty, \qquad (3.8.12)$$

and $g(\infty, Z) = 0$. Then

$$M^g_t = \int_\Omega I_{\{s \le t\}} g \, dq$$

$$= I_{\{t \ge T\}} \left[g(T, Z) - \int_{]0,t]} \int_E g(s, z) F^{-1}_{s-} P(ds, dz) \right]$$

$$- I_{\{t < T\}} \int_{]0,t]} \int_E g(s, z) F^{-1}_{s-} P(ds, dz). \qquad (3.8.13)$$

From (3.8.11) and (3.8.13) we see that $M_t = M^g_t$ if

$$h(t, z) = g(t, z) - \int_{]0,t]} \int_E g(s, z) F^{-1}_{s-} P(ds, dz). \qquad (3.8.14)$$

However, if g is given by (3.8.12) and $t < c$,

$$\int_{]0,t]} \int_E g(s,z) F_{s-}^{-1} P(ds, dz) = \int_{]0,t]} \int_E h(s,z) F_{s-}^{-1} P(ds, dz)$$

$$- \int_{]0,t]} F_s^{-1} F_{s-}^{-1} \int_{]0,s]} \int_E h(u,z) P(du, dz) dF_s$$

$$= \int_{]0,t]} \int_E h(s,z) F_{s-}^{-1} P(ds, dz)$$

$$+ \int_{]0,t]} \int_E \left(- \int_{]u,t]} F_s^{-1} F_{s-}^{-1} dF_s \right) h(u,z) P(du, dz)$$

$$= \int_{]0,t]} \int_E h(s,z) F_{s-}^{-1} P(ds, dz)$$

$$+ \int_{]0,t]} \int_E \left(F_t^{-1} F_{u-}^{-1} \right) h(u,z) P(du, dz)$$

$$= F_t^{-1} \int_{]0,t]} \int_E h(u,z) P(du, dz).$$

Therefore, (3.8.14) is satisfied if $t < c$. A similar calculation shows that the coefficients of $I_{t<T}$ agree in (3.8.11) and (3.8.13), so $M_t = M_t^g$ if $t < c$. However, both M_t and M_t^g are stopped at T and $P(T > c) = 0$ so it remains only to show that $M_c = M_c^g$ when $T(\omega) = c$. This is verified by a similar calculation to that above.

We now check that $g \in L^1(\mu)$. Because $\{M_t\}$ is uniformly integrable

$$\int_\Omega |h| dp < \infty.$$

Therefore

$$\int_\Omega |h| dp \le \int_\Omega |h| dp - \int_{]0,c[} F_t^{-1} \int_{]0,t]} \int_E |h| P(ds, dz) dF_t$$

$$\le \int_\Omega |h| dp - F_{c-}^{-1} \int_{]0,c[} \int_{]0,t]} \int_E |h| P(ds, dz) dF_t$$

$$= \int_\Omega |h| dp + F_{c-}^{-1} \int_{]0,c[} \int_E |h| (F_t - F_{c-} P(ds, dz)$$

$$\le (1 + F_{c-}^{-1}) \int |h| dp < \infty.$$

Consequently, $g \in L^1(\mu)$.

2. Now suppose $c = \infty$, or $c < \infty$ and $F_{c-} = 0$. Then from Lemma 3.8.9 $\{M_t\}$ is a martingale on $[0, c[$, and so uniformly integrable on $[0, t]$ for any $t < c$. Therefore M_t is of the form (3.8.11) for some h satisfying

$$\int_{]0,t]} \int_E |h| P(ds, dz) < \infty,$$

for all $t < c$. Calculations as in (1.) above show that, for g given by (3.8.12) and $t < c$, $M_t = M_t^g$. Also

$$\int_{]0,t]} \int_E |g| P(\mathrm{d}s, \mathrm{d}z) \le \int_{]0,t]} \int_E |h| \mathrm{d}p - \int_{]0,t[} F_s^{-1} \int_{]0,s]} \int_E |h| P(\mathrm{d}s, \mathrm{d}z) \mathrm{d}F_s$$

$$\le \int_{]0,t]} \int_E |h| \mathrm{d}p \left(1 - \int_{]0,t[} F_s^{-1} \mathrm{d}F_s \right) < \infty \quad \text{if } t < \infty.$$

Therefore $g \in L^1_{loc}(P)$ and the proof is complete. ∎

3.9 Random measures

Definition 3.9.1 *A measure μ on $(\mathbb{R}^+, \mathcal{B}(\mathbb{R}^+))$ is a* counting measure *if*

1. $\mu(B) \in \{0, 1, \ldots, +\infty\} \overset{\triangle}{=} \mathbb{N} \cup \{\infty\}$ *for every $B \in \mathcal{B}(\mathbb{R}^+)$,*
2. $\mu([a, b]) < \infty$ *for all bounded intervals $[a, b] \subset \mathbb{R}^+$.*

In other words a counting measure μ is just a countable subset $D \subset \mathbb{R}^+$, and for any given $B \in \mathcal{B}(\mathbb{R}^+)$, $\mu(B)$ is the number of points in D which belong to B and we write

$$\mu(\mathrm{d}x) = \sum_{x \in D} \delta_x(\mathrm{d}x).$$

Here $\delta_x(\mathrm{d}x)$ denotes the unit mass at x.

Integration with respect to a counting measure μ is reduced to discrete time summation, i.e. for any real valued function f, we have

$$\int_{\mathbb{R}} f(x)\mu(\mathrm{d}x) = \sum_{x \in D} f(x).$$

Definition 3.9.2 *Let $(E, \mathcal{E})$ be a measurable space. Let $D \subset \mathbb{R}^+$ be a countable set. A function $\mathbf{p}$ from D to E is called a* point function. *A point function $\mathbf{p}$ defines a counting measure $\mu_{\mathbf{p}}(\mathrm{d}t, \mathrm{d}x)$ on $\mathcal{B}(\mathbb{R}^+) \otimes \mathcal{E}$ by*

$$\mu_{\mathbf{p}}((0, t] \times A) = \#\{s \in D; s \le t, \mathbf{p}(s) \in A\}, \quad t > 0, A \in \mathcal{E}. \tag{3.9.1}$$

The right hand side of (3.9.1) stands for the number of times s up to time t when $\mathbf{p}(s)$ landed in A.

Definition 3.9.3 *Let $(\Omega, \mathcal{F}, P)$ be a given probability space and $(E, \mathcal{E})$ be a measurable space. A nonnegative kernel $\mu(\omega, \mathrm{d}t, \mathrm{d}x)$ is called a* random measure *(on E) if*

1. $\mu(., A)$ *is $\mathcal{F}$-measurable for each fixed $A \in \mathcal{B}(\mathbb{R}^+) \otimes \mathcal{E}$,*
2. $\mu(\omega, .)$ *is a σ-finite measure for each ω. Such a random measure is said to be* integer valued *if also*
3. $\mu(\omega, A) \in \{0, 1, \ldots, +\infty\} \overset{\triangle}{=} \mathbb{N} \cup \{\infty\}$ *for every $A \in \mathcal{B}(\mathbb{R}^+) \otimes \mathcal{E}$,*
4. $\mu(\omega, (0, t] \times E) - \mu(\omega, (0, t) \times E) = \mu(\omega, \{t\} \times E) \le 1$ *for all (ω, t), that is to say the counting done by μ cannot increase by more than one at any isolated single time t.*

Remarks 3.9.4

1. If μ is an integer valued random measure write

$$D = \{(\omega, t) : \mu(\omega, \{t\} \times E) = 1\}.$$

It follows from Definition 3.9.3(3, 4) that, for each fixed ω, the set

$$D_\omega = \{t \in \mathbb{R}^+ : \mu(\omega, \{t\} \times E) = 1\}, \tag{3.9.2}$$

which is the set of times when $\mu(\omega, .)$ jumps, is at most countable.

2. For $t \in D_\omega$ write $\{t\} \times E = \bigcup_{n=1}^\infty \{t\} \times A_n$, where $\{A_n\}$ is a partition of E into proper nonempty subsets. Then there exists one and only one subset $A_{n_t}^\omega \in \bigcup_{n=1}^\infty A_n$, say, such that $\mu(\omega, \{t\} \times A_{n_t}^\omega) = 1$, which implies that there exists a single point $\varepsilon_t(\omega) \in A_{n_t}^\omega$ such that $\mu(\omega, \{t\} \times \varepsilon_t(\omega)) = 1$. In summary, for each $(\omega, t) \in D$ there is a unique point $\varepsilon_t(\omega) \in E$ such that

$$\mu(\omega, \{t\} \times dx) = \delta_{\varepsilon_t(\omega)}(dx).$$

Here $\delta_{\varepsilon_t(\omega)}(dx)$ denotes the unit mass at $\varepsilon_t(\omega)$ and we can write

$$\mu(\omega, dt, dx) = \sum_{(\omega,s)\in D} \delta_{(s,\varepsilon_s(\omega))}(dt, dx)$$

$$= \sum_{s \geq 0} I_{(\varepsilon_s \in E)} \delta_{(s,\varepsilon_s(\omega))}(dt, dx). \tag{3.9.3}$$

Note that the set given by (3.9.2) can be written

$$D_\omega = \{t \in \mathbb{R}^+ : \varepsilon_t(\omega) \in E)\}. \tag{3.9.4}$$

3. If $\{\varepsilon_t\}$ is an E-valued stochastic process such that for each fixed ω the t-function $\varepsilon(\omega, .)$ takes on at most a countable number of values in E, then the above expression (3.9.3) defines an integer valued random measure. These random measures are sometimes called *point processes*, since for each ω the sample path $\{\varepsilon_t(\omega)\}$ consists of the countable set of points $\{t, \varepsilon_t(\omega)\}$, that is, $\{\varepsilon_t(\omega)\}$ is a point function as given by Definition 3.9.2.

4. Since an integer valued random measure process μ satisfies the assumptions of Doob–Meyer Theorem 2.6.9, then there exists a predictable increasing process μ_p, the compensator of μ, such that

$$\mu_p(\omega, \{t\} \times E) \leq 1,$$

and $\mu - \mu_p$ is a local martingale. $\square$

Random measures associated with jump processes

When dealing with stochastic processes which are not continuous everywhere but with sample paths which are right-continuous with left limits, the notion of random measures enters naturally into the scene. For instance, the process

$$\mu^B(\omega, (0, t], B) = \sum_{0<s\leq t} I(\Delta X_s \in B), \qquad B \in \mathcal{B}(\mathbb{R} - \{0\}), \tag{3.9.5}$$

is called the measure of jumps of the process X and it counts the increments of X which fall in B up to time t. Note that, since X is right-continuous with left limits, the series given by (3.9.5) is finite (a.s.) for every finite t and any subset B that is bounded away from zero. However, the number of jumps of X_t need not be finite on finite intervals of time.

If we eliminate the randomness parameter ω from (3.9.5) we obtain a σ-finite measure on the product σ-field generated by $(0, \infty) \times (\mathbb{R} - \{0\})$.

Remark 3.9.5 Since the process (3.9.5) is an integer valued random measure, by Remark 3.9.4(4) there exists a predictable increasing process μ_p^B, the compensator of μ^B, such that $\mu^B - \mu_p^B$ is a local martingale. □

Examples 3.9.6

1. Counting processes. Since all the jumps are of size $+1$, in formula (3.9.5) the only set of interest is $B = \{+1\}$, that is,

$$\mu^B(\omega, (0, t], B) = \sum_{0<s\leq t} I(\Delta X_s = 1) = X_t.$$

2. Finite state processes. Suppose that for all $t \geq 0$, $X_t \in \{0, 1, \ldots, N\}$. Then the possible jumps are the integers
   $\{-N, -N+1, \ldots, -1, +1, \ldots, N-1, N\}$,
   and in formula (3.9.5) the sets B of interest are all subsets of $\{-N, -N+1, \ldots, -1, +1, \ldots, N-1, N\}$.
3. Let Z_t, $t \in \mathbb{R}_+$, be a finite state space process with right-constant sample paths on the state space $S = \{e_1, e_2, \ldots, e_N\}$. Here e_i is the standard basis (column) vector in $\mathbb{R}^N$ with unity in the i-th position and zero elsewhere.

   Let $T_k(\omega)$ be the k-th jump time of Z, $\delta_{T_k(\omega)}(dr)$ be the unit mass at time $T_k(\omega)$ and $\delta_{Z_{T_k(\omega)}}(e_i)$ be the unit mass at $Z_{T_k(\omega)}(\omega)$.
   Since Z_t is a jump process taking values in the vector space $\mathbb{R}^N$ we can write

$$Z_t = Z_0 + \sum_{0<r\leq t} \Delta Z_r.$$

Here

$$\Delta Z_r = Z_r - Z_{r-} = \sum_{i=1}^N (e_i - Z_{r-}) \sum_{k=1}^\infty \delta_{T_k(\omega)}(dr)\delta_{Z_{T_k(\omega)}}(e_i)$$

$$\stackrel{\triangle}{=} \sum_{i=1}^N (e_i - Z_{r-})\mu^Z(dr, e_i).$$

We assume that each Z_t has almost surely finitely many jumps in any finite interval so that the random measure μ^Z is σ-finite. Let $\tilde{\mu}^Z(dr, e_i)$ be the predictable compensator of μ^Z so that

$$Z_t = Z_0 + \sum_{i=1}^N \int_0^t (e_i - Z_{r-})\tilde{\mu}^Z(dr, e_i) + W_t,$$

where $W_t \stackrel{\triangle}{=} \sum_{i=1}^N \int_0^t (e_i - Z_{r-})(\mu^Z(dr, e_i) - \tilde{\mu}^Z(dr, e_i))$. □

Definition 3.9.7 *An integer valued random measure $\mu(\omega, dt, dx)$ is a* Poisson *random measure if*

1. *for each $A \in \mathcal{B}(\mathbb{R}^+) \otimes \mathcal{E}$ the random variable $\mu(., A)$ is Poisson distributed with parameter $\lambda(A) = E[\mu(., A)]$, i.e.*

$$P(\mu(., A) = k) = \exp\{-\lambda(A)\}\frac{(\lambda(A))k}{k!},$$

and

2. *if $A_1, A_2, \ldots, A_n$ are disjoint subsets of $\mathcal{B}(\mathbb{R}^+) \otimes \mathcal{E}$, then the random variables $\mu(., A_1), \mu(., A_2), \ldots, \mu(., A_n)$ are mutually independent.*

More of the differentiation rule

Suppose $X = \{X_t, \mathcal{F}_t\}$ is a real local martingale. Let $X_t = X_0 + X^c + X^d$ be the unique decomposition given in Lemma 3.2.14, where X^c is the continuous local martingale part of X and X^d is the purely discontinuous local martingale part of X. Suppose

$$\mu^X(\omega, dt, dx) = \sum_{s>0} I_{(\Delta X_s \neq 0)}\delta_{(s, \Delta X_s)}(dt, dx).$$

with predictable compensator $\mu_p^X(\omega, dt, dx)$.
   We also have from [18]

$$X_t = X_0 + X_t^c + \int_0^t \int_{\mathbb{R}} x(\mu^X(dt, dx) - \mu_p^X(dt, dx)).$$

Suppose $f \in C^{1,2}$, the space of functions continuously differentiable in t and twice continuously differentiable in x. Then the differentiation rule gives (see [18])

$$f(t, X_t) = f(0, X_0) + \int_0^t \frac{\partial f(s, X_{s-})}{\partial s}ds$$

$$+ \int_0^t \frac{\partial f(s, X_{s-})}{\partial x}dX_s^c$$

$$+ \frac{1}{2}\int_0^t \frac{\partial^2 f(s, X_{s-})}{\partial x^2}d\langle X^c, X^c\rangle_s$$

$$+ \int_0^t \int_{\mathbb{R}} [f(s, X_{s-} + x) - f(s, X_{s-})](\mu^X(ds, dx) - \mu_p^X(ds, dx))$$

$$+ \int_0^t \int_{\mathbb{R}} \left[f(s, X_{s-} + x) - f(s, X_{s-}) - \frac{\partial f(s, X_{s-})}{\partial x}\right]\mu_p^X(ds, dx).$$

Example 3.9.8 Suppose that the scalar process $\{X_t\}$ is described by the stochastic differential equation

$$dX_t = f(t, X_t)dt + \sigma(t, X_t)dB_t + \int_{\mathbb{R}} \gamma(t, X_{t-}, x)(\mu(dt, dx) - \mu_p(dt, dx)).$$

Here B is a standard Brownian motion and μ is a random measure with compensator μ_p.

Suppose $f \in C^{1,2}$. Then the differentiation rule gives (see [18])

$$f(t, X_t) = f(0, X_0) + \int_0^t \frac{\partial f(s, X_{s-})}{\partial s} ds$$

$$+ \int_0^t \frac{\partial f(s, X_{s-})}{\partial x} (f(s, X_s)ds + \sigma(s, X_{s-})dB_s)$$

$$+ \int_0^t \int_R \frac{\partial f(s, X_{s-})}{\partial s} \gamma(s, X_{s-}, x)(\mu(ds, dx) - \mu_p(ds, dx))$$

$$+ \frac{1}{2} \int_0^t \frac{\partial^2 f(s, X_{s-})}{\partial x^2} \sigma^2(s, X_{s-})ds$$

$$+ \int_0^t \int_R [f(s, X_{s-} + \gamma(s, X_{s-}, x)) - f(s, X_{s-})](\mu(dt, dx) - \mu_p^X(dt, dx))$$

$$+ \int_0^t \int_R \left[f(s, X_{s-} + \gamma(s, X_{s-}, x)) - f(s, X_{s-}) \right.$$

$$\left. - \gamma(s, X_{s-}, x) \frac{\partial f(s, X_{s-})}{\partial x} \right] \mu_p^X(dt, dx). \qquad \square$$

3.10 Problems

1. Let $X_1, X_2, \ldots$ be a sequence of i.i.d. $N(0, 1)$ random variables and consider the process $Z_0 = 0$ and $Z_n = \sum_{k=1}^n X_k$.
   Show that

$$[Z, Z]_n = \sum_{k=1}^n X_k^2,$$

$$\langle Z, Z \rangle_n = n,$$

$$E([Z, Z]_n) = E(\sum_{k=1}^n X_k^2) = n.$$

2. Show that if X and Y are (square integrable) martingales, then $XY - \langle X, Y \rangle$ is a martingale.

3. Establish the identity

$$[X, Y]_n = \frac{1}{2}([X + Y, X + Y]_n - [X, X]_n - [Y, Y]_n).$$

4. Show that for any processes X, Y,

$$X_n Y_n = \sum_{k=1}^n X_{k-1} \Delta Y_k + \sum_{k=1}^n Y_{k-1} \Delta X_k + [X, Y]_n.$$

5. Show that for a real valued differentiable function f and a stochastic process X we have the discrete time version of the Itô formula,

$$f(X_n) = f(X_0) + \sum_{k=1}^n f'(X_{k-1}) \Delta X_k$$

$$+ \sum_{k=1}^n [f(X_k) - f(X_{k-1}) - f'(X_{k-1}) \Delta X_k].$$

6. Show that if $\{X_n\}$ is a square integrable martingale then $X^2 - \langle X, X \rangle$ is a martingale.

7. Find $[B + N, B + N]_t$ and $\langle B + N, B + N \rangle_t$ for a Brownian motion process $\{B_t\}$ and a Poisson process $\{N_t\}$.

8. Show that $\lim_{\delta_n \to 0} S_n \overset{L^2}{=} B_t^2/2 + (\alpha - \frac{1}{2})t$, where S_n is given by (3.4.1) where $\tau_k^n = (1 - \alpha)t_k + \alpha t_{k-1}^n$, $0 \leq \alpha \leq 1$.

9. Let f be a deterministic square integrable function and B_t a Brownian motion. Show that the stochastic integral

$$\int_0^t f(s) dB_s$$

is a normally distributed random variable with distribution $N(0, \int_0^t f^2(s) ds)$.

10. Show that if

$$\int_0^t E[f(s)]^2 ds < \infty,$$

the Itô process

$$I_t = \int_0^t f(s) dB_s$$

has orthogonal increments, i.e., for $0 \leq r \leq s \leq t \leq u$,

$$E[(I_u - I_t)(I_s - I_r)] = 0.$$

11. Show that

$$\int_0^t (B_s^2 - s) dB_s = \frac{B_t^3}{3} - t B_t.$$

12. Prove the second part of Lemma 3.4.3.

13. Show that the process $B_t^2/2 + (\alpha - \frac{1}{2})t$ is an $\mathcal{F}_t$-martingale if and only if $\alpha = 0$.

14. Using the Itô formula, show that the Doob–Meyer decomposition of B_t^4 is given by

$$B_t^4 = 4 \int_0^t B_s^3 dB_s + 6 \int_0^t B_s^2 ds,$$

where B is the Brownian motion process.

15. Using the Itô formula, show that

$$d(B_t)^n = n B_t^{n-1} dB_t + \frac{n(n - 1)}{2} B_t^{n-2} dt.$$

16. If N is a standard Poisson process show that the stochastic integral

$$\int_0^t N_s d(N_s - s)$$

is not a martingale. However, show that

$$\int_0^t N_{s-} d(N_s - s)$$

is a martingale. Here N_t is a Poisson process. (Note that at any jump time s, $N_{s-} = N_s - 1$.)

17. Prove that

$$\int_0^t 2^{N_{s-}} dN_s = 2^{N_t} - 1.$$

Here N_t is a Poisson process.

18. Show that the unique solution of

$$x_t = 1 + \int_0^t x_{s-} dy_s$$

is given by $x_t = e^{y_t^c} \prod_{0 \le s \le t} (1 + \Delta y_s)$. Here y_t is a finite-variation deterministic function.

19. Show that the unique solution of

$$x_t = 1 + \int_0^t x_{s-} dN_s$$

is given by $x_t = 2^{N_t}$. Here N_t is a Poisson process.

20. Show that given two adapted, measurable processes x_t and y_t, such that

$$E\left[\int_0^t (x_s)^2 ds\right] < \infty,$$

and

$$E\left[\int_0^t (y_s)^2 ds\right] < \infty,$$

we have for $0 \le r \le t$,

$$E\left[\int_0^t x_s dB_s \int_0^t y_s dB_s \mid \mathcal{F}_r\right] = E\left[\int_0^t x_s y_s ds \mid \mathcal{F}_r\right],$$

where B_t is a Brownian motion process and $\mathcal{F}_t$ is its natural filtration.

21. Show that the linear stochastic differential equation

$$dX_t = F(t)X_t dt + G(t)dt + H(t)dB_t, \qquad (3.10.1)$$

with $X_0 = \xi$ has the solution

$$X_t = \Phi(t)\left[\xi + \int_0^t \Phi^{-1}(s)G(s)ds + \int_0^t \Phi^{-1}(s)H(s)dB_s\right]. \qquad (3.10.2)$$

Here $F(t)$ is an $n \times n$ bounded measurable matrix, $H(t)$ is an $n \times m$ bounded measurable matrix, B_t is an m-dimensional Brownian motion and $G(t)$ is an $\mathbb{R}^n$-valued bounded measurable function. $\Phi(t)$ is the fundamental matrix solution of the deterministic equation

$$dX_t = F(t)X_t dt.$$

See [1].

22. Show that the solution (3.10.2) of the stochastic differential equation (3.10.1) with $E|X_0|^2 = E|\xi|^2 < \infty$ has mean

$$\mu_t = E[X_t] = \Phi(t)\left[E[\xi] + \int_0^t \Phi^{-1}(s)G(s)ds\right],$$

satisfying the deterministic differential equation

$$d\mu_t = F(t)\mu_t dt + G(t)dt, \qquad \mu_0 = E[\xi],$$

and covariance matrix $P(t)$ satisfying the deterministic matrix differential equation

$$dp(t) = F(t)P(t)dt + P(t)F(t)'dt + H(t)H(t)'dt, \qquad \mu_0 = E[\xi],$$

with initial value $P(0) = E[\xi - E\xi][\xi - E\xi]'$.

23. Show that the solution (3.10.2) of the stochastic differential equation (3.10.1) is a Gaussian stochastic process if and only if X_0 is normally distributed or constant.

24. Show that the linear stochastic differential equation

$$dX_t = -\alpha X_t dt + \sigma dB_t,$$

with $E|X_0|^2 = E|\xi|^2 < \infty$ has the solution

$$X_t = e^{-\alpha t}\xi + \sigma \int_0^t e^{-\alpha(t-s)}dB_s,$$

and

$$\mu_t = E[X_t] = e^{-\alpha t}E[\xi],$$

$$P(t) = \text{Var}(X_t) = e^{-2\alpha t}\text{Var}(\xi) + \frac{\sigma^2(1 - e^{-2\alpha t})}{2\alpha}.$$

25. Show that the sequence of stopping times given by Remark 3.4.5 is indeed a localizing sequence of stopping times, i.e. a nondecreasing sequence, converging to ∞ with probability 1.

26. Suppose for $\theta \in \mathbb{R}$,

$$X_t^\theta = e^{\theta M_t - \frac{1}{2}\theta^2 A_t}$$

is a martingale, and suppose there is an open neighborhood I of $\theta = 0$ such that for all $\theta \in I$ and all t (P- a.s.),

1. $|X_t^\theta| \le a$,
2. $|\dfrac{dX_t^\theta}{d\theta}| \le b$,
3. $|\dfrac{d^2 X_t^\theta}{d\theta^2}| \le c$.

Here a, b, c are nonrandom constants which depend on I, but not on t. Show that then the processes $\{M_t\}$ and $\{M_t^2 - A_t\}$ are martingales.

27. Prove the result of Example 3.6.12.

4

Change of measures

4.1 Introduction

We begin by giving a conditional form of Bayes' Theorem. The result relates conditional expectations under two different measures.

Consider first a simple situation like, for instance, the throwing of a die. Here

$$\Omega = \{\omega_1, \omega_2, \omega_3, \omega_4, \omega_5, \omega_6\} = \{1, 2, 3, 4, 5, 6\}.$$

Suppose $P(\omega_i) = p_i \neq 1/6$.

Let $\overline{P}$ be another probability measure such that

$$\overline{P}(\omega_i) = \frac{1}{6}.$$

Then the two measures are related by the Radon–Nikodym derivative

$$\frac{\overline{P}}{P}(\omega) = \Lambda(\omega) = \sum_i \frac{1}{6p_i} I_{\{\omega_i\}}(\omega).$$

Write $\Lambda_j = \Lambda(\omega_j) = 1/(6p_j)$.

Consider the sub-σ-field

$$\mathcal{G} = \{\{\text{odd}\}, \{\text{even}\}, \Omega, \emptyset\}.$$

Consider a set of real numbers $\{x_1, x_2, \ldots, x_6\}$ and an associated random variable $X(\omega) \to \mathbb{R}$ given by:

$$X(\omega_i) = x_i, \quad i = 1, \ldots, 6.$$

The $\mathcal{G}$-measurable random variable $E[X\Lambda \mid \mathcal{G}](\omega)$ is constant on the two atoms of $\mathcal{G}$ and is given by the following expression:

$$E[X\Lambda \mid \mathcal{G}](\omega) = \sum_j x_j \Lambda_j P[\omega_j \mid \mathcal{G}](\omega)$$

$$= \sum_j x_j \Lambda_j P[\omega_j \mid \{\text{even}\}] I_{\{\text{even}\}}(\omega) + \sum_j x_j \Lambda_j P[\omega_j \mid \{\text{odd}\}] I_{\{\text{odd}\}}(\omega)$$

$$= \frac{x_2\Lambda_2 p_2 + x_4\Lambda_4 p_4 + x_6\Lambda_6 p_6}{P(\{\text{even}\})} I_{\{\text{even}\}}(\omega)$$

$$+ \frac{x_1\Lambda_1 p_1 + x_3\Lambda_3 p_3 + x_5\Lambda_5 p_5}{P(\{\text{odd}\})} I_{\{\text{odd}\}}(\omega)$$

$$= \frac{x_2 + x_4 + x_6}{6(p_2 + p_4 + p_6)} I_{\{\text{even}\}}(\omega) + \frac{x_1 + x_3 + x_5}{6(p_1 + p_3 + p_5)} I_{\{\text{odd}\}}(\omega).$$

Similarly,

$$E[\Lambda \mid \mathcal{G}](\omega) = \sum \Lambda_j P[\omega_j \mid \mathcal{G}](\omega)$$

$$= \sum \Lambda_j \{P[\omega_j \mid \{\text{even}\}] I_{\{\text{even}\}}(\omega) + P[\omega_j \mid \{\text{odd}\}] I_{\{\text{odd}\}}(\omega)\}$$

$$= \frac{\Lambda_2 p_2 + \Lambda_4 p_4 + \Lambda_6 p_6}{P(\{\text{even}\})} I_{\{\text{even}\}}(\omega) + \frac{\Lambda_1 p_1 + \Lambda_3 p_3 + \Lambda_5 p_5}{P(\{\text{odd}\})} I_{\{\text{odd}\}}(\omega)$$

$$= \frac{1}{2(p_2 + p_4 + p_6)} I_{\{\text{even}\}}(\omega) + \frac{1}{2(p_1 + p_3 + p_5)} I_{\{\text{odd}\}}(\omega).$$

Now with $\overline{E}$ denoting expectation under $\overline{P}$,

$$\overline{E}[X \mid \mathcal{G}](\omega) = \sum x_j \overline{P}[X = x_j \mid \mathcal{G}](\omega)$$

$$= \sum x_j \{\overline{P}[X = x_j \mid \{\text{even}\}] I_{\{\text{even}\}}(\omega) + \overline{P}[X = x_j \mid \{\text{odd}\}] I_{\{\text{odd}\}}(\omega)\}$$

$$= \frac{x_2\overline{p}_2 + x_4\overline{p}_4 + x_6\overline{p}_6}{\overline{P}(\{\text{even}\})} I_{\{\text{even}\}}(\omega) + \frac{x_1\overline{p}_1 + x_3\overline{p}_3 + x_5\overline{p}_5}{\overline{P}(\{\text{odd}\})} I_{\{\text{odd}\}}(\omega)$$

$$= \frac{x_2 + x_4 + x_6}{3} I_{\{\text{even}\}}(\omega) + \frac{x_1 + x_3 + x_5}{3} I_{\{\text{odd}\}}(\omega).$$

We, therefore, see that

$$\overline{E}[X \mid \mathcal{G}](\omega) = \frac{E[\Lambda X \mid \mathcal{G}]}{E[\Lambda \mid \mathcal{G}]} I_{\{\text{even}\}}(\omega) + \frac{E[\Lambda X \mid \mathcal{G}]}{E[\Lambda \mid \mathcal{G}]} I_{\{\text{odd}\}}(\omega).$$

We now prove this result in full generality. Recall that ϕ is integrable if $E|\phi| < \infty$.

Theorem 4.1.1 *(Conditional Bayes' Theorem) Suppose $(\Omega, \mathcal{F}, P)$ is a probability space and $\mathcal{G} \subset \mathcal{F}$ is a sub-σ-field. Suppose $\overline{P}$ is another probability measure absolutely continuous with respect to P ($\overline{P} \ll P$) and with a Radon–Nikodym derivative*

$$\frac{d\overline{P}}{dP} = \Lambda.$$

Then if ϕ is any integrable $\mathcal{F}$-measurable random variable,

$$\overline{E}[\phi \mid \mathcal{G}] = \begin{cases} \dfrac{E[\Lambda\phi \mid \mathcal{G}]}{E[\Lambda \mid \mathcal{G}]} & \text{if } E[\Lambda \mid \mathcal{G}] > 0, \\ 0 & \text{otherwise.} \end{cases}$$

Proof We must show that for any $A \in \mathcal{G}$,

$$\int_A \overline{E}[\phi \mid \mathcal{G}] d\overline{P} = \int_A \alpha d\overline{P},$$

where

$$\alpha = \begin{cases} \dfrac{E\big[\Lambda\phi \mid \mathcal{G}\big]}{E\big[\Lambda \mid \mathcal{G}\big]} & \text{if } E\big[\Lambda \mid \mathcal{G}\big] > 0, \\ 0 \text{ otherwise.} \end{cases}$$

Write

$$G = \{\omega : E\big[\Lambda \mid \mathcal{G}\big] = 0\},$$

so $G \in \mathcal{G}$. Then

$$\int_G E\big[\Lambda \mid \mathcal{G}\big]\mathrm{d}P = 0 = \int_G \Lambda \mathrm{d}P,$$

and $\Lambda \geq 0$ a.s. So either $P(G) = 0$, or the restriction of Λ to G is 0 a.s. In either case, $\Lambda = 0$ a.s. on G.

Now

$$G^c = \{\omega : E\big[\Lambda \mid \mathcal{G}\big] > 0\}.$$

Suppose $A \in \mathcal{G}$; then $A = B \cup C$, where $B = A \cap G^c$ and $C = A \cap G$. Further,

$$\int_A \overline{E}[\phi \mid \mathcal{G}]\mathrm{d}\overline{P} = \int_A \phi \mathrm{d}\overline{P} = \int_A \phi\Lambda\mathrm{d}P$$

$$= \int_B \phi\Lambda\mathrm{d}P + \int_C \phi\Lambda\mathrm{d}P. \tag{4.1.1}$$

Of course, $\Lambda = 0$ a.s. on $C \subset G$, so

$$\int_C \phi\Lambda\mathrm{d}P = 0 = \int_C \alpha\mathrm{d}\overline{P}, \tag{4.1.2}$$

by definition.

Now

$$\int_B \alpha\mathrm{d}\overline{P} = \int_B (E\big[\Lambda\phi \mid \mathcal{G}\big]/E\big[\Lambda \mid \mathcal{G}\big])\mathrm{d}\overline{P}$$

$$= \overline{E}\Big[I_B \frac{E\big[\Lambda\phi \mid \mathcal{G}\big]}{E\big[\Lambda \mid \mathcal{G}\big]}\Big]$$

$$= E\Big[I_B\Lambda \frac{E\big[\Lambda\phi \mid \mathcal{G}\big]}{E\big[\Lambda \mid \mathcal{G}\big]}\Big]$$

$$= E\Big[E[I_B\Lambda \frac{E\big[\Lambda\phi \mid \mathcal{G}\big]}{E\big[\Lambda \mid \mathcal{G}\big]} \mid \mathcal{G}]\Big]$$

$$= E\Big[I_B E[\Lambda \mid \mathcal{G}]\frac{E\big[\Lambda\phi \mid \mathcal{G}\big]}{E\big[\Lambda \mid \mathcal{G}\big]}\Big]$$

$$= E\big[I_B\Lambda\phi\big].$$

That is

$$\int_B \Lambda\phi\mathrm{d}P = \int_B \alpha\mathrm{d}\overline{P}. \tag{4.1.3}$$

From (4.1.1), adding (4.1.2) and (4.1.3), we see that

$$\int_C \Lambda\phi\,\mathrm{d}P + \int_B \Lambda\phi\,\mathrm{d}P = \int_A \Lambda\phi\,\mathrm{d}P$$
$$= \int_A \overline{E}[\phi \mid \mathcal{G}]\mathrm{d}\overline{P} = \int_A \alpha\mathrm{d}\overline{P},$$

and the result follows.   ∎

Another useful version of the preceding theorem is the following result.

Theorem 4.1.2 *Suppose* $(\Omega, \mathcal{F}, P)$ *is a probability space with a filtration* $\{\mathcal{F}_t, t \geq 0\}$. *Suppose* $\overline{P}$ *is another probability measure absolutely continuous with respect to* P $(\overline{P} \ll P)$ *on* $\mathcal{F}$ *and with Radon–Nikodym derivative*

$$\frac{\mathrm{d}\overline{P}}{\mathrm{d}P} = \Lambda.$$

Define the martingale

$$E\big[\Lambda \mid \mathcal{F}_t\big] \stackrel{\Delta}{=} \Lambda_t.$$

Then if $\{\phi_t\}$ *is any* $\{\mathcal{F}_t\}$-*adapted process,*

$$\overline{E}\big[\phi_t \mid \mathcal{F}_s\big] = \begin{cases} \dfrac{E\big[\Lambda_t\phi_t \mid \mathcal{F}_s\big]}{E\big[\Lambda_t \mid \mathcal{F}_s\big]} & \text{if } E\big[\Lambda_t \mid \mathcal{F}_s\big] > 0, \\ 0 \text{ otherwise.} \end{cases}$$

4.2 Measure change for discrete time processes

Example 4.2.1   Let $\{b_n\}$ be a sequence of i.i.d. Bernouilli random variables on a probability space $(\Omega, \mathcal{F}, P)$ such that $P(b_k = 1) = p_1$ and $P(b_k = 2) = p_2, p_1 + p_2 = 1$. Consider the filtration $\{\mathcal{F}_k\} = \sigma\{b_1, \ldots, b_k\}$. Suppose that we wish to define a new probability measure $\overline{P}$ on $(\Omega, \bigvee \mathcal{F}_k\}$ such that $\overline{P}(b_k = 1) = \overline{P}(b_k = 2) = 1/2$. For $1 \leq k \leq N$ define a positive $\{\mathcal{F}_k, P\}$-martingale $\{\Lambda_k\}$ with P-mean 1 and put

$$\frac{\mathrm{d}\overline{P}}{\mathrm{d}P}(\omega)\bigg|_{\mathcal{F}_N} = \Lambda_N(\omega). \tag{4.2.1}$$

Let $\Lambda_0 = 1$. Since Λ_1 is $\mathcal{F}_1 = \sigma\{b_1\}$-measurable we have

$$\Lambda_1(\omega) = \frac{\overline{P}(b_1 = 1)}{P(b_1 = 1)}I_{(b_1=1)}(\omega) + \frac{\overline{P}(b_1 = 2)}{P(b_1 = 2)}I_{(b_1=2)}(\omega),$$

or

$$\Lambda_1(\omega) = \frac{1}{2p_1}I_{(b_1=1)}(\omega) + \frac{1}{2p_2}I_{(b_1=2)}(\omega).$$

Similarly,

$$\Lambda_2(\omega) = \sum_{i,j=1}^{2} \frac{\overline{P}(b_i = j, b_j = i)}{P(b_i = j, b_j = i)} I_{(b_i=j,b_j=i)}$$

$$= \sum_{i,j=1}^{2} \frac{1}{4 p_i p_j} I_{(b_i=j,b_j=i)}.$$

Define

$$\lambda_k(\omega) = \sum_{i=1}^{2} \frac{1}{2 p_i} I_{(b_k=i)}(\omega),$$

$$\Lambda_N(\omega) = \prod_{k=1}^{N} \lambda_k(\omega).$$

Now

$$E[\Lambda_k \mid \mathcal{F}_{k-1}] = \Lambda_{k-1} E[\lambda_k \mid \mathcal{F}_{k-1}]$$

$$= \Lambda_{k-1} E\Big[\sum_{i=1}^{2} \frac{1}{2 p_i} I_{(b_k=i)}(\omega) \mid \mathcal{F}_{k-1}\Big]$$

$$= \Lambda_{k-1} \sum_{i=1}^{2} \Big(\frac{1}{2 p_i}\Big) p_i = \Lambda_{k-1}.$$

Hence for $1 \leq k \leq N$, $\{\Lambda_k\}$ is a martingale and since $\Lambda_0 = 1$, $E[\Lambda_k] = 1$.

Lemma 4.2.2 *Under the probability measure $\overline{P}$ defined by (4.2.1), $\{b_n\}$ is a sequence of i.i.d. Bernouilli random variables such that $\overline{P}(b_n = 1) = \overline{P}(b_n = 2) = 1/2$.*

Proof Using Bayes' Theorem 4.1.1 write

$$\overline{P}[b_n = \ell \mid \mathcal{F}_{n-1}] = \overline{E}[I_{(b_n=\ell)} \mid \mathcal{F}_{n-1}] = \frac{E[I_{(b_n=\ell)}\Lambda_n \mid \mathcal{F}_{n-1}]}{E[\Lambda_n \mid \mathcal{F}_{n-1}]}$$

$$= \frac{\Lambda_{n-1} E[I_{(b_n=\ell)}\lambda_n \mid \mathcal{F}_{n-1}]}{\Lambda_{n-1} E[\lambda_n \mid \mathcal{F}_{n-1}]} = \frac{E[I_{(b_n=\ell)}\lambda_n]}{E[\lambda_n]}$$

$$= E[I_{(b_n=\ell)}\lambda_n].$$

Here $\lambda_n = \sum_{i=1}^{2} = \frac{1}{2}\frac{1}{2 p_i} I_{(b_n=i)}(\omega)$ and $E[\lambda_n] = 1$ so that

$$\overline{P}[b_n = \ell \mid \mathcal{F}_{n-1}] = \frac{1}{2 p_\ell} P[b_n = \ell]$$

$$= \frac{1}{2 p_\ell} p_\ell = \frac{1}{2},$$

which shows that under $\overline{P}$ $\{b_n\}$ is a sequence of i.i.d. Bernouilli random variables such that $P(b_n = 1) = P(b_n = 2) = 1/2$. ∎

□

Example 4.2.3 Let $\{X_n\}$ be a sequence of random variables with positive probability density functions ϕ_n on some probability space $(\Omega, \mathcal{F}, P)$. Consider the filtration $\{\mathcal{F}_n\} = \sigma\{X_1, \ldots, X_n\}$. Suppose that we wish to define a new probability measure $\overline{P}$ on $(\Omega, \bigvee \mathcal{F}_n\}$ such that X_n are i.i.d. with positive probability density function α. Let $\lambda_0 = 1$ and for $k \geq 1$,

$$\lambda_k = \frac{\alpha(X_k)}{\phi_k(X_k)},$$

$$\Lambda_n = \prod_{k=0}^{n} \lambda_k,$$

and

$$\frac{\mathrm{d}\overline{P}}{\mathrm{d}P}(\omega)\bigg|_{\mathcal{F}_n} = \Lambda_n(\omega).$$

Lemma 4.2.4 *The sequence of random variables* $\{\Lambda_n\}$, $n \geq 0$ *is an* $\{\mathcal{F}_n, P\}$*-martingale with P-mean 1. Moreover, under $\overline{P}$, $\{X_n\}$ is a sequence of i.i.d. random variables with probability density function α.*

Proof   We have to show that

$$E[\Lambda_n \mid \mathcal{F}_{n-1}] = \Lambda_{n-1}.$$

However, $\Lambda_n = \Lambda_{n-1}\lambda_n$ and since Λ_{n-1} is $\mathcal{F}_{n-1}$-measurable we must show that $E[\lambda_n \mid \mathcal{F}_{n-1}] = 1$. In view of the definition of λ_n we have

$$E[\lambda_n \mid \mathcal{F}_{n-1}] = E\left[\frac{\alpha(X_n)}{\phi_k(X_n)} \mid \mathcal{F}_{n-1}\right] = E\left[\int_{\mathbb{R}} \frac{\alpha(x)}{\phi_k(x)}\phi_k(x)\mathrm{d}x \mid \mathcal{F}_{n-1}\right] = 1.$$

Since $\{\Lambda_n\}$ is a martingale, for all n, $E[\Lambda_n] = E[\lambda_0] = 1$.

Let f be any integrable real-valued "test" function. Using Bayes' Theorem 4.1.1,

$$\overline{E}[f(x_n) \mid \mathcal{F}_{n-1}] = \frac{E[f(x_n)\Lambda_n \mid \mathcal{F}_{n-1}]}{E[\Lambda_n \mid \mathcal{F}_{n-1}]} = E[f(x_n)\lambda_n \mid \mathcal{F}_{n-1}].$$

Using the form of λ_n we have

$$E\left[\int_{\mathbb{R}} f(x)\frac{\alpha(x)}{\phi_k(x)}\phi_k(x) \mid \mathcal{F}_{n-1}\right] = \int_{\mathbb{R}} f(x)\alpha(x)\mathrm{d}x,$$

which finishes the proof.   ∎
                                                                                    □

The next example is a generalization of Example 4.2.1. Some dependence between the random variables b_n is introduced.

Example 4.2.5 Let $\{\eta_n\}, 1 \leq n \leq N$ be a Markov chain with state space $\{1, 2\}$ on a probability space $(\Omega, \mathcal{F}, P)$ such that $P(\eta_n = j \mid \eta_{n-1} = i) = p_{ij}$ and let $\{p_1^0, p_2^0\}$ be the distribution of η_0. Consider the filtration $\{\mathcal{F}_n\} = \sigma\{\eta_0, \eta_1, \ldots, \eta_n\}$. Suppose that we wish to define a new probability measure $\overline{P}$ on $(\Omega, \bigvee \mathcal{F}_n)$ such that $\overline{P}(\eta_n = j \mid \eta_{n-1} = i) = \overline{p}_{ij}$. Let $\Lambda_0 = 1$. Since Λ_1 is $\mathcal{F}_1 = \sigma\{\eta_0, \eta_1\}$-measurable we have that

$$\Lambda_1(\omega) = \frac{\overline{p}_{11}}{p_{11}} I_{(\eta_0=1, \eta_1=1)}(\omega) + \frac{\overline{p}_{12}}{p_{12}} I_{(\eta_0=1, \eta_1=2)}(\omega)$$

$$+ \frac{\overline{p}_{21}}{p_{21}} I_{(\eta_0=2, \eta_1=1)}(\omega) + \frac{\overline{p}_{22}}{p_{22}} I_{(\eta_0=2, \eta_1=2)}(\omega).$$

Define

$$\lambda_n(\omega) = \sum_{ij} \frac{\overline{p}_{ji}}{p_{ji}} I_{(\eta_{n-1}=i, \eta_n=j)}(\omega),$$

$$\Lambda_N = \prod_{n=1}^{N} \lambda_n.$$

Lemma 4.2.6 $\{\Lambda_n\}$ is an $\{\mathcal{F}_n, P\}$-martingale and under $\overline{P}$ the Markov chain η has transition probabilities $\overline{p}_{ij}$.

Proof Using the fact that Λ_{n-1} is $\mathcal{F}_{n-1}$-measurable and the Markov property of $\{\eta_n\}$ under P we can write

$$E[\Lambda_n \mid \mathcal{F}_{n-1}] = \Lambda_{n-1} \sum_{ij} \frac{\overline{p}_{ji}}{p_{ij}} E[I_{(\eta_{n-1}=i, \eta_n=j)} \mid \eta_{n-1}]$$

$$= \Lambda_{n-1} \sum_{ij} \frac{\overline{p}_{ji}}{p_{ji}} p_{ij} I_{(\eta_{n-1}=i)}$$

$$= \Lambda_{n-1} \sum_{i} I_{(\eta_{n-1}=i)} \sum_{j} \overline{p}_{ji}$$

$$= \Lambda_{n-1}.$$

Hence $\{\Lambda_n\}$ is a martingale and since $\Lambda_0 = 1$, $E[\Lambda_n] = 1$ for all $n \geq 0$. Using Bayes' Theorem 4.1.1 write

$$\overline{P}[\eta_n = \ell \mid \mathcal{F}_{n-1}] = \overline{E}[I_{(\eta_n=\ell)} \mid \mathcal{F}_{n-1}] = \frac{E[I_{(\eta_n=\ell)}\Lambda_n \mid \mathcal{F}_{n-1}]}{E[\Lambda_n \mid \mathcal{F}_{n-1}]}$$

$$= \frac{\Lambda_{n-1} E[I_{(\eta_n=\ell)}\lambda_n \mid \mathcal{F}_{n-1}]}{\Lambda_{n-1} E[\lambda_n \mid \mathcal{F}_{n-1}]}$$

$$= \frac{E[I_{(\eta_n=\ell)}\lambda_n \mid \mathcal{F}_{n-1}]}{E[\lambda_n \mid \mathcal{F}_{n-1}]}$$

$$= E[I_{(\eta_n=\ell)}\lambda_n \mid \mathcal{F}_{n-1}].$$

Here $\lambda_n(\omega) = \sum_{ij} \dfrac{\overline{p}_{ji}}{p_{ji}} I_{(\eta_{n-1}=i,\eta_n=j)}(\omega)$ and $E[\lambda_n] = 1]$ so that:

$$\overline{P}[\eta_n = \ell \mid \mathcal{F}_{n-1}] = \sum_i \frac{\overline{p}_{\ell i}}{p_{\ell i}} I_{(\eta_{n-1}=i)} P[\eta_n = \ell \mid \mathcal{F}_{n-1}]$$

$$= \sum_i \frac{\overline{p}_{\ell i}}{p_{\ell i}} I_{(\eta_{n-1}=i)} P[\eta_n = \ell \mid \eta_{n-1}]$$

$$= \sum_i \frac{\overline{p}_{\ell i}}{p_{\ell i}} p_{\ell i} I_{(\eta_{n-1}=i)}$$

$$= \overline{p}_{X_{n-1},\ell},$$

which shows that under $\overline{P}$, $\{\eta_n\}$ is a Markov chain with transition probabilities $\overline{p}_{ij}$.   ■

□

Example 4.2.7 Let $\{\eta_n\}$ be a Markov chain with state space
$S = \{e_1, \ldots, e_M\}$, where e_i are unit vectors in $\mathbb{R}^M$ with unity as the i-th element and zeros elsewhere.

Write $\mathcal{F}_n^0 = \sigma\{\eta_0, \ldots, \eta_n\}$ for the σ-field generated by $\eta_0, \ldots, \eta_n$, and $\{\mathcal{F}_n\}$ for the complete filtration generated by the $\mathcal{F}_n^0$; this augments $\mathcal{F}_n^0$ by including all subsets of events of probability zero. The Markov property implies here that

$$P(\eta_{n+1} = e_j \mid \mathcal{F}_n) = P(\eta_{n+1} = e_j \mid \eta_n).$$

Write

$$\Pi = (p_{ji}) \in \mathbb{R}^{M \times M},$$

so that $E[\eta_{k+1} \mid \mathcal{F}_k] = E[\eta_{k+1} \mid \eta_k] = \Pi \eta_k$.

From (2.4.3) we have the semimartingale

$$\eta_{n+1} = \Pi \eta_n + V_{n+1}. \tag{4.2.2}$$

The Markov chain is a simple kind of stochastic process on S. However, a more simple process would be one in which η is independently and uniformly distributed over its state space S at each time n.

This is modeled by supposing there is a probability measure $\overline{P}$ on $(\Omega, \mathcal{F})$ such that at time n, $\overline{P}(\eta_{n+1} = j \mid \eta_n = i) = 1/M$.

Given such a simple process, and its probability $\overline{P}$, we shall construct a new probability P so that under P, η is a Markov chain with transition matrix Π.

Recall that, if $\Pi = (p_{ji})$ is a transition matrix, then $(p_{ji}) \geq 0$ and $\sum_{j=1}^M p_{ji} = 1$.

Suppose Π is any transition matrix. Suppose $\{\eta_n\}$, $n \geq 0$, is a process on the finite state space S such that, under a probability $\overline{P}$,

$$\overline{P}(\eta_n = j \mid \eta_{n-1} = i) = \frac{1}{M}.$$

That is, the probability distribution of η is independent, and uniformly distributed at each time n.

Lemma 4.2.8 *Define*

$$\bar{\lambda}_\ell = M \sum_{j=1}^{M} (\langle \Pi \eta_{\ell-1}, e_j \rangle \langle \eta_\ell, e_j \rangle),$$

and $\bar{\Lambda}_n = \prod_{\ell=1}^{n} \bar{\lambda}_\ell$.

A new probability measure P is defined by putting $\dfrac{dP}{d\bar{P}}\Big|_{\mathcal{F}_n} = \bar{\Lambda}_n$, and under P, η is a Markov chain with transition matrix Π.

Proof Note first that

$$\bar{E}[\bar{\lambda}_\ell \mid \mathcal{F}_{\ell-1}] = M \bar{E}\left[\sum_{j=1}^{M} (\langle \Pi \eta_{\ell-1}, e_j \rangle \langle \eta_\ell, e_j \rangle) \mid \mathcal{F}_{\ell-1} \right]$$

$$= M \frac{1}{M} \sum_{j=1}^{M} \langle \Pi \eta_{\ell-1}, e_j \rangle$$

$$= \sum_{i=1}^{M} \sum_{j=1}^{M} \langle \eta_{\ell-1}, e_i \rangle p_{ji} = 1.$$

Then, using Bayes' Theorem 4.1.1,

$$P(\eta_n = e_j \mid \mathcal{F}_{n-1}) = E[\langle X_n, e_j \rangle \mid \mathcal{F}_{n-1}]$$

$$= \frac{\bar{E}[\langle X_n, e_j \rangle \bar{\Lambda}_n \mid \mathcal{F}_{n-1}]}{\bar{E}[\bar{\Lambda}_n \mid \mathcal{F}_{n-1}]}.$$

Because $\bar{\Lambda}_n = \bar{\Lambda}_{n-1} \bar{\lambda}_n$ and $\bar{\Lambda}_{n-1}$ is $\mathcal{F}_{n-1}$-measurable this is

$$\frac{\bar{E}[\langle X_n, e_j \rangle \bar{\lambda}_n \mid \mathcal{F}_{n-1}]}{\bar{E}[\bar{\lambda}_n \mid \mathcal{F}_{n-1}]} = M \bar{E}[\langle \Pi \eta_{n-1}, e_j \rangle \langle \eta_n, e_j \rangle) \mid \mathcal{F}_{n-1}]$$

$$= \langle \Pi \eta_{n-1}, e_j \rangle,$$

and, as this depends on η_{n-1} this equals $P(\eta_n = e_j \mid \eta_{n-1})$. If $\eta_{n-1} = e_i$ we see that $P(\eta_n = e_j \mid \eta_{n-1} = e_i) = p_{ji}$ and so, under P, η is a Markov chain with transition matrix Π. ∎

□

Example 4.2.9 In this example we discuss the filtering of a partially observed discrete-time, finite-state Markov chain, that is, the Markov chain is not observed directly; rather there is a discrete-time, finite-state observation process $\{Y_k\}$, $k \in \mathbb{N}$, which is a "noisy" function of the chain.

All processes are defined initially on a probability space $(\Omega, \mathcal{F}, P)$; below a new probability measure $\bar{P}$ is defined.

A system is considered whose state is described by a finite-state, homogeneous, discrete-time Markov chain X_k, $k \in \mathbb{N}$. We suppose X_0 is given, or its distribution known. If the state space of X_k has N elements it can be identified, without loss of generality, with the set

$$S_X = \{e_1, \ldots, e_N\},$$

where e_i are unit vectors in $\mathbb{R}^N$ with unity as the i-th element and zeros elsewhere.

Write $\mathcal{F}_k = \sigma\{X_0, \ldots, X_k\}$, for the complete filtration generated by $X_0, \ldots, X_k$. The Markov property implies here that

$$P(X_{k+1} = e_j \mid \mathcal{F}_k) = P(X_{k+1} = e_j \mid X_k).$$

Write

$$a_{ji} = P(X_{k+1} = e_j \mid X_k = e_i), \quad A = (a_{ji}) \in \mathbb{R}^{N \times N}, \tag{4.2.3}$$

so that $E[X_{k+1} \mid \mathcal{F}_k] = E[X_{k+1} \mid X_k] = AX_k$ and $X_{k+1} = AX_k + V_{k+1}$.

The state process X is not observed directly. We suppose there is a function $c(., .)$ with finite range and we observe the values

$$Y_{k+1} = c(X_k, w_{k+1}), \quad k \in \mathbb{N}, \tag{4.2.4}$$

where the w_k are a sequence of independent, identically distributed (i.i.d.) random variables.

We shall write $\{\mathcal{G}_k\}$ for the complete filtration generated by X and Y, and $\{\mathcal{Y}_k\}$ for the complete filtration generated by Y. Suppose the range of $c(., .)$ consists of M points. Then we can identify the range of $c(., .)$ with the set of unit vectors

$$S_Y = \{f_1, \ldots, f_M\}, \quad f_j = (0, \ldots, 1, \ldots, 0)' \in \mathbb{R}^M,$$

where the unit element is the j-th element.

Now (4.2.4) implies

$$P(Y_{k+1} = f_j \mid \mathcal{G}_k) = P(Y_{k+1} = f_j \mid X_k).$$

Write

$$C = (c_{ji}) \in \mathbb{R}^{M \times N}, \quad c_{ji} = P(Y_{k+1} = f_j \mid X_k = e_i), \tag{4.2.5}$$

so that $\sum_{j=1}^M c_{ji} = 1$ and $c_{ji} \geq 0$, $1 \leq j \leq M$, $1 \leq i \leq N$. Note that, for simplicity, we assume that the c_{ji} are independent of k.

We have, therefore, $E[Y_{k+1} \mid X_k] = CX_k$.

If $W_{k+1} := Y_{k+1} - CX_k$ then, taking the conditional expectation and noting $E[CX_k \mid X_k] = CX_k$, we have

$$\begin{aligned} E[W_{k+1} \mid \mathcal{G}_k] &= E[Y_{k+1} - CX_k \mid X_k] \\ &= CX_k - CX_k = 0, \end{aligned}$$

so W_k is a $(P, \mathcal{G}_k)$ martingale increment and

$$Y_{k+1} = CX_k + W_{k+1},$$

Write $Y_k^i = \langle Y_k, f_i \rangle$ so $Y_k = (Y_k^1, \ldots, Y_k^M)'$, $k \in \mathbb{N}$. For each $k \in \mathbb{N}$, exactly one component is equal to 1, the remainder being 0.

Note $\sum_{i=1}^M Y_k^i = 1$. Write

$$c_{k+1}^i = E[Y_{k+1}^i \mid \mathcal{G}_k] = \sum_{j=1}^N c_{ij} \langle e_j, X_k \rangle,$$

and $c_{k+1} = (c_{k+1}^1, \ldots, c_{k+1}^M)'$. Then

$$c_{k+1} = E[Y_{k+1} \mid \mathcal{G}_k] = CX_k.$$

We shall suppose initially that $c_k^i > 0$, $1 \leq i \leq M$, $k \in \mathbb{N}$. (See, however, Remark 4.2.12.) Note $\sum_{i=1}^M c_k^i = 1$, $k \in \mathbb{N}$.

In summary then, we have under P,

$$X_{k+1} = AX_k + V_{k+1} \tag{4.2.6}$$

$$Y_{k+1} = CX_k + W_{k+1}, \quad k \in \mathbb{N}, \tag{4.2.7}$$

where $X_k \in S_X$, $Y_k \in S_Y$, A and C are matrices of transition probabilities given in (4.2.3), (4.2.5). The entries satisfy

$$\sum_{j=1}^{N} a_{ji} = 1, \quad a_{ji} \geq 0, \tag{4.2.8}$$

$$\sum_{j=1}^{M} c_{ji} = 1, \quad c_{ji} \geq 0. \tag{4.2.9}$$

We assume, for this measure change, $c_\ell^i > 0$, $1 \leq i \leq M$, $\ell \in \mathbb{N}$. This assumption, in effect, is that given any $\mathcal{G}_k$, the observation noise is such that there is a nonzero probability that $Y_{k+1}^i > 0$ for all i. This assumption is later relaxed to achieve the main results of this section. Define

$$\lambda_\ell = \sum_{i=1}^{M} \frac{M^{-1}}{c_\ell^i} \langle Y_\ell, f_i \rangle, \quad \Lambda_k = \prod_{\ell=1}^{k} \lambda_\ell.$$

Lemma 4.2.10 *With the above definitions* $E[\lambda_{k+1} \mid \mathcal{G}_k] = 1$.

Proof

$$E[\lambda_{k+1} \mid \mathcal{G}_k] = \frac{1}{M} \sum_{i=1}^{M} \frac{1}{c_{k+1}^i} P(Y_{k+1}^i = 1 \mid \mathcal{G}_k)$$

$$= \frac{1}{M} \sum_{i=1}^{M} \frac{1}{c_{k+1}^i} \cdot c_{k+1}^i = 1.$$

∎

We now define a new probability measure $\overline{P}$ on $\left(\Omega, \bigvee_{\ell=1}^{\infty} \mathcal{G}_\ell \right)$ by putting the restriction of the Radon–Nikodym derivative $\dfrac{d\overline{P}}{dP}$ to the σ-field $\mathcal{G}_k$ equal to Λ_k. Thus $\dfrac{d\overline{P}}{dP}\Big|_{\mathcal{G}_k} = \Lambda_k$. This means that, for any set $B \in \mathcal{G}_k$,

$$\overline{P}(B) = \int_B \Lambda_k \, dP.$$

Equivalently, for any $\mathcal{G}_k$- measurable random variable ϕ,

$$\overline{E}[\phi] = \int \phi \, d\overline{P} = \int \phi \frac{d\overline{P}}{dP} \, dP = \int \phi \Lambda_k \, dP = E[\Lambda_k \phi],$$

where $\overline{E}$ and E denote expectations under $\overline{P}$ and P, respectively.

Lemma 4.2.11 *Under* $\overline{P}$, $\{Y_k\}$, $k \in \mathbb{N}$, *is a sequence of i.i.d. random variables each having the uniform distribution which assigns probability* $1/M$ *to each point* f_i, $1 \leq i \leq M$, *in its range space.*

Proof   Using Lemma 4.2.10 and Bayes' Theorem 4.1.1 we have

$$
\begin{aligned}
\overline{P}(Y_{k+1}^j = 1 \mid \mathcal{G}_k) &= \overline{E}[\langle Y_{k+1}, f_j \rangle \mid \mathcal{G}_k] \\
&= E[\Lambda_{k+1} \langle Y_{k+1}, f_j \rangle \mid \mathcal{G}_k] / E[\Lambda_{k+1} \mid \mathcal{G}_k] \\
&= \Lambda_k E[\lambda_{k+1} \langle Y_{k+1}, f_j \rangle \mid \mathcal{G}_k] / \Lambda_k E[\lambda_{k+1} \mid \mathcal{G}_k] \\
&= E[\lambda_{k+1} \langle Y_{k+1}, f_j \rangle \mid \mathcal{G}_k] \\
&= E\left[\sum_{i=1}^{M} \frac{1}{M c_{k+1}^i} \langle Y_{k+1}, f_i \rangle \langle Y_{k+1}, f_j \rangle \mid \mathcal{G}_k \right] \\
&= \frac{1}{M c_{k+1}^j} E\left[Y_{k+1}^j \mid \mathcal{G}_k \right] \\
&= \frac{1}{M c_{k+1}^j} c_{k+1}^j = \frac{1}{M},
\end{aligned}
$$

a quantity independent of $\mathcal{G}_k$, which finishes the proof.   ∎

Note that

$$
\overline{E}[X_{k+1} \mid \mathcal{G}_k] = \frac{E[\Lambda_{k+1} X_{k+1} \mid \mathcal{G}_k]}{E[\Lambda_{k+1} \mid \mathcal{G}_k]} = E[\lambda_{k+1} X_{k+1} \mid \mathcal{G}_k] = A X_k,
$$

so that under $\overline{P}$, X remains a Markov chain with transition matrix A.

A reverse measure change

What we wish to do now is start with a probability measure $\overline{P}$ on $\left(\Omega, \bigvee_{n=1}^{\infty} \mathcal{G}_n \right)$ such that

1. the process X is a finite-state Markov chain with transition matrix A and
2. $\{Y_k\}$, $k \in \mathbb{N}$, is a sequence of i.i.d. random variables and

$$
\overline{P}(Y_{k+1}^j = 1 \mid \mathcal{G}_k) = \overline{P}(Y_{k+1}^j = 1) = 1/M.
$$

Suppose $C = (c_{ji})$, $1 \leq j \leq M$, $1 \leq i \leq N$ is a matrix such that $c_{ji} \geq 0$ and $\sum_{j=1}^{M} c_{ji} = 1$.

We shall now construct a new measure P on $\left(\Omega, \bigvee_{n=1}^{\infty} \mathcal{G}_n \right)$ such that under P, (4.2.7) still holds and $E[Y_{k+1} \mid \mathcal{G}_k] = C X_k$. We again write

$$
c_{k+1} = C X_k,
$$

and $c_{k+1}^i = \langle c_{k+1}, f_i \rangle = \langle C X_k, f_i \rangle$, so that $\sum_{i=1}^{K} c_{k+1}^i = 1$.

Remark 4.2.12   We do not divide by the c_k^i in the construction of P from $\overline{P}$. Therefore, we no longer require the c_k^i to be strictly positive.   □

The construction of P from $\overline{P}$ is inverse to that of $\overline{P}$ from P. Write

$$
\overline{\Lambda}_\ell = M \sum_{i=1}^{M} c_\ell^i \langle Y_\ell, f_i \rangle, \qquad \overline{\Lambda}_k = \prod_{\ell=1}^{k} \overline{\lambda}_\ell.
$$

Lemma 4.2.13 *With the above definitions* $\overline{E}[\bar{\lambda}_{k+1} \mid \mathcal{G}_k] = 1.$

Proof Following the proof of Lemma 4.2.13,

$$\overline{E}[\bar{\lambda}_{k+1} \mid \mathcal{G}_k] = M \sum_{i=1}^{M} c_{k+1}^i \overline{P}(Y_{k+1}^i = 1 \mid \mathcal{G}_k)$$

$$= M \sum_{i=1}^{M} \frac{c_{k+1}^i}{M} = \sum_{i=1}^{M} c_{k+1}^i = 1. \qquad \blacksquare$$

This time set $\left. \dfrac{\mathrm{d}P}{\mathrm{d}\overline{P}} \right|_{\mathcal{G}_k} = \overline{\Lambda}_k.$ (The existence of P follows from Kolmogorov's Extension Theorem.)

Lemma 4.2.14 *Under P,*

$$E[Y_{k+1} \mid \mathcal{G}_k] = C X_k.$$

Proof The proof is left as an exercise. $\qquad \blacksquare$

Write $q_k(e_r)$, $1 \le r \le N$, $k \in \mathbb{N}$, for the unnormalized, conditional probability distribution such that

$$\overline{E}[\bar{\Lambda}_k \langle X_k, e_r \rangle \mid \mathcal{Y}_k] = q_k(e_r).$$

Now $\sum_{i=1}^{N} \langle X_k, e_i \rangle = 1$, so

$$\sum_{i=1}^{N} q_k(e_i) = \overline{E}\Big[\bar{\Lambda}_k \sum_{i=1}^{N} \langle X_k, e_i \rangle \mid \mathcal{Y}_k\Big] = \overline{E}[\bar{\Lambda}_k \mid \mathcal{Y}_k].$$

Therefore, the normalized conditional probability distribution

$$p_k(e_r) = E[\langle X_k, e_r \rangle \mid \mathcal{Y}_k]$$

is given by

$$p_k(e_r) = \frac{q_k(e_r)}{\sum\limits_{j=1}^{k} q_k(e_j)}.$$

Theorem 4.2.15 *For $k \in \mathbb{N}$, and $1 \le r \le N$, we have the recursive estimate*

$$q_{k+1} = A \operatorname{diag}(q_k) M \prod_{i=1}^{M} c_{ij}^{Y_k^i}.$$

Proof   Using the independence assumptions under $\overline{P}$ and the fact that $\sum_{j=1}^{N} \langle X_k, e_j \rangle = 1$, we have

$$\overline{E}[\langle X_{k+1}, e_r \rangle \overline{\Lambda}_{k+1} \mid \mathcal{Y}_{k+1}] = \overline{E}\left[\langle AX_k + V_{k+1}, e_r \rangle \overline{\Lambda}_k \overline{\Lambda}_{k+1} \mid \mathcal{Y}_{k+1} \right]$$

$$= M \sum_{j=1}^{N} \overline{E}[\langle X_k, e_j \rangle a_{rj} \overline{\Lambda}_k \mid \mathcal{Y}_k] \prod_{i=1}^{M} c_{ij}^{Y_{k+1}^i}$$

$$= M \sum_{j=1}^{N} q_k(e_j) a_{rj} \prod_{i=1}^{M} c_{ij}^{Y_{k+1}^i},$$

and the result follows.   ∎

□

Example 4.2.16 (Change of measure for linear systems). Consider a system whose state at times $k = 1, 2, \ldots$ is $x_k \in \mathbb{R}$.

Let $(\Omega, \mathcal{F}, P)$ be a probability space upon which $\{v_k\}$, $k \in \mathbb{N}$ is a sequences of $N(0, 1)$ Gaussian random variables, having zero means and variances 1. Let $\{\mathcal{F}_k\}$, $k \in \mathbb{N}$ be the complete filtration (that is, $\mathcal{F}_0$ contains all the P-null events) generated by $\{x_0, x_1, \ldots, x_k\}$.

The state of the system satisfies the linear dynamics

$$x_{k+1} = ax_k + bv_{k+1}. \tag{4.2.10}$$

Note that $E[v_{k+1} \mid \mathcal{F}_k] = 0$.

Initially we suppose all processes are defined on an "ideal" probability space $(\Omega, \mathcal{F}, \overline{P})$; then under a new probability measure P, to be defined, the model dynamics (4.2.10) will hold.

Suppose that under $\overline{P}$, $\{x_k\}$, $k \in \mathbb{N}$, is an i.i.d. $N(0, 1)$ sequence with density function ϕ. For each $l = 0, 1, 2, \ldots$ define

$$\overline{\lambda}_l = \frac{\phi(b^{-1}(x_l - ax_{l-1}))}{b\phi(x_l)},$$

$$\overline{\Lambda}_k = \prod_{l=0}^{k} \overline{\lambda}_l.$$

Lemma 4.2.17 *The process $\{\overline{\Lambda}_k\}$, $k \in \mathbb{N}$, is a $\overline{P}$-martingale with respect to the filtration $\{\mathcal{F}_k\}$.*

Proof   Since $\overline{\Lambda}_k$ is $\mathcal{F}_k$-measurable,

$$\overline{E}[\overline{\Lambda}_{k+1} \mid \mathcal{F}_k] = \overline{\Lambda}_k \overline{E}[\overline{\lambda}_{k+1} \mid \mathcal{F}_k].$$

So that it is enough to show that $\overline{E}[\overline{\lambda}_{k+1} \mid \mathcal{F}_k] = 1$:

$$\overline{E}[\overline{\lambda}_{k+1} \mid \mathcal{F}_k] = \overline{E}[\frac{\phi(b^{-1}(x_{k+1} - ax_k))}{b\phi(x_{k+1})} \mid \mathcal{F}_k]$$

$$= \int_{\mathbb{R}} \frac{\phi(b^{-1}(x - ax_k))}{b\phi(x)} \phi(x) dx.$$

Using the change of variable $b^{-1}(x - ax_k) = u$,

$$\int_{\mathbb{R}} \phi(u)du = 1,$$

and the result follows. ∎

Define P on $\{\Omega, \mathcal{F}\}$ by setting the restriction of the Radon–Nykodim derivative $\dfrac{dP}{d\overline{P}}$ to $\mathcal{G}_k$ equal to $\overline{\Lambda}_k$. Then:

Lemma 4.2.18 *On $\{\Omega, \mathcal{F}\}$ and under P, $\{v_k\}$, $k \in \mathbb{N}$, is a sequence of i.i.d. $N(0, 1)$ random variables, where*

$$v_{k+1} \stackrel{\triangle}{=} b^{-1}(x_{k+1} - ax_k).$$

Proof Suppose $f : \mathbb{R} \rightarrow \mathbb{R}$ is a "test" function (i.e. measurable function with compact support). Then with E (resp. $\overline{E}$) denoting expectation under P (resp. $\overline{P}$) and using Bayes' Theorem 4.1.1,

$$E[f(v_{k+1}) \mid \mathcal{F}_k] = \frac{\overline{E}[\overline{\Lambda}_{k+1} f(v_{k+1}) \mid \mathcal{F}_k]}{\overline{E}[\overline{\Lambda}_{k+1} \mid \mathcal{F}_k]}$$

$$= \overline{E}[\overline{\lambda}_{k+1} f(v_{k+1}) \mid \mathcal{F}_k],$$

where the last equality follows from Lemma 4.2.17. Consequently

$$E[f(v_{k+1}) \mid \mathcal{F}_k] = \overline{E}[\overline{\lambda}_{k+1} f(v_{k+1}) \mid \mathcal{F}_k]$$

$$= \overline{E}\left[\frac{\phi(b^{-1}(x_{k+1} - ax_k))}{b\phi(x_{k+1})} f(b^{-1}(x_{k+1} - ax_k)) \mid \mathcal{F}_k \right].$$

Using the independence assumption under $\overline{P}$ this is

$$\int_{\mathbb{R}} \frac{\phi(b^{-1}(x - ax_k))}{b\phi(x)} f(b^{-1}(x - ax_k))\phi(x)dx = \int_{\mathbb{R}} \phi(u)f(u)du,$$

and the lemma is proved. ∎ □

4.3 Girsanov's Theorem

In this section we investigate how martingales, and in particular, Brownian motion, are changed when a new, absolutely continuous, probability measure is introduced. We need first the following results.

Theorem 4.3.1 *Suppose $(\Omega, \mathcal{F}, P)$ is a probability space with a filtration $\{\mathcal{F}_t, t \geq 0\}$. Suppose $\overline{P}$ is another probability measure equivalent to P ($\overline{P} \ll P$ and $P \ll \overline{P}$) and with Radon–Nikodym derivative*

$$\frac{d\overline{P}}{dP} = \Lambda.$$

Define the martingale

$$E[\Lambda \mid \mathcal{F}_t] \triangleq \Lambda_t$$

Then

1. $\{X_t \Lambda_t\}$ *is a local martingale under P if and only if $\{X_t\}$ is a local martingale under $\overline{P}$.*
2. *Every $\overline{P}$-semimartingale is a P-semimartingale.*

Proof

1. We prove the result for martingales. The extension to local martingales can be found in Proposition 3.3.8 of Jacod and Shiryayev [19].
   Let $\{X_t\}$ be a $\overline{P}$ martingale and $F \in \mathcal{F}_s$, $s \leq t$. We have

$$\int_F X_t \mathrm{d}\overline{P} = \int_F X_s \mathrm{d}\overline{P} = \int_F X_s \Lambda_s \mathrm{d}P,$$

and

$$\int_F X_t \mathrm{d}\overline{P} = \int_F X_t \Lambda_t \mathrm{d}P,$$

that is

$$\int_F X_t \Lambda_t \mathrm{d}P = \int_F X_s \Lambda_s \mathrm{d}P.$$

   Hence ΛX is a P-martingale. The proof of the converse is identical.
2. By definition, a semimartingale is the sum of a local martingale and a process of finite variation. We need only prove the theorem in one direction and we can suppose $X_0 = 0$. If $\{X_t\}$ is a semimartingale under P, then by the product rule $\{X_t \Lambda_t\}$ is a semimartingale under P, which has a decomposition

$$X_t \Lambda_t = N_t + V_t,$$

   where N a local martingale and V is a process of finite variation. Therefore

$$X_t = N_t \Lambda_t^{-1} + V_t \Lambda_t^{-1},$$

   since, by the equivalence of P and $\overline{P}$, Λ_t^{-1} exists and is a $\overline{P}$-martingale. By the first part of this theorem $N_t \Lambda_t^{-1}$ is a local martingale under $\overline{P}$, and the second term is the product of the $\overline{P}$-semimartingale V of finite variation and the $\overline{P}$-martingale Λ_t^{-1}. ∎

Theorem 4.3.2 *Suppose Λ_t and $\overline{P}$ are as mentioned in Theorem 4.3.1 above. Suppose $\{X_t\}$ is a local martingale under P with $X_0 = 0$,*

(i) *$\{X_t\}$ is a special semimartingale under $\overline{P}$ if the process $\{\langle X, \Lambda\rangle_t\}$ exists and then under $\overline{P}$,*

$$X_t = \left(X_t - \int_0^t \Lambda_{s-}^{-1} \mathrm{d}\langle X, \Lambda\rangle_s \right) + \int_0^t \Lambda_{s-}^{-1} \mathrm{d}\langle X, \Lambda\rangle_s.$$

Here, the first term is a local martingale under $\overline{P}$, and the second is a predictable process of finite variation.

(ii) In general, the process

$$X_t - \int_0^t \Lambda_{s-}^{-1} d[X, \Lambda]_s$$

is a local martingale under $\overline{P}$.

Proof See [11] page 162. ∎

The following important theorem is an extension of the following rather simple situation. Let $X_1, \ldots, X_n$ be i.i.d. normal random variables with mean $E(X) = 0$ and variance $E(X^2) = \sigma^2 \neq 0$ under probability measure P and with mean $E(X) = \mu$ and variance $E(X^2) = \sigma^2 \neq 0$ under probability measure P^μ. Then it is clear that $P^\mu \ll P$ (and $P \ll P^\mu$) and that

$$\frac{dP^\mu}{dP}(\omega) = \exp\left(\sum_{i=1}^n \mu_i X_i(\omega) - \frac{1}{2} \sum_{i=1}^n \mu_i^2\right).$$

Theorem 4.3.3 (Girsanov) *Suppose B_t, $t \in [0, T]$, is an m-dimensional Brownian motion on a filtered space $\{\Omega, \mathcal{F}, \mathcal{F}_t, P\}$. Let $f = (f_1, \ldots, f_m) : \Omega \times [0, T] \to \mathbb{R}^m$ be a predictable process such that*

$$\int_0^T |f_t|^2 dt < \infty \quad a.s.$$

Write

$$\Lambda_t(f) = \exp\left(\sum_{i=1}^m \int_0^t f_s^i dB_s^i - \frac{1}{2} \int_0^t |f_s|^2 ds\right),$$

and suppose

$$E[\Lambda_T(f)] = 1,$$

(which holds if Novikov's condition $E\left[e^{\frac{1}{2}\int_0^T |f_t|^2 dt}\right] < \infty$ holds). (See [11].) If P^f is the probability measure on $\{\Omega, \mathcal{F}\}$ defined by $\dfrac{dP^f}{dP} = \Lambda_T(f)$, then W_t is an m-dimensional Brownian motion on $\{\Omega, \mathcal{F}, \mathcal{F}_t, P^f\}$, where

$$W_t^i \stackrel{\triangle}{=} B_t^i - \int_0^t f_s^i ds. \tag{4.3.1}$$

Proof We prove here the scalar case. To show W is a standard Brownian motion we verify the conditions of Theorem 2.7.1. That is, we show that (i) it is continuous a.s., (ii) it is a (local) martingale, and (iii) $\{W_t^2, t \geq 0\}$ is a (local) martingale. By definition W is a continuous process a.s. (B_t is continuous a.s. and an indefinite integral is a continuous process.) For (ii) we must show W is a local $(\mathcal{F}_t)$-martingale under measure P^f. Equivalently, from

Lemma 4.3.1 we must show that $\{\Lambda_t W_t\}$ is a local martingale under P. Using the Itô rule we see, as in Example 3.6.11, that

$$\Lambda_t(f) = 1 + \int_0^t \Lambda_s(f) f_s \mathrm{d}B_s. \tag{4.3.2}$$

Applying the Itô rule to (4.3.2) and W,

$$\begin{aligned}
\Lambda_t W_t &= W_0 + \int_0^t \Lambda_s \mathrm{d}W_s + \int_0^t W_s \mathrm{d}\Lambda_s + \int_0^t \mathrm{d}\langle \Lambda, W \rangle_s \\
&= W_0 + \int_0^t \Lambda_s \mathrm{d}B_s - \int_0^t \Lambda_s f_s \mathrm{d}s + \int_0^t W_s \Lambda_s f_s \mathrm{d}B_s + \int_0^t \Lambda_s f_s \mathrm{d}s \\
&= W_0 + \int_0^t \Lambda_s (1 + W_s f_s) \mathrm{d}B_s,
\end{aligned}$$

and, as a stochastic integral with respect to B, $\{\Lambda_t W_t, t \geq 0\}$ is a (local) martingale under P. Property (iii) is established similarly,

$$W_t^2 = 2 \int_0^t W_s \mathrm{d}W_s + \langle W, W \rangle_t = 2 \int_0^t W_s \mathrm{d}W_s + t,$$

or

$$W_t^2 - t = 2 \int_0^t W_s \mathrm{d}W_s,$$

which, from (ii), is a (local) martingale under P^f and the result follows.   ∎

Example 4.3.4 As a simple application of Girsanov's theorem, let us derive the distribution of the first passage time, $\alpha = \inf\{t, B_t = b\}$, for Brownian motion with drift to a level $b \in \mathbb{R}$ (see Example 2.2.5).

Suppose that under probability measure P, $\{B_t, \mathcal{F}_t^B\}$ is a standard Brownian motion. Write

$$\Lambda_t = \exp\left(\mu B_t - \frac{1}{2}\mu^2 t\right),$$

and set

$$\frac{\mathrm{d}P^\mu}{\mathrm{d}P} = \Lambda_t.$$

Using Girsanov's theorem, the process $B_t^\mu \overset{\triangle}{=} B_t - \mu t$ is a standard Brownian motion under probability measure P^μ. That is, under probability measure P^μ, $B_t = \mu t + B_t^\mu$ is a Brownian motion with drift μt.

Now

$$P^\mu(\alpha \le t) = E^\mu[I(\alpha \le t)] = E[\Lambda_t I(\alpha \le t)]$$
$$= E[I(\alpha \le t)E[\Lambda_t \mid \mathcal{F}_\alpha]]$$
$$\text{(see (2.2.2) and (2.2.3) for the definition of } \mathcal{F}_\alpha)$$
$$= E[I(\alpha \le t)\Lambda_\alpha]$$
$$= E[I(\alpha \le t)\exp\left(\mu b - \frac{1}{2}\mu^2\alpha\right)]$$
$$= \int_0^t \frac{|b|}{\sqrt{2\pi s^3}} \exp\left(\mu b - \frac{1}{2}\mu^2 s - b/2s\right) ds.$$

See Problem 10, Chapter 2 for the density function of α under P.   □

Remark 4.3.5 Equation (4.3.1) is equivalent to saying that the original Brownian motion process $\{B_t\}$ is a weak solution of the stochastic differential equation

$$dX_t = f(t, \omega)dt + d\overline{B}_t, \quad X_0 = 0,$$

where $\{\overline{B}_t\}$ is a Brownian motion. That is, we have constructed a probability measure $\overline{P}$ on $(\Omega, \mathcal{F})$ and a new Brownian motion process $\{\overline{B}_t\}$ such that

$$dB_t = f(t, \omega)dt + d\overline{B}_t.$$

□

Remark 4.3.6 Let X_t be a special semimartingale; then (see Example 3.6.11)

$$\Lambda_t = 1 + \int_0^t \Lambda_{s-} dX_s, \tag{4.3.3}$$

has the unique solution ($\Lambda_0 = 1$)

$$\Lambda_t = e^{X_t - \frac{1}{2}\langle X^c, X^c \rangle_t} \Pi_{s \le t}(1 + \Delta X_s)e^{-\Delta X_s},$$

which is called the stochastic exponential of the semimartingale $\{X_t\}$. If Λ_t is a uniformly integrable positive martingale then $\Lambda_\infty = \lim_{t \to \infty} \Lambda_t$ exists and

$$E[\Lambda_\infty \mid \mathcal{F}_t] = \Lambda_t \quad \text{(a.s.)}.$$

Consequently,

$$E[\Lambda_\infty] = E[\Lambda_0] = 1,$$

so that a new probability measure $\overline{P}$ can be defined on $(\Omega, \mathcal{F})$ by putting

$$\frac{d\overline{P}}{dP} = \Lambda_\infty.$$

$\overline{P}$ is equivalent to P if and only if $\Lambda_\infty > 0$ a.s. More precisely, we have the following form of Girsanov's theorem. (See [11] page 165.)   □

Theorem 4.3.7 *Suppose the exponential* Λ_t *and* $\overline{P}$ *are as mentioned in (4.3.3) and Remark 4.3.6.*

If $\{M_t\}$ *is a local martingale under probability measure* P, *and the predictable covariation process* $\{\langle M, X\rangle_t\}$ *exists under probability measure* P, *then* $\overline{M}_t = M_t - \langle M, X\rangle_t$ *is a local martingale under probability measure* $\overline{P}$.

Proof   First note that Λ_t plays the role of Λ_t in part (i) of Theorem 4.3.2. However,

$$\Lambda_t = 1 + \int_0^t \Lambda_{s-} dX_s,$$

so

$$\langle M, \Lambda \rangle_t = \int_0^t \Lambda_{s-} d\langle M, X\rangle_s$$

and

$$\int_0^t \Lambda_{s-}^{-1} d\langle M, \Lambda \rangle_s = \langle M, X\rangle_t.$$

That is, from part (i) of Theorem 4.3.2,

$$\overline{M}_t = M_t - \langle M, X\rangle_t$$

is a local martingale under probability measure $\overline{P}$.   ∎

More generally, we have the following result which is proven in [11].

Theorem 4.3.8 *Suppose for a continuous local martingale* $\{X_t\}$ *the exponential* Λ_t *and* $\overline{P}$ *are as mentioned in Remark 4.3.6.*

Let $\{M_t\} = \{M_t^1, \ldots, M_t^m\}$ *be an* $\mathbb{R}^m$*-valued continuous local martingale under probability measure* P. *Then* $\{\overline{M}_t\} = \{\overline{M}_t^1, \ldots, \overline{M}_t^m\}$ *is a continuous local martingale under probability measure* $\overline{P}$, *where* $\overline{M}_t^i = M_t^i - \langle M^i, X\rangle_t$, *and the predictable covariation under probability measure* $\overline{P}$ *of* $\{\overline{M}_t\}$ *is equal to the predictable covariation under probability measure* P *of* $\{M_t\}$, *that is*

$$\langle \overline{M}^i, \overline{M}^j\rangle_t^{\overline{P}} = \langle M^i, M^j\rangle_t^P.$$

4.4 The single jump process

In this section we investigate Radon–Nikodym derivatives relating probability measures that describe when the jump happens and where it goes for a single jump process. Recall a few facts from Chapters 2 and 3.

Consider a stochastic process $\{X_t\}$, $t \geq 0$, which takes its values in some measurable space $\{E, \mathcal{E}\}$ and which remains at its initial value $z_0 \in E$ until a random time T, when it jumps to a random position Z. A sample path of the process is

$$X_t(\omega) = \begin{cases} z_0 \text{ if } t < T(\omega), \\ Z(\omega) \text{ if } t \geq T(\omega). \end{cases}$$

The underlying probability space can be taken to be $\Omega = [0, \infty] \times E$, with the σ-field $\mathcal{B} \times \mathcal{E}$. A probability measure P is given on $(\Omega, \mathcal{B} \times \mathcal{E})$. Write $F_t = P[T > t, Z \in E]$, $c = \inf\{t : F_t = 0\}$ and

$$d\Lambda(t) = P(T \le t, Z \in E \mid T > t - \epsilon) = \frac{-dF_t}{F_{t-}},$$

for the rate of the jump of the process X.

Write $F_t^A = P[T > t, Z \in A]$, then there is a Radon–Nikodym derivative $\lambda(A, s)$ such that

$$F_t^A - F_0^A = \int_{]0,t[} \lambda(A, s) dF_s.$$

There is a bijection between probability measures P on $(\Omega, \mathcal{B} \times \mathcal{E})$ and Lévy systems (λ, Λ). For $A \in \mathcal{E}$ define

$$P(]0, t] \times A) = -\int_{]0,t]} \lambda(A, s) dF_s.$$

For $t \ge 0$ define $\mu(t, A) = I_{T \le t} I_{Z \in A}$. The predictable compensator of μ is given by

$$\mu_p(t, A) = -\int_{]0, T \wedge t]} \frac{dF_s^A}{F_{s-}}.$$

Write $\mathcal{F}_t$ for the completed σ-field generated by $\{X_s\}$, $s \le t$, then

$$q(t, A) = \mu(t, A) - \mu_p(t, A)$$

is an $\mathcal{F}_t$-martingale.

Suppose $\overline{P}$ is absolutely continuous with respect to P. Then there is a Radon–Nikodym derivative $L = \dfrac{d\overline{P}}{dP}$. Write $L_t = E[L \mid \mathcal{F}_t]$. From Lemma 3.8.8,

$$L_t = L(T, Z) I_{\{T \le t\}} + I_{\{T > t\}} \frac{1}{F_t} \int_{]t,\infty]} \int_E L(s, z) P(ds, dz).$$

However, the $P(ds, dz)$-integral is equivalent to

$$\overline{P}(T > t, Z \in E) \overset{\triangle}{=} \overline{F}_t,$$

so that

$$L_t = L(T, Z) I_{\{T \le t\}} + I_{\{T > t\}} \frac{\overline{F}_t}{F_t}.$$

If we substitute the mean 0 martingale $L_t - 1$ for M_t in Theorem 3.8.10 we have the stochastic integral representation

$$L_t - 1 = \int_\Omega I_{\{s \le t\}} g(s, x) q(ds, dx),$$

where $g(s, x) = L(s, x) - I_{\{s < c\}} \overline{F}_s / F_s$. With $\overline{c} = \inf\{t : \overline{F}_t = 0\}$ the absolute continuity of $\overline{P}$ with respect to P implies that $\overline{c} \le c$ and $g(s, x) = 0$ for $s > \overline{c}$.

In order to use the exponential formula given in Example 3.6.11 we write

$$L_t = 1 + \int_0^t L_{s-} dM_s. \qquad (4.4.1)$$

Here

$$M_t = \int_\Omega I_{\{s \le t\}} g(s, x) L_{s-}^{-1} q(ds, dx).$$

The unique solution of (4.4.1) is the stochastic exponential ($L_0 = 1$)

$$L_t = e^{M_t} \prod_{s \le t} (1 + \Delta M_s) e^{-\Delta M_s}.$$

At the discontinuity of F_s,

$$\Delta M_s = \int_E g(s, z) L_{s-}^{-1} \lambda(dz, s) \frac{\Delta F_s}{F_{s-}},$$

and at the jump time T,

$$\Delta M_T = g(T, z) L_{T-}^{-1} + \int_E g(T, z) L_{T-}^{-1} \lambda(dz, T) \frac{\Delta F_T}{F_{T-}}.$$

Hence

$$L_t = \exp\left\{ -\int_\Omega I_{\{s \le t\}} g(s, x) L_{s-}^{-1} d\mu_p \right\}$$

$$\times \left[1 + g(T, z) L_{T-}^{-1} I_{\{T \le t\}} + I_{\{T \ge t\}} \int_E g(T, z) L_{T-}^{-1} \lambda(dz, T) \frac{\Delta F_T}{F_{T-}} \right]$$

$$\times \prod_{s \le t \wedge T, u \ne T} \left[1 + \int_E g(s, z) L_{s-}^{-1} \lambda(dz, s) \frac{\Delta F_s}{F_{s-}} \right].$$

We can relate the Lévy system $(\overline{\lambda}, \overline{\Lambda})$ of probability measure $\overline{P}$ to that of probability measure P. This is given in the next theorem (see [11]).

Theorem 4.4.1 *Suppose $(\overline{\lambda}, \overline{\Lambda})$ is the Lévy system of probability measure $\overline{P}$. Then $d\overline{F}$-a.s.:*

$$\overline{\lambda}(A, s) = \frac{\int_A \left[1 + g(s, z) L_{s-}^{-1} + \frac{\Delta F_s}{F_{s-}} \int_E g(s, z) L_{s-}^{-1} d\lambda \right] d\lambda}{\int_E \left[1 + g(s, z) L_{s-}^{-1} + \frac{\Delta F_s}{F_{s-}} \int_E g(s, z) L_{s-}^{-1} d\lambda \right] d\lambda},$$

and

$$\overline{\Lambda}_t = \int_{]0,t]} \int_E \left[1 + g(s, z) L_{s-}^{-1} + \frac{\Delta F_s}{F_{s-}} \int_E g(s, z) L_{s-}^{-1} \right] \lambda(dz, s) d\Lambda_s.$$

Proof For $t > 0$ and $A \in \mathcal{E}$,

$$\overline{F}_t^A = \overline{P}(]t, \infty] \times A) = \int_{]t,\infty] \times A} L dP = -\int_{]t,\infty]} \int_A L(s, z) \lambda(dz, ds) dF_s.$$

However,

$$\overline{F}_t^A = -\int_{]t,\infty]} \overline{\lambda}(A, s) d\overline{F}_s = -\int_{]t,\infty]} \overline{\lambda}(A, s) \frac{d\overline{F}_s}{dF_s} dF_s.$$

so dF_s-a.s.:

$$\bar\lambda(A, s)\frac{d\bar F_s}{dF_s} = \int_A L(s, z)\lambda(dz, ds) = \int_A \left(\frac{\bar F_{s-}}{F_{s-}}g(s, z)L_{s-}^{-1} + \frac{\bar F_s}{F_s}\right)\lambda(dz, ds).$$

Therefore, for $s < \bar c$, and if $\bar F_{\bar c-} \neq 0$, for $s \leq \bar c$,

$$\begin{aligned}
\bar\lambda(A, s)\frac{F_s}{\bar F_{s-}}\frac{d\bar F_s}{dF_s} &= \int_A \left(\frac{F_s}{F_{s-}}g(s, z)L_{s-}^{-1} + \frac{F_s}{\bar F_{s-}}\right)\lambda(dz, ds) \quad d\bar F_s\text{-a.s.}\\
&= \int_A \left(\left(1 + \frac{\Delta F_s}{F_{s-}}\right)g(s, z)L_{s-}^{-1} + \left(1 + \frac{\Delta\bar F_s}{\bar F_{s-}}\right)\right)\lambda(dz, ds).
\end{aligned}$$

(4.4.2)

Now if s is a point of continuity of F then it is also a point of continuity of $\bar F$, and $\Delta F_s = \Delta\bar F_s = 0$. If $\Delta F_s \neq 0$ then the Radon–Nikodym derivative $\frac{d\bar F_s}{dF_s} = \frac{\Delta\bar F_s}{\Delta F_s}$, and the left hand side above is

$$\bar\lambda(A, s)\frac{(F_{s-} + \Delta F_s)}{\bar F_s}\frac{\Delta\bar F_s}{\Delta F_s} = \bar\lambda(A, s)\frac{\Delta\bar F_s}{\bar F_{s-}}\left(1 + \frac{F_{s-}}{\Delta F_s}\right).$$

Evaluating (4.4.2) when $A = E$, so $\bar\lambda(E, s) = 1 = \lambda(E, s)$,

$$\frac{\Delta\bar F_s}{\bar F_{s-}} = \frac{\Delta F_s}{F_{s-}}\int_E \left(1 + g(s, z)L_{s-}^{-1} + \frac{\Delta F_s}{F_{s-}}g(s, z)L_{s-}^{-1}\right)\lambda(dz, s),$$

if $\Delta F_s \neq 0$, and we have

$$\frac{F_s}{\bar F_{s-}}\frac{d\bar F_s}{dF_s} = \int_E (1 + g(s, z)L_{s-}^{-1})\lambda(dz, s),$$

if $\Delta F_s = 0$. Substituting in (4.4.2) we have if $(1 + \frac{\Delta F_s}{F_{s-}}) \neq 0$,

$$\bar\lambda(A, s) = \frac{\int_A \left[1 + g(s, z)L_{s-}^{-1} + \frac{\Delta F_s}{F_{s-}}\int_E g(s, z)L_{s-}^{-1}d\lambda\right]d\lambda}{\int_E \left[1 + g(s, z)L_{s-}^{-1} + \frac{\Delta F_s}{F_{s-}}\int_E g(s, z)L_{s-}^{-1}d\lambda\right]d\lambda}$$

$d\bar F_s$-a.s. for $s < \bar c$, and for $s \leq \bar c$ if $\bar F_{\bar c-} \neq 0$.

Now $(1 + \Delta F_s/F_{s-}) = 0$ only if $s = c$, $c < \infty$ and $F_{c-} \neq 0$. This situation is only of interest here if also $\bar c = c$ and $\bar F_{c-} \neq 0$. However, in this case it is easily seen that substituting

$$g(c, z)L_{c-}^{-1} = \frac{F_{c-}}{\bar F_{c-}}L(c, z)$$

in (4.4.2) gives the correct expression for $\bar\lambda(A, c) = \lambda(A, c)$, because $L(c, z) = \frac{\Delta\bar F_c}{\Delta F_c}\frac{d\bar\lambda}{d\lambda}$.

Now

$$\bar\Lambda_t = -\int_{]0, t]}\frac{d\bar F_s}{\bar F_s} = \int_{]0, t]}\frac{F_s}{\bar F_{s-}}\frac{d\bar F_s}{dF_s}d\Lambda_s.$$

If F_t is continuous at s, again $\Delta \bar{F}_s = \Delta F_s = 0$ and evaluating (4.4.2) for $A = E$,

$$\frac{d\bar{\Lambda}_s}{d\Lambda_s} = \frac{F_s \, d\bar{F}_s}{\bar{F}_s \, dF_s} = \int_E (1 + g(s, z)L_{s-}^{-1})\lambda(dz, s).$$

That is

$$\bar{\Lambda}_t = \int_{]0,t]} \int_E \left[1 + g(s, z)L_{s-}^{-1} + \frac{\Delta F_s}{F_{s-}} \int_E g(s, z)L_{s-}^{-1} \right] \lambda(dz, s)d\Lambda_s.$$

∎

Notation 4.4.2 *Denote by $\mathcal{A}$ the set of right-continuous, monotonic increasing (deterministic) functions Λ_t, $t \geq 0$, such that*

(Λ1) $\Lambda_0 = 0$,
(Λ2) $\Delta\Lambda_u = \Lambda_u - \Lambda_{u-} \leq 1$ *for all points of discontinuity u,*
(Λ3) *if $\Delta\Lambda_u = 1$ then $\Lambda_t = \Lambda_u$ for $t \geq u$.*

Remark 4.4.3 If $\Lambda_t \in \mathcal{A}$ then $\Lambda_t = \Lambda_t^c + \Lambda_t^d$, where $\Lambda_t^d = \sum_{s \leq t} \Delta\Lambda_s$ and Λ_t^c is continuous. The decomposition is unique and both Λ_t^d and Λ_t^c are in $\mathcal{A}$.

If $\Lambda_t^d = 0$ and Λ_t^c is absolutely continuous with respect to Lebesgue measure, there is a measurable function r_s such that

$$\Lambda_t^c = \int_0^t r_s ds.$$

The function r_s is often called the "rate" of the jump process.

Note that Λ might equal $+\infty$ for finite t.   □

Lemma 4.4.4 *The formulae*

$$F_t = 1 - G_t,$$
$$F_t = \exp(-\Lambda_t^c) \prod_{u \leq t}(1 - \Delta\Lambda_u), \tag{4.4.3}$$

$$\Lambda_t = -\int_{]0,t]} F_{s-}^{-1} dF_s, \tag{4.4.4}$$

define a bijection between the set $\mathcal{A}$ and the set of all probability distributions $\{G\}$ on $]0, \infty]$.

Proof   Clearly if $\Lambda_t \in \mathcal{A}$ then F_t, defined by (4.4.3), is monotonic decreasing, right-continuous, $F_0 = 0$ and $0 \leq F_t \leq 1$. Therefore $G_t = 1 - F_t$ is a probability distribution on $]0, \infty]$. Conversely, if G_t is a probability distribution, if $F_t = 1 - G_t$ and Λ_t is given by (4.4.4), then Λ_t is in $\mathcal{A}$.

From Example 3.6.11 (taking Ω to be a single point), F_t defined by (4.4.3) is the unique solution of the equation

$$dF_t = -F_{t-}d\Lambda_t, \qquad F_0 = 1.$$

This shows the correspondence is a bijection.   ∎

Lemma 4.4.5 *Suppose $\overline{\Lambda}_t \in \mathcal{A}$ is a second process whose associated Stieltjes measure $\mathrm{d}\overline{\Lambda}_t$ is absolutely continuous with respect to $\mathrm{d}\Lambda_t$, that is*

$$\frac{\mathrm{d}\overline{\Lambda}_t}{\mathrm{d}\Lambda_t} = \alpha_t.$$

Then the associated $\overline{F}_t$ has the form

$$\overline{F}_t = F_t \prod_{s \le t} \frac{(1 - \alpha(s)\Delta\Lambda_s^d)}{(1 - \Delta\Lambda_s^d)} \exp\left[-\int_0^t (\alpha(s) - 1)\mathrm{d}\Lambda_s^c\right],$$

where F_t is defined by (4.4.3). Furthermore, $\alpha(s)\Delta\Lambda_s^d \le 1$, and if $\alpha(s)\Delta\Lambda_s^d = 1$ then $\alpha(t) = 0$ for $t \ge s$.

Proof By hypothesis

$$\overline{\Lambda}_t = \int_0^t \alpha(s)\mathrm{d}\Lambda_s^c + \sum_{s \le t} \alpha(s)\Delta\Lambda_s^d,$$

so from (4.4.3)

$$\overline{F}_t = \mathrm{e}^{-\overline{\Lambda}_t^c} \prod_{u \le t}(1 - \Delta\overline{\Lambda}_u)$$

$$= \exp\left[-\int_0^t \alpha(s)\mathrm{d}\Lambda_s^c\right] \prod_{u \le t}(1 - \alpha(s)\Delta\Lambda_s^d)$$

$$= F_t \prod_{s \le t} \frac{(1 - \alpha(s)\Delta\Lambda_s^d)}{(1 - \Delta\Lambda_s^d)} \exp\left[-\int_0^t (\alpha(s) - 1)\mathrm{d}\Lambda_s^c\right].$$

The conditions on α follow from Lemma 4.4.4 and the definition of $\mathcal{A}$. ∎

If $\lambda(.,.)$ is such that

(λ1) $\lambda(A, s) \ge 0$ for $A \in \mathcal{E}, s > 0$,
(λ2) for each $A \in \mathcal{E}$ $\lambda(A, .)$ is Borel measurable,
(λ3) for all $s \in]0, c[$, (except perhaps on a set of $\mathrm{d}\Lambda$-measure 0), $\lambda(., s)$ is a probability measure on $(E, \mathcal{E})$, and if $c < \infty$ and $\Lambda_{c-} < \infty$ then $\lambda(., c)$ is a probability measure.

Then:

Lemma 4.4.6 *There is a bijection between probability measures P on $(\Omega, \mathcal{B} \times \mathcal{E})$ and Lévy systems (λ, Λ).*

Proof In Example 2.1.4 we saw how a Lévy system is determined by a measure P. Conversely, given a pair (λ, Λ), because $\Lambda \in \mathcal{A}$ we can determine a function F_t by (4.4.3). For $A \in \mathcal{E}$ define

$$P(]0, t] \times A) = -\int_{]0,t]} \lambda(A, s)\mathrm{d}F_s.$$

∎

Now the converse of theorem 4.4.1 is given. (Theorem 17.12 of [11].)

Theorem 4.4.7 *Suppose P, $\bar{P}$ have Lévy systems (λ, Λ) and $(\bar{\lambda}, \bar{\Lambda})$. Write*

$$\bar{c} = \inf\{t : \bar{F}_t = 0\},$$

and suppose $\bar{c} \leq c$, $d\bar{\Lambda}_t \ll d\Lambda$ on $]0, \bar{c}]$ and $\bar{\lambda}(., t) \ll \lambda(., t)$ $d\Lambda$-a.e. Then $\bar{P} \ll P$ with Radon–Nikodym derivative

$$L(t, z) = \alpha(t)\beta(t, z)\Pi_{t-}\exp\left[-\int_0^t (\alpha(s) - 1)d\Lambda_s^c\right]I_{\{t \leq \bar{c}\}}. \tag{4.4.5}$$

Here $\Pi_t = \displaystyle\prod_{s \leq t} \frac{\left(1 + \dfrac{\Delta F_s}{F_{s-}}\alpha(s)\right)}{\left(1 + \dfrac{\Delta F_s}{F_{s-}}\right)}$,

$$\frac{d\bar{\Lambda}_t}{d\Lambda_t} = \alpha(t), \text{ and } \frac{d\bar{\lambda}_t}{d\lambda_t} = \beta(t, z).$$

Proof Define $L(t, Z)$ by (4.4.5) and write

$$\eta(t) = \exp\left(-\int_0^t (\alpha(s) - 1)d\Lambda_s^c\right).$$

Then, because $\displaystyle\int_E \beta(t, z)d\lambda = 1$ a.s.

$$E[L(t, Z)] = -\int_{]0,\bar{c}]} \alpha(t)\eta(t)\Pi_{t-}dF_t.$$

From Lemma 4.4.5 and Equations (4.4.3) and (4.4.4),

$$\eta(t)\Pi_{t-} = \frac{\bar{F}_{t-}}{F_{t-}}.$$

As measures on $[0, \infty]$,

$$d\bar{\Lambda}_t = \frac{d\bar{F}_t}{\bar{F}_{t-}} = -\alpha(t)\frac{dF_t}{F_{t-}} = \alpha(t)d\Lambda_t,$$

so

$$E[L(t, Z)] = -\int_{]0,\bar{c}]} \alpha(t)\bar{F}_{t-}\frac{dF_t}{F_{t-}} = -\int_{]0,\bar{c}]} \bar{F}_{t-}\frac{d\bar{F}_{t-}}{\bar{F}_{t-}} = \bar{F}_0 - \bar{F}_{\bar{c}} = 1.$$

A probability measure $P^* \ll P$ can, therefore, be defined on $(\Omega, \mathcal{B} \times \mathcal{E})$ by putting $\dfrac{dP^*}{dP} = L$. For $t < c$ we have

$$L_t = E[L \mid \mathcal{F}] = L(T, Z)I_{\{t \geq T\}} + I_{\{t < T\}}\int_{]t,\bar{c}] \times E} L(s, z)dP.$$

By similar calculations to those above the later term is

$$F_t^{-1}\int_{]t,\bar{c}]} \alpha(s)\eta(s)\Pi_{s-}dF_s = \frac{\bar{F}_t}{F_t} = \eta(t)\Pi_t,$$

so $L_t = \alpha(T)\beta(T, Z)\eta(T)\Pi_{T-}I_{\{t\geq T\}} + I_{\{t<T\}}\eta(t)\Pi_t.$

Write $\phi(t, z) = \dfrac{F_{t-}}{\bar{F}_{t-}}\left(L(t, z) - \dfrac{\bar{F}_t}{F_t}\right).$

(1)

$$\phi(t, z) = [\alpha(t)\beta(t, Z)\eta(t)\Pi_{t-}I_{\{t\leq\bar{c}\}} - \eta(t)\Pi_t]\eta(t)^{-1}\Pi_{t-}^{-1}$$

$$= \alpha(t)\beta(t, Z)I_{\{t\leq\bar{c}\}} - \frac{\left(1 + \dfrac{\Delta F_t}{F_{t-}}\alpha(t)\right)}{\left(1 + \dfrac{\Delta F_t}{F_{t-}}\right)},$$

for $t < \bar{c}$.

(2) $\phi(t, z) = 0$ for $t > \bar{c}$, and for $t = \bar{c}$ if $\bar{c} = \infty$ or $\bar{F}_{\bar{c}-} = 0$.

(3) $\phi(\bar{c}, z) = L(\bar{c}, z)$ if $\bar{c} < \infty$ and $\bar{F}_{\bar{c}-} \neq 0$, that is, substituting, in this case $\phi(\bar{c}, z) = \alpha(\bar{c})\beta(\bar{c}, z).$

The Lévy system (λ^*, Λ^*) associated with P^* is then defined by

$$\lambda^*(A, s) = \frac{\displaystyle\int_A\left[1 + \phi + \frac{\Delta F_s}{F_{s-}}\int_E \phi d\lambda\right]d\lambda}{\displaystyle\int_E\left[1 + \phi + \frac{\Delta F_s}{F_{s-}}\int_E \phi d\lambda\right]d\lambda},$$

and

$$\Lambda_t^* = \int_{]0,t]}\int_E\left[1 + \phi + \frac{\Delta F_s}{F_{s-}}\int_E \phi\right]\lambda(dz, s)d\Lambda_s.$$

Substituting the above expression for ϕ we have

$$\int_E \phi d\lambda = \alpha(t)I_{\{t\leq\bar{c}\}} - \frac{(1 + (\Delta F_t/F_{t-})\alpha(t))}{(1 + \Delta F_t/F_{t-})},$$

and $(1 + (\Delta F_t/F_{t-}))\displaystyle\int_E \phi d\lambda = 1.$

The above expression gives

$$\frac{d\Lambda_t^*}{d\Lambda_t} = \alpha(t), \text{ and } \frac{d\lambda_t^*}{d\lambda_t} = \beta(t, z),$$

so

$$\Lambda_t^* = \bar{\Lambda}, \text{ and } \lambda_t^* = \bar{\lambda}.$$

By Lemma 4.4.6, $\bar{P} = P^* \ll P$ and the result is proved. ∎

4.5 Change of parameter in Poisson processes

Let N_t be a Poisson process with constant parameter λ on a filtered probability space $(\Omega, \mathcal{F}, \mathcal{F}_t, P)$ and suppose that we wish to define a new probability $\bar{P}$ such that N_t is a

Poisson process with constant parameter $\hat{\lambda}$. Define the stochastic process

$$\Lambda_t = \left(\frac{\hat{\lambda}}{\lambda}\right)^{N_t} e^{(\lambda-\hat{\lambda})t}. \tag{4.5.1}$$

Lemma 4.5.1 *The process (4.5.1) is a martingale under probability measure P.*

Proof Let $0 \le s \le t$ and recall that N_t is an independent increment process adapted to $\mathcal{F}_t$ so that

$$E[\Lambda_t \mid \mathcal{F}_s] = \left(\frac{\hat{\lambda}}{\lambda}\right)^{N_s} e^{(\lambda-\hat{\lambda})s} E\left[\left(\frac{\hat{\lambda}}{\lambda}\right)^{(N_t-N_s)} e^{(\lambda-\hat{\lambda})(t-s)}\right]$$

$$= \Lambda_s e^{(\lambda-\hat{\lambda})(t-s)} E\left[\left(\frac{\hat{\lambda}}{\lambda}\right)^{(N_t-N_s)}\right]$$

$$= \Lambda_s e^{(\lambda-\hat{\lambda})(t-s)} \sum_k \left(\frac{\hat{\lambda}}{\lambda}\right)^k e^{-\lambda(t-s)} \frac{(\lambda(t-s))^k}{k!}$$

$$= \Lambda_s,$$

which finishes the proof. ∎

Lemma 4.5.2 *The exponential martingale $\{\Lambda_t\}$ given by (4.5.1) is the unique solution of*

$$\Lambda_t = 1 - \int_0^t \Lambda_{s-} \lambda^{-1}(\lambda - \hat{\lambda})(dN_s - \lambda ds). \tag{4.5.2}$$

Proof Write

$$\Lambda_t = e^{(\lambda-\hat{\lambda})t} Y_t, \tag{4.5.3}$$

where $Y_t = \left(\frac{\hat{\lambda}}{\lambda}\right)^{N_t}$, and $\Lambda_t = f(t, Y_t)$. Using rule (3.6.9),

$$f(t, Y_t) = 1 + \int_0^t \frac{\partial f(s, Y_{s-})}{\partial s} ds + \int_0^t \frac{\partial f(s, Y_{s-})}{\partial Y} dY_s$$

$$+ \sum_{0 < s \le t} \left(f(s, Y_s) - f(s, Y_{s-}) - \frac{\partial f(s, Y_{s-})}{\partial Y} \Delta Y_s\right). \tag{4.5.4}$$

Because Y_t is a purely discontinuous process and of bounded variation, the second integral in (4.5.4) is equal to $\sum_{0 < s \le t} e^{(\lambda-\hat{\lambda})s} \Delta Y_s$.

In expression (4.5.4) we have

$$f(s, Y_s) - f(s, Y_{s-}) = \Lambda_s - \Lambda_{s-}$$

$$= e^{(\lambda - \hat{\lambda})s} \prod_{r \leq s-} \left(\frac{\hat{\lambda}}{\lambda}\right)^{\Delta N_r} \left(\frac{\hat{\lambda}}{\lambda}\right)^{\Delta N_s}$$

$$-e^{(\lambda - \hat{\lambda})s} \prod_{r \leq s-} \left(\frac{\hat{\lambda}}{\lambda}\right)^{\Delta N_r}$$

$$= \Lambda_{s-} \left(\left(\frac{\hat{\lambda}}{\lambda}\right)^{\Delta N_s} - 1\right)$$

$$= \Lambda_{s-} \left(\frac{\hat{\lambda}}{\lambda} - 1\right) \Delta N_s.$$

Putting all these results together gives

$$\Lambda_t = 1 + \int_0^t (\lambda - \hat{\lambda}) \Lambda_{s-} ds + \sum_{0 < s \leq t} e^{(\lambda - \hat{\lambda})s} \Delta Y_s$$

$$+ \sum_{0 < s \leq t} \left(\Lambda_{s-} \left(\frac{\hat{\lambda}}{\lambda} - 1\right) \Delta N_s - e^{(\lambda - \hat{\lambda})s} \Delta Y_s\right)$$

$$= 1 + \int_0^t (\lambda - \hat{\lambda}) \Lambda_{s-} ds + \sum_{0 < s \leq t} e^{(\lambda - \hat{\lambda})s} \Delta Y_s$$

$$- \sum_{0 < s \leq t} e^{(\lambda - \hat{\lambda})s} \Delta Y_s - \int_0^t (\lambda - \hat{\lambda}) \Lambda_{s-} ds$$

$$- \int_0^t \Lambda_{s-} \lambda^{-1} (\lambda - \hat{\lambda})(dN_s - \lambda ds),$$

which, after simplification, is (4.5.2). ∎

Now define a new probability measure $\overline{P}$ by setting

$$E\left[\frac{d\overline{P}}{dP} \Big| \mathcal{F}_t\right] = \Lambda_t.$$

Lemma 4.5.3 *Under probability measure $\overline{P}$ the process N_t is a Poisson process with parameter $\hat{\lambda}$.*

Proof By the characterization theorem of Poisson processes (see Theorem 3.6.14) we must show that $M_t \stackrel{\Delta}{=} N_t - \hat{\lambda}t$ and $M_t^2 - \hat{\lambda}t$ are $(\overline{P}, \mathcal{F}_t)$-martingales.

By Bayes' Theorem 4.1.1, for $t \geq s \geq 0$,

$$\overline{E}[M_t \mid \mathcal{F}_s] = \frac{E[\Lambda_t M_t \mid \mathcal{F}_s]}{E[\Lambda_t \mid \mathcal{F}_s]} = \frac{E[\Lambda_t M_t \mid \mathcal{F}_s]}{\Lambda_s}.$$

Therefore, M_t is a $(\overline{P}, \mathcal{F}_t)$-martingale if and only if $\Lambda_t M_t$ is a $(P, \mathcal{F}_t)$-martingale. Now

$$\Lambda_t M_t = \int_0^t \Lambda_{s-} dM_s + \int_0^t M_{s-} d\Lambda_s + [\Lambda, M]_t.$$

Recall

$$[\Lambda, M]_t = \sum_{0 < s \le t} \Delta \Lambda_s \Delta M_s$$

$$= \sum_{0 < s \le t} \Delta \left(\int_0^t \Lambda_{s-} \lambda^{-1} (\lambda - \hat{\lambda}) dN_s \right) \Delta N_s$$

$$= - \int_0^t \Lambda_{s-} \lambda^{-1} (\lambda - \hat{\lambda}) d[N, N]_s$$

$$= - \int_0^t \Lambda_{s-} \lambda^{-1} (\lambda - \hat{\lambda}) dN_s.$$

Therefore

$$\Lambda_t M_t = \int_0^t \Lambda_{s-} (dN_s - \hat{\lambda} ds) + \int_0^t M_{s-} d\Lambda_s - \int_0^t \Lambda_{s-} \lambda^{-1} (\lambda - \hat{\lambda}) dN_s. \qquad (4.5.5)$$

The second integral on the right of (4.5.5) is a $(P, \mathcal{F}_t)$-martingale. (Recall that $N_t - \lambda t$ is a $(P, \mathcal{F}_t)$-martingale.) The other two integrals are written as

$$\int_0^t \Lambda_{s-} (dN_s - \hat{\lambda} ds) = \int_0^t \Lambda_{s-} (dN_s - \lambda ds + \lambda ds - \hat{\lambda} ds)$$

$$= \int_0^t \Lambda_{s-} (dN_s - \lambda ds) + \int_0^t \Lambda_{s-} \lambda ds - \int_0^t \Lambda_{s-} \hat{\lambda} ds, \qquad (4.5.6)$$

and

$$\int_0^t \Lambda_{s-} \lambda^{-1} (\lambda - \hat{\lambda}) dN_s = \int_0^t \Lambda_{s-} \lambda^{-1} (\lambda - \hat{\lambda}) (dN_s - \lambda ds + \lambda ds)$$

$$= \int_0^t \Lambda_{s-} (dN_s - \lambda ds) + \int_0^t \Lambda_{s-} \lambda^{-1} (\lambda - \hat{\lambda}) \lambda ds. \qquad (4.5.7)$$

Substituting (4.5.6) and (4.5.7) in (4.5.5) yields the desired result and it remains to show that $M_t^2 - \hat{\lambda} t$ is also a $(\overline{P}, \mathcal{F}_t)$-martingale.

Now

$$M_t^2 = 2 \int_0^t M_{s-} dM_s + [M, M]_t = 2 \int_0^t M_{s-} dM_s + N_t. \qquad (4.5.8)$$

Subtracting $\hat{\lambda} t$ from both sides of (4.5.8) makes the last term on the right of (4.5.8) a $(\overline{P}, \mathcal{F}_t)$-martingale and since the dM integral is a $(\overline{P}, \mathcal{F}_t)$-martingale the result follows.   ∎

4.6 Poisson process with drift

Let N_t be a Poisson process with parameter λ on a filtered probability space $(\Omega, \mathcal{F}, \mathcal{F}_t, P)$ and suppose that we have the following process:

$$X_t = \mu t + \sigma N_t = \mu t + \sigma(N_t - \lambda t + \lambda t)$$
$$= (\mu + \sigma \lambda)t + \sigma M_t. \tag{4.6.1}$$

Here μ and σ are constants and $\{M_t\} = \{N_t - \lambda t\}$ is an $(\mathcal{F}_t, P)$-martingale. The dynamics given by (4.6.1) could describe the evolution of a system with a linear trend perturbated by random jumps of size σ given by the Poisson process N.

We wish to define a new probability $\overline{P}$ such that X_t has dynamics

$$X_t = \sigma \overline{M}_t, \tag{4.6.2}$$

where $\{\overline{M}_t\}$ is an $(\mathcal{F}_t, \overline{P})$-martingale.

Define the stochastic process

$$\Lambda_t = \exp\left\{-\int_0^t \frac{\mu}{\lambda\sigma} dM_r\right\} \prod_{0<s\leq t} (1 - \frac{\mu}{\lambda\sigma}\Delta N_s) e^{\frac{\mu}{\lambda\sigma}\Delta N_s}$$

$$= e^{-\frac{\mu}{\lambda\sigma}M_t} \prod_{0<s\leq t} (1 - \frac{\mu}{\lambda\sigma}\Delta N_s) e^{\frac{\mu}{\lambda\sigma}\Delta N_s}$$

$$= e^{\frac{\mu}{\sigma}t} \prod_{0<s\leq t} (1 - \frac{\mu}{\lambda\sigma}\Delta N_s), \tag{4.6.3}$$

where the last expression is obtained if we recall that $M_t = N_t - \lambda t$ and that

$$\prod_{0<s\leq t} e^{\frac{\mu}{\lambda\sigma}\Delta N_s} = \exp\left\{\sum_{0<s\leq t} \frac{\mu}{\lambda\sigma}\Delta N_s\right\}$$

$$= e^{\frac{\mu}{\lambda\sigma}N_t}.$$

Lemma 4.6.1 *The process (4.6.3) is a martingale under probability measure P.*

Proof Let $0 \leq s \leq t$ and recall that N_t is an independent increment process and is adapted to $\mathcal{F}_t$ so that

$$E[\Lambda_t \mid \mathcal{F}_s] = e^{\frac{\mu}{\sigma}t} E\left[\prod_{0<r\leq t} (1 - \frac{\mu}{\lambda\sigma}\Delta N_r) \mid \mathcal{F}_s\right]$$

$$= \Lambda_s e^{\frac{\mu}{\sigma}(t-s)} E\left[\prod_{s<r\leq t} (1 - \frac{\mu}{\lambda\sigma}\Delta N_r)\right]$$

$$= \Lambda_s e^{\frac{\mu}{\sigma}(t-s)} E\left[\left[1 - \frac{\mu}{\lambda\sigma}\right]^{N_t - N_s}\right]$$

$$= \Lambda_s e^{\frac{\mu}{\sigma}(t-s)} \sum_k \left(1 - \frac{\mu}{\lambda\sigma}\right)^k e^{-\lambda(t-s)} \frac{(\lambda(t-s))^k}{k!}$$

$$= \Lambda_s.$$

∎

Lemma 4.6.2 *The exponential martingale* $\{\Lambda_t\}$ *given by (4.6.3) is the unique solution of*

$$\Lambda_t = 1 - \int_0^t \Lambda_{s-} \frac{\mu}{\lambda\sigma} (dN_s - \lambda ds). \tag{4.6.4}$$

Proof Write $\Lambda_t = e^{\frac{\mu}{\sigma}t} Y_t$, where $Y_t = \prod_{0<r\le t}(1 - \frac{\mu}{\lambda\sigma}\Delta N_r)$, and $\Lambda_t = f(t, Y_t)$. Using rule (3.6.9),

$$f(t, Y_t) = 1 + \int_0^t \frac{\partial f(s, Y_{s-})}{\partial s} ds + \int_0^t \frac{\partial f(s, Y_{s-})}{\partial Y} dY_s \tag{4.6.5}$$

$$+ \sum_{0<s\le t} \left[f(s, Y_s) - f(s, Y_{s-}) - \frac{\partial f(s, Y_{s-})}{\partial Y} \Delta Y_s \right]. \tag{4.6.6}$$

Because Y_t is a purely discontinuous process and of bounded variation the second integral in (4.6.5) is equal to $\sum_{0<s\le t} e^{\frac{\mu}{\sigma}s} \Delta Y_s$.

In expression (4.6.6) we have

$$f(s, Y_s) - f(s, Y_{s-}) = \Lambda_s - \Lambda_{s-}$$

$$= e^{\frac{\mu}{\sigma}s} \prod_{r\le s-} \left(1 - \frac{\mu}{\lambda\sigma}\Delta N_r\right) \left(1 - \frac{\mu}{\lambda\sigma}\Delta N_s\right)$$

$$- e^{\frac{\mu}{\sigma}s} \prod_{r\le s-} \left(1 - \frac{\mu}{\lambda\sigma}\Delta N_r\right)$$

$$= \Lambda_{s-} \left(1 - \frac{\mu}{\lambda\sigma}\Delta N_s - 1\right)$$

$$= -\Lambda_{s-} \left(\frac{\mu}{\lambda\sigma}\Delta N_s\right).$$

Combining these results together gives

$$\Lambda_t = 1 + \int_0^t \left(\frac{\mu}{\sigma}\right) \Lambda_{s-} ds + \sum_{0<s\le t} e^{(\frac{\mu}{\sigma})s} \Delta Y_s$$

$$+ \sum_{0<s\le t} \left[\Lambda_{s-} \left(-\frac{\mu}{\lambda\sigma}\Delta N_s\right) - e^{(\frac{\mu}{\sigma})s} \Delta Y_s \right]$$

$$= 1 + \int_0^t \left(\frac{\mu}{\sigma}\right) \Lambda_{s-} ds + \sum_{0<s\le t} e^{(\frac{\mu}{\sigma})s} \Delta Y_s$$

$$- \sum_{0<s\le t} e^{(\frac{\mu}{\sigma})s} \Delta Y_s - \int_0^t \left(\frac{\mu}{\sigma}\right) \Lambda_{s-} ds$$

$$- \int_0^t \Lambda_{s-} \frac{\mu}{\lambda\sigma} (dN_s - \lambda ds),$$

which, after simplification, is (4.6.4). ∎

Now define a new probability measure $\overline{P}$ by setting

$$E\left[\frac{d\overline{P}}{dP} \Big| \mathcal{F}_t\right] = \Lambda_t.$$

Lemma 4.6.3 *Under probability measure $\overline{P}$ the process X_t has dynamics given by (4.6.2).*

Proof   Let $\overline{M}_t = N_t - \overline{\lambda}t$, where $\overline{\lambda} = \lambda - \mu/\sigma$ which is assumed to be positive. We claim that $\overline{M}_t$ is a $\overline{P}$-martingale. To see this we need only show that $\Lambda \overline{M}$ is a P-martingale. Using the differentiation rule,

$$
\begin{aligned}
d(\Lambda_t \overline{M}_t) &= \Lambda_{t-}d\overline{M}_t + \overline{M}_{t-}d\Lambda_t + d[\Lambda, \overline{M}]_t \\
&= \Lambda_{t-}(dN_t - \overline{\lambda}dt) + \overline{M}_{t-}\Lambda_{t-}\frac{\mu}{\lambda\sigma}dM_t - \Lambda_{t-}\frac{\mu}{\lambda\sigma}dN_t \\
&= \Lambda_{t-}\left(dN_t - \lambda dt + \frac{\mu}{\sigma}dt + \overline{M}_{t-}\frac{\mu}{\lambda\sigma}dM_t - \frac{\mu}{\lambda\sigma}dN_t\right) \\
&= \Lambda_{t-}\left(dM_t + \overline{M}_{t-}\frac{\mu}{\lambda\sigma}dM_t - \frac{\mu}{\lambda\sigma}dM_t\right),
\end{aligned}
$$

so that $\Lambda \overline{M}$ is a dM stochastic integral, and hence is a P-martingale.
So we can write $X_t = \sigma \overline{M}_t$, under the new measure $\overline{P}$.   ∎

Remark 4.6.4   The results of this section hold if (4.6.1) is replaced with the more general dynamics

$$
dX_t = \mu(t, X_{t-})dt + \sigma(t, X_{t-})dN_t.
$$

The stochastic exponential martingale (4.6.3) takes the form

$$
\Lambda_t = \exp\left\{\int_0^t \frac{\mu_s}{\sigma_s}ds\right\}\prod_{0<s\leq t}(1 - \frac{\mu_s}{\lambda\sigma_s}\Delta N_s).
$$

□

4.7 Continuous-time Markov chains

Consider again the finite-state Markov process $\{X_t\}$ on the set of standard unit vectors of $\mathbb{R}^N$ (see Example 2.6.17 and Section 3.8).

Write $\mathcal{F}_t$ for the right-continuous, complete filtration $\sigma\{X_r : 0 \leq r \leq t\}$.
We saw in Lemma 2.6.18 that X_t has the semimartingale representation

$$
X_t = X_0 + \int_0^t A_r X_r dr + V_t.
$$

Recall that $\mathcal{J}_t$ denotes the total number of jumps (of all kinds) of the process X up to time t and

$$
\mathcal{J}_t = -\sum_{i=1}^N \int_0^t \langle X_s, e_i\rangle a_{ii}(s)ds + Q_t.
$$

Write

$$
\lambda_t = -\sum_{i=1}^N \langle X_t, e_i\rangle a_{ii}(t).
$$

Suppose that we wish to define a new probability $\overline{P}$ such that $\mathcal{J}_t$ is a standard Poisson process with parameter 1. Define the P-martingale

$$\Lambda_t = \exp\{-\int_0^t \log \lambda_r d\mathcal{J}_r + \int_0^t (\lambda_r - 1)dr\}.$$

Since $\{\Lambda_t, \mathcal{F}_t\}$ is a P-martingale such that

$$\Lambda_t = 1 - \int_0^t \Lambda_{r-}\lambda_r^{-1}(\lambda_r - 1)(d\mathcal{J}_r - \lambda_r dr),$$

we can define a new probability measure $\overline{P}$ by setting

$$E\left[\frac{d\overline{P}}{dP}\Big|\mathcal{F}_t\right] = \Lambda_t.$$

Lemma 4.7.1 *Under probability measure $\overline{P}$ the process $\mathcal{J}_t$ is a Poisson process with parameter* 1.

Proof   By the characterization theorem of Poisson processes (see Theorem 3.6.14) we must show that $\overline{Q}_t \overset{\Delta}{=} \mathcal{J}_t - t$ and $\overline{Q}_t^2 - t$ are $(\overline{P}, \mathcal{F}_t)$-martingales.
    By Bayes' Theorem 4.1.1, for $t \geq s \geq 0$,

$$\overline{E}[\overline{Q}_t \mid \mathcal{F}_s] = \frac{E[\Lambda_t \overline{Q}_t \mid \mathcal{F}_s]}{E[\Lambda_t \mid \mathcal{F}_s]} = \frac{E[\Lambda_t \overline{Q}_t \mid \mathcal{F}_s]}{\Lambda_s}.$$

Therefore, $\overline{Q}_t$ is a $(\overline{P}, \mathcal{F}_t)$-martingale if and only if $\Lambda_t \overline{Q}_t$ is a $(P, \mathcal{F}_t)$-martingale. Now

$$\Lambda_t \overline{Q}_t = \int_0^t \Lambda_{s-} d\overline{Q}_s + \int_0^t \overline{Q}_{s-} d\Lambda_s + [\Lambda, \overline{Q}]_t,$$

and

$$\begin{aligned}
[\Lambda, \overline{Q}]_t &= \left[1 - \int_0^t \Lambda_{s-}\lambda_s^{-1}(\lambda_s - 1)(d\mathcal{J}_s - \lambda_s ds), \int_0^t d\mathcal{J}_s - t\right] \\
&= -\int_0^t \Lambda_{s-}\lambda_s^{-1}(\lambda_s - 1)d[\mathcal{J}, \mathcal{J}]_s \\
&= \int_0^t \Lambda_{s-}(\lambda_s^{-1} - 1)d\mathcal{J}_s.
\end{aligned}$$

Therefore

$$\Lambda_t \overline{Q}_t = \int_0^t \Lambda_{s-}(d\mathcal{J}_s - ds) + \int_0^t \overline{Q}_{s-} d\Lambda_s + \int_0^t \Lambda_{s-}(\lambda_s^{-1} - 1)d\mathcal{J}_s. \qquad (4.7.1)$$

The second integral on the right of (4.7.1) is a $(P, \mathcal{F}_t)$-martingale. However, $\mathcal{J}_t - \lambda_t$ is a $(P, \mathcal{F}_t)$-martingale so that the other two integrals are written as

$$\int_0^t \Lambda_{s-}(d\mathcal{J}_s - ds) = \int_0^t \Lambda_{s-}(d\mathcal{J}_s - \lambda_s ds + \lambda_s ds - ds)$$

$$= P\text{-martingale} + \int_0^t \Lambda_{s-}\lambda_s ds - \int_0^t \Lambda_{s-} ds, \qquad (4.7.2)$$

and

$$\int_0^t \Lambda_{s-}(\lambda_s^{-1} - 1)\mathrm{d}\mathcal{J}_s = \int_0^t \Lambda_{s-}(\lambda_s^{-1} - 1)(\mathrm{d}\mathcal{J}_s - \lambda_s \mathrm{d}s + \lambda_s \mathrm{d}s)$$

$$= P\text{-martingale} + \int_0^t \Lambda_{s-}(\lambda_s^{-1} - 1)\lambda_s \mathrm{d}s. \qquad (4.7.3)$$

Substituting (4.7.2) and (4.7.3) in (4.7.1) yields the desired result.

To finish the proof we have to show that $\overline{Q}_t^2 - t$ is also a $(\overline{P}, \mathcal{F}_t)$-martingale.

Now

$$\overline{Q}_t^2 = 2\int_0^t \overline{Q}_{s-}\mathrm{d}\overline{Q}_s + [\overline{Q}, \overline{Q}]_t = 2\int_0^t \overline{Q}_{s-}\mathrm{d}\overline{Q}_s + \mathcal{J}_t. \qquad (4.7.4)$$

Subtracting t from both sides of (4.7.4) makes the last term on the right of (4.7.4) a $(\overline{P}, \mathcal{F}_t)$-martingale and since the $\mathrm{d}M$ integral is a $(\overline{P}, \mathcal{F}_t)$-martingale the result follows. ∎

Remark 4.7.2 The above counting processes generate the same information as the Markov chain $\{X_t\}$. ☐

If we wish to change the intensity matrix A to a another one, $\overline{A}$, under a new probability measure $\overline{P}$ we change the intensities in the counting processes

$$\mathcal{J}_t^{ij} = \int_0^t \langle X_s, e_i \rangle a_{ji}(s)\mathrm{d}s + V_t^{ij} \triangleq \int_0^t \lambda_s^{ij}\mathrm{d}s + V_t^{ij}.$$

In order to define a new probability $\overline{P}$ such that

$$\mathcal{J}_t^{ij} = \int_0^t \overline{\lambda}_s^{ij}\mathrm{d}s + \overline{V}_t^{ij},$$

define the P-martingale

$$\Lambda_t = \prod_{i \neq j} \exp\left\{ \int_0^t \log \frac{\overline{\lambda}_s^{ij}}{\lambda_s^{ij}}\mathrm{d}\mathcal{J}_r^{ij} - \int_0^t (\overline{\lambda}_s^{ij} - \lambda_s^{ij})\mathrm{d}r \right\}, \qquad (4.7.5)$$

and set

$$E\left[\frac{\mathrm{d}\overline{P}}{\mathrm{d}P}\bigg|\mathcal{F}_t \right] = \Lambda_t.$$

Lemma 4.7.3 *Under probability measure $\overline{P}$ the processes $\mathcal{J}_t^{ij}$ have intensities $\overline{\lambda}_t^{ij}$ respectively.*

4.8 Problems

1. Consider the probability space $([0, 1], \mathcal{B}([0, 1]), \lambda)$, where λ is the Lebesgue measure on the Borel σ-field $\mathcal{B}([0, 1])$. Let P be another probability measure carried by the singleton $\{0\}$, i.e. $P(\{0\}) = 1$. Let

$$\pi_1 = \{[0, \tfrac{1}{2}], (\tfrac{1}{2}, 1]\},$$
$$\pi_2 = \{[0, \tfrac{1}{4}], (\tfrac{1}{4}, \tfrac{3}{4}], (\tfrac{3}{4}, 1]\}, \dots,$$
$$\pi_n = \{[0, 1/2^n], \dots, (1 - 1/2^n, 1]\}.$$

Define the random variable

$$\Lambda_n([0, \frac{1}{2^n}]) = \frac{P([0, 2^{-n}])}{\lambda([0, 2^{-n}])} = \begin{cases} \dfrac{1}{2^n}, \\ 0 \text{ elswhere in } [0, 1]. \end{cases}$$

Show that the sequence Λ_n is a positive martingale (with respect to the filtration generated by the partitions π_n) such that $E_\lambda[\Lambda_n] = 1$ for all n but $\lim \Lambda_n = 0$ λ-almost surely.

2. Prove Lemma 4.2.14.

3. Consider the order-2 Markov chain $\{X_n\}$, $1 \le n \le N$ discussed in Example 2.4.6. Define a new probability measure $\overline{P}$ on $(\Omega, \bigvee \mathcal{F}_n\}$ such that $\overline{P}(X_n = e_k \mid X_{n-2} = e_i, X_{n-1} = e_j) = \overline{p}_{ij,k}$.

4. On a probability space $(\Omega, \mathcal{F}, P)$ consider the stochastic process X_n with a finite state space and transition probabilities $P(X_{n+1} = k \mid X_{n-1} = i, X_n = j) = p_{ij,k}$.
   Transform the process X into a Markov chain Y with an appropriate state space and do a change of measure under which the process Y becomes a sequence of i.i.d. uniform random variables on the state space of the Markov chain. (Hint: show that the process $(X_{n-1}, X_n), (X_n, X_{n+1}), \dots$ is a Markov chain.)

5. Let N_t be a Poisson process with parameter λ on a filtered probability space $(\Omega, \mathcal{F}, \mathcal{F}_t, P)$ and suppose that we have the following process:

$$X_t = \mu t + \sigma N_t.$$

Here μ and σ are constants.
Define a new probability $\overline{P}$ such that X_t is a Poisson process with parameter $\lambda = 1$.

6. Show that the exponential martingale $\{\Lambda_t\}$ given by (4.7.5) is the unique solution of

$$\Lambda_t = 1 + \sum_{i,j} \int_0^t \Lambda_{s-}(\lambda_s^{ij})^{-1}(\overline{\lambda}_s^{ij} - \lambda_s^{ij})(d\mathcal{J}_s^{ij} - \lambda_s^{ij}\,ds).$$

7. Prove Lemma 4.7.3.

Part II
Applications

5

Kalman filtering

5.1 Introduction

This chapter discusses the filtering of partially observed linear (and nonlinear) dynamics using the tools developed in Chapter 4. The chapter starts with simple applications and evolves into less easy situations.

5.2 Discrete-time scalar dynamics

Consider a system whose state at time k is $x_k \in \mathbb{R}$. The time index k of the state evolution will be discrete and identified with $\mathbb{N} = \{0, 1, 2, \dots\}$.

Let $(\Omega, \mathcal{F}, P)$ be a probability space upon which $\{v_k\}, \{w_k\}, k \in \mathbb{N}$ are independent and identically distributed (i.i.d.) sequences of Gaussian random variables, having zero means and variances 1 $(N(0, 1))$, respectively; x_0 is normally distributed. Let $\{\mathcal{F}_k\}, k \in \mathbb{N}$ be the complete filtration (that is, $\mathcal{F}_0$ contains all the P-null events) generated by $\{x_0, x_1, \dots, x_k\}$.

The state of the system satisfies the linear dynamics

$$x_{k+1} = ax_k + bv_{k+1}. \tag{5.2.1}$$

Note that $E[v_{k+1} \mid \mathcal{F}_k] = 0$.

A useful and simple model for a noisy observation of x_k is to suppose it is given as a linear function of x_k plus a random "noise" term. That is, we suppose that for some real numbers c and d our observations have the form

$$y_k = cx_k + dw_k. \tag{5.2.2}$$

We shall also write $\{\mathcal{Y}_k\}, k \in \mathbb{N}$ for the complete filtration generated by $\{y_0, y_1, \dots, y_k\}$.

Using measure change techniques we shall derive a recursive expression for the conditional distribution of x_k given $\mathcal{Y}_k$.

5.3 Recursive estimation

Initially we suppose all processes are defined on an "ideal" probability space $(\Omega, \mathcal{F}, \overline{P})$; then under a new probability measure P, to be defined, the model dynamics (5.2.1) and (5.2.2) will hold.

Suppose that under $\overline{P}$:

1. $\{x_k\}$, $k \in \mathbb{N}$ is an i.i.d. $N(0, 1)$ sequence with density function $\phi(x) = \dfrac{1}{\sqrt{2\pi}} e^{-x^2/2}$,

2. $\{y_k\}$, $k \in \mathbb{N}$ is an i.i.d. $N(0, 1)$ sequence with density function $\psi(y) = \dfrac{1}{\sqrt{2\pi}} e^{-y^2/2}$.

For $l = 0$, $\overline{\lambda}_0 = \dfrac{\psi(d^{-1}(y_0 - cx_0))}{d\psi(y_0)}$ and for $l = 1, 2, \ldots$ define

$$\overline{\lambda}_l = \frac{\phi(b^{-1}(x_l - ax_{l-1}))\psi(d^{-1}(y_l - cx_l))}{b\,d\phi(x_l)\psi(y_l)}, \tag{5.3.1}$$

$$\overline{\Lambda}_k = \prod_{l=0}^{k} \overline{\lambda}_l. \tag{5.3.2}$$

Let $\mathcal{G}_k$ be the complete σ-field generated by $\{x_0, x_1, \ldots, x_k, y_0, y_1, \ldots, y_k\}$ for $k \in \mathbb{N}$.

Lemma 5.3.1 *The process* $\{\overline{\Lambda}_k\}$, $k \in \mathbb{N}$ *is an* $\overline{P}$*-martingale with respect to the filtration* $\{\mathcal{G}_k\}$, $k \in \mathbb{N}$.

Proof   Since $\overline{\Lambda}_k$ is $\mathcal{G}_k$-measurable,

$$\overline{E}[\overline{\Lambda}_{k+1} \mid \mathcal{G}_k] = \overline{\Lambda}_k \overline{E}[\overline{\lambda}_{k+1} \mid \mathcal{G}_k].$$

Now

$$\overline{E}\left[\overline{\lambda}_{k+1} \mid \mathcal{G}_k\right] = \overline{E}\left[\frac{\phi(b^{-1}(x_{k+1} - ax_k))\psi(d^{-1}(y_{k+1} - cx_{k+1}))}{bd\phi(x_{k+1})\psi(y_{k+1})} \mid \mathcal{G}_k\right]$$

$$= \overline{E}\left[\frac{\phi(b^{-1}(x_{k+1} - ax_k))}{b\phi(x_{k+1})}\overline{E}\left[\frac{\psi(d^{-1}(y_{k+1} - cx_{k+1}))}{d\psi(y_{k+1})} \mid \mathcal{G}_k, x_{k+1}\right] \mid \mathcal{G}_k\right].$$

Now,

$$\overline{E}\left[\frac{\psi(d^{-1}(y_{k+1} - cx_{k+1}))}{d\psi(y_{k+1})} \mid \mathcal{G}_k, x_{k+1}\right] = \int_{\mathbb{R}} \frac{\psi(d^{-1}(y - cx_{k+1}))}{d\psi(y)}\psi(y)dy = 1,$$

and

$$\overline{E}\left[\frac{\phi(b^{-1}(x_{k+1} - ax_k))}{b\phi(x_{k+1})} \mid \mathcal{G}_k\right] = \int_{\mathbb{R}} \frac{\phi(b^{-1}(x - ax_k))}{b\phi(x)}\phi(x)dx = 1,$$

using the change of variable $d^{-1}(y - cx_{k+1}) = u$ in the first integral and a similar change of variable in the second integral.   ∎

Define P on $\{\Omega, \mathcal{F}\}$ by setting the restriction of the Radon–Nikodym derivative $\dfrac{dP}{d\overline{P}}$ to $\mathcal{G}_k$ equal to $\overline{\Lambda}_k$. Then:

Lemma 5.3.2 *On* $\{\Omega, \mathcal{F}\}$ *and under* P, $\{v_k\}$, $k \in \mathbb{N}$, $\{w_k\}$, $k \in \mathbb{N}$ *are i.i.d.* $N(0, 1)$ *sequences of random variables, where*

$$v_{k+1} \stackrel{\triangle}{=} b^{-1}(x_{k+1} - ax_k),$$

$$w_k \stackrel{\triangle}{=} d^{-1}(y_k - cx_k).$$

Proof Suppose f, $g : \mathbb{R} \to \mathbb{R}$ are "test" functions (i.e. measurable functions with compact support). Then with E (resp. $\overline{E}$) denoting expectation under P (resp. $\overline{P}$) and using Bayes' Theorem (4.1.1)

$$E[f(v_{k+1})g(w_{k+1}) \mid \mathcal{G}_k] = \frac{\overline{E}[\overline{\Lambda}_{k+1} f(v_{k+1})g(w_{k+1}) \mid \mathcal{G}_k]}{\overline{E}[\overline{\Lambda}_{k+1} \mid \mathcal{G}_k]}$$

$$= \overline{E}[\overline{\lambda}_{k+1} f(v_{k+1})g(w_{k+1}) \mid \mathcal{G}_k],$$

where the last equality follows from Lemma 5.3.1. Consequently

$$E[f(v_{k+1})g(w_{k+1}) \mid \mathcal{G}_k] = \overline{E}[\overline{\lambda}_{k+1} f(v_{k+1})g(w_{k+1}) \mid \mathcal{G}_k]$$

$$= \overline{E}\left[\frac{\phi(b^{-1}(x_{k+1} - ax_k))\psi(d^{-1}(y_{k+1} - cx_{k+1}))}{bd\phi(x_{k+1})\psi(y_{k+1})} \right.$$

$$\left. \times f(b^{-1}(x_{k+1} - ax_k))g(d^{-1}(y_{k+1} - cx_{k+1})) \mid \mathcal{G}_k \right]$$

$$= \overline{E}\left[\frac{\phi(b^{-1}(x_{k+1} - ax_k))}{b\phi(x_{k+1})} f(b^{-1}(x_{k+1} - ax_k)) \right.$$

$$\left. \times \overline{E}\left[\frac{\psi(d^{-1}(y_{k+1} - cx_{k+1}))}{d\psi(y_{k+1})} g(d^{-1}(y_{k+1} - cx_{k+1})) \mid \mathcal{G}_k, x_{k+1} \right] \mid \mathcal{G}_k \right].$$

Now

$$\overline{E}\left[\frac{\psi(d^{-1}(y_{k+1} - cx_{k+1}))}{d\psi(y_{k+1})} g(d^{-1}(y_{k+1} - cx_{k+1})) \mid \mathcal{G}_k, x_{k+1} \right]$$

$$= \int_{\mathbb{R}} \frac{\psi(d^{-1}(y - cx_{k+1}))}{d\psi(y)} \psi(y)g(d^{-1}(y - cx_{k+1}))dy$$

$$= \int_{\mathbb{R}} \psi(u)g(u)du.$$

Similarly

$$\overline{E}\left[\frac{\phi(b^{-1}(x_{k+1} - ax_k))}{b\phi(x_{k+1})} f(b^{-1}(x_{k+1} - ax_k)) \mid \mathcal{G}_k \right] = \int_{\mathbb{R}} \phi(u)f(u)du.$$

Therefore

$$E[f(v_{k+1})g(w_{k+1}) \mid \mathcal{G}_k] = \int_{\mathbb{R}} \phi(u)f(u)du \int_{\mathbb{R}} \psi(u)g(u)du,$$

and the lemma is proved. ■

Let $g : \mathbb{R} \to \mathbb{R}$ be a "test" function. Using Bayes' Theorem 4.1.1,

$$E[g(x_k) \mid \mathcal{Y}_k] = \frac{\overline{E}[\overline{\Lambda}_k g(x_k) \mid \mathcal{Y}_k]}{\overline{E}[\overline{\Lambda}_k \mid \mathcal{Y}_k]}, \tag{5.3.3}$$

where $\overline{E}$ (resp. E) denotes expectations with respect to $\overline{P}$ (resp. P). Consider the unnormalized, conditional expectation which is the numerator of (5.3.3),

$$\overline{E}[\overline{\Lambda}_k g(x_k) \mid \mathcal{Y}_k].$$

This is a measure-valued process. Write $\alpha_k(.), k \in \mathbb{N}$ for its density so that

$$\overline{E}[\overline{\Lambda}_k g(x_k) \mid \mathcal{Y}_k] = \int_{\mathbb{R}} g(x)\alpha_k(x)\mathrm{d}x. \qquad (5.3.4)$$

If $p_k(.)$ denotes the normalized conditional density, such that

$$E[g(x_k) \mid \mathcal{Y}_k] = \int_{\mathbb{R}} g(x)p_k(x)\mathrm{d}x,$$

then from (5.3.3) we see that

$$p_k(x) = \alpha_k(x)\left[\int_{\mathbb{R}} \alpha_k(z)\mathrm{d}z\right]^{-1}, \qquad (5.3.5)$$

for $x \in \mathbb{R}, k \in \mathbb{N}$.

Then we have the following result.

Theorem 5.3.3

$$\alpha_{k+1}(x) = \frac{\psi(d^{-1}(y_{k+1} - cx))}{db\psi(y_{k+1})} \int \phi(b^{-1}(x - az))\alpha_k(z)\mathrm{d}z. \qquad (5.3.6)$$

Proof   For any "test" function g and in view of (5.3.1) and (5.3.2),

$$\int_{\mathbb{R}} g(x)\alpha_{k+1}(x)\mathrm{d}x = \overline{E}[\overline{\Lambda}_{k+1} g(x_{k+1}) \mid \mathcal{Y}_{k+1}]$$

$$= \overline{E}[\overline{\Lambda}_k \overline{\lambda}_{k+1} g(x_{k+1}) \mid \mathcal{Y}_{k+1}]$$

$$= \overline{E}\left[\overline{\Lambda}_k \frac{\phi(b^{-1}(x_{k+1} - ax_k))\psi(d^{-1}(y_{k+1} - cx_{k+1}))}{bd\phi(x_{k+1})\psi(y_{k+1})} g(x_{k+1}) \mid \mathcal{Y}_{k+1}\right]$$

$$= \frac{1}{db\psi(y_{k+1})} \overline{E}\left[\overline{\Lambda}_k \int \frac{\phi(b^{-1}(x - ax_k))\psi(d^{-1}(y_{k+1} - cx))}{\phi(x)} \right.$$

$$\left. \times \phi(x)g(x)\mathrm{d}x \mid \mathcal{Y}_{k+1}\right].$$

The last equality follows from the fact that x_{k+1} has distribution ϕ and is independent of everything else under $\overline{P}$. Also, note that given y_{k+1} we condition only on $\mathcal{Y}_k$ to get an expression similar to notation (5.3.4), that is

$$\int_{\mathbb{R}} g(x)\alpha_{k+1}(x)\mathrm{d}x = \frac{1}{db\psi(y_{k+1})}$$

$$\times \int_{\mathbb{R}} \int_{\mathbb{R}} \psi(d^{-1}(y_{k+1} - cx))\phi(b^{-1}(x - az))g(x)\alpha_k(z)\mathrm{d}x\mathrm{d}z.$$

This holds for all "test" functions g so we can conclude that (5.3.6) holds.   ∎

Remark 5.3.4 The linearity of (5.2.1) and (5.2.2) implies that (5.3.5) is also normally distributed with mean $\hat{x}_{k|k} = E[x_k \mid \mathcal{Y}_k]$ and variance $\Sigma_{k|k} = E[(x_k - \hat{x}_{k|k})^2 \mid \mathcal{Y}_k]$. Our purpose now is to give recursive estimates of $\hat{x}_{k|k}$ and $\Sigma_{k|k}$ using the recursion for $\alpha_k(x)$. $\qquad\square$

Theorem 5.3.5 *For the linear model described by (5.2.1) and (5.2.2) the conditional mean and variance of the state process x_k are given by the following recursions:*

$$\Sigma_{k+1|k+1} = A_k,$$
$$\hat{x}_{k+1|k+1} = A_k B_k,$$

where

$$A_k = \left[\frac{c^2}{d^2} + \frac{1}{b^2} - \frac{a^2 \overline{\Sigma}_{k|k}}{b^4} \right]^{-1},$$

$$B_k = \left[\frac{c y_{k+1}}{d^2} + \frac{a \hat{x}_{k|k} \overline{\Sigma}_{k|k}}{b^2 \Sigma_{k|k}} \right],$$

and $\overline{\Sigma}_{k|k}^{-1} = \left(\dfrac{a^2}{b^2} + \dfrac{1}{\Sigma_{k|k}} \right)$.

Proof Recall that $\phi(.)$ and $\psi(.)$ are normal densities with zero means and variances 1, and using the fact that $\alpha_n(x)$ is proportional to a normal density with mean $\hat{x}_{k|k}$ and variance $\Sigma_{k|k}$ we write

$$\alpha_{k+1}(x) = \frac{\psi(d^{-1}(y_{k+1} - cx))}{db\psi(y_{k+1})} \int \phi(b^{-1}(x - az))\alpha_k(z)dz$$

$$= \frac{\psi(d^{-1}(y_{k+1} - cx))}{db\psi(y_{k+1})} \int \exp\left(-\frac{1}{2b^2}(x - az)^2 - \frac{1}{2\Sigma_{k|k}}(z - \hat{x}_{k|k})^2 \right) dz$$

$$= \frac{\psi(d^{-1}(y_{k+1} - cx))}{db\psi(y_{k+1})} \exp\left(\frac{-x^2}{2b^2} - \frac{\hat{x}_{k|k}^2}{2\Sigma_{k|k}} \right)$$

$$\times \int \exp\left\{ \frac{-1}{2} \left[z^2 \left(\frac{a^2}{b^2} + \frac{1}{\Sigma_{k|k}} \right) - 2z \left(\frac{ax}{b^2} + \frac{\hat{x}_{k|k}}{\Sigma_{k|k}} \right) \right] \right\} dz.$$

Let

$$K(x) \triangleq \frac{\psi(d^{-1}(y_{k+1} - cx))}{db\psi(y_{k+1})} \exp\left(\frac{-x^2}{2b^2} - \frac{\hat{x}_{k|k}^2}{2\Sigma_{k|k}} \right),$$

$$\overline{\Sigma}_{k|k}^{-1} \triangleq \frac{a^2}{b^2} + \frac{1}{\Sigma_{k|k}},$$

and

$$\beta_k(x) \triangleq \frac{ax}{b^2} + \frac{\hat{x}_{k|k}}{\Sigma_{k|k}}.$$

$$\alpha_{k+1}(x) = K(x) \int \exp\left\{\frac{-1}{2}[z^2 \overline{\Sigma}_{k|k}^{-1} - 2z\beta_k(x)]\right\} dz$$

$$= K(x) \int \exp\left\{\frac{-1}{2\overline{\Sigma}_{k|k}}[z^2 - 2z\overline{\Sigma}_{k|k}\beta_k(x) + (\overline{\Sigma}_{k|k}\beta_k(x))^2 - (\overline{\Sigma}_{k|k}\beta_k(x))^2]\right\} dz$$

$$= K(x)\exp\left\{\frac{(\overline{\Sigma}_{k|k}\beta_k(x))^2}{2}\right\} \int \exp\left\{\frac{-1}{2\overline{\Sigma}_{k|k}}(z - \overline{\Sigma}_{k|k}\beta_k(x))^2\right\} dz$$

$$= K(x)\exp\left\{\frac{(\overline{\Sigma}_{k|k}\beta_k(x))^2}{2}\right\} \sqrt{2\pi\overline{\Sigma}_{k|k}}.$$

The last expression follows from the fact the integrand is proportional to a normal density with mean $\overline{\Sigma}_{k|k}\beta_k(x)$ and variance $\overline{\Sigma}_{k|k}$ and hence the integral is 1 when properly normalized. The final step is to group together all the terms containing x and then completing the square with respect to the variable x:

$$\alpha_{k+1}(x) = K_1 \exp\left\{\frac{-1}{2}\left[x^2\left(\frac{c^2}{d^2} + \frac{1}{b^2} - \frac{a^2\overline{\Sigma}_{k|k}}{b^4}\right) - 2x\left(\frac{cy_{k+1}}{d^2} + \frac{a\hat{x}_{k|k}\overline{\Sigma}_{k|k}}{b^2\Sigma_{k|k}}\right)\right]\right\}$$

$$= K_2 \exp\left\{\left(\frac{-1}{2A_k}\right)(x - A_k B_k)^2\right\},$$

where K_1 and K_2 are constants independent of x and

$$A_k = \left(\frac{c^2}{d^2} + \frac{1}{b^2} - \frac{a^2\overline{\Sigma}_{k|k}}{b^4}\right)^{-1},$$

$$B_k = \left(\frac{cy_{k+1}}{d^2} + \frac{a\hat{x}_{k|k}\overline{\Sigma}_{k|k}}{b^2\Sigma_{k|k}}\right),$$

that is

$$\Sigma_{k+1|k+1} = A_k,$$

$$\hat{x}_{k+1|k+1} = A_k B_k,$$

which finishes the proof. ∎

The Kalman filter is usually presented in terms of the one-step ahead prediction

$$\hat{x}_{k+1|k} = E[x_{k+1} \mid \mathcal{Y}_k] = a\hat{x}_{k|k}.$$

Similarly,

$$\Sigma_{k+1|k} = E[(x_{k+1} - \hat{x}_{k+1|k})^2 \mid \mathcal{Y}_k] = a^2\Sigma_{k|k} + b^2.$$

Then, with $K_{k+1} \triangleq \dfrac{c\Sigma_{k+1|k}}{c^2\Sigma_{k+1|k} + d^2}$,

$$\hat{x}_{k+1|k+1} = a\hat{x}_{k+1|k} + K_{k+1}(y_k - c\hat{x}_{k+1|k}),$$

$$\Sigma_{k+1|k+1} = \Sigma_{k+1|k} - \frac{\Sigma_{k+1|k}^2 c^2}{c^2\Sigma_{k+1|k} + d^2}.$$

5.4 Vector dynamics

Consider a system whose state at time $k = 0, 1, 2, \dots$ is $X_k \in \mathbb{R}^m$ and which can be observed only indirectly through another process $Y_k \in \mathbb{R}^d$.

Let $(\Omega, \mathcal{F}, P)$ be a probability space upon which V_k and W_k are normally distributed with means 0 and respective covariance identity matrices $I_{m \times m}$ and $I_{d \times d}$. Assume that D_k is nonsingular, B_k is nonsingular and symmetric (for notational convenience). X_0 is a Gaussian random variable with zero mean and covariance matrix B_0^2 (of dimension $m \times m$). Let $\{\mathcal{F}_k\}$, $k \in \mathbb{N}$ be the complete filtration (that is $\mathcal{F}_0$ contains all the P-null events) generated by $\{X_0, X_1, \dots, X_k\}$.

The state and observations of the system satisfies the linear dynamics

$$X_{k+1} = A_{k+1} X_k + B_{k+1} V_{k+1} \in \mathbb{R}^m, \tag{5.4.1}$$
$$Y_k = C_k X_k + D_k W_k \in \mathbb{R}^d. \tag{5.4.2}$$

A_k, C_k are matrices of appropriate dimensions.

We shall also write $\{\mathcal{Y}_k\}$, $k \in \mathbb{N}$ for the complete filtration generated by $\{Y_0, Y_1, \dots Y_k\}$.

Using measure change techniques we shall derive a recursive expression for the conditional distribution of X_k given $\mathcal{Y}_k$.

Recursive estimation

Initially we suppose all processes are defined on an "ideal" probability space $(\Omega, \mathcal{F}, \overline{P})$; then under a new probability measure P, to be defined, the model dynamics (5.4.1) and (5.4.2) will hold.

Suppose that under $\overline{P}$:

1. $\{X_k\}$, $k \in \mathbb{N}$ is an i.i.d. $N(0, I_{m \times m})$ sequence with density function $\phi(x) = \frac{1}{(2\pi)^{m/2}} e^{-x'x/2}$,
2. $\{Y_k\}$, $k \in \mathbb{N}$ is an i.i.d. $N(0, I_{d \times d})$ sequence with density function $\psi(y) = \frac{1}{(2\pi)^{d/2}} e^{-y'y/2}$.

For any square matrix B write $|B|$ for the absolute value of its determinant.

For $l = 0$, $\overline{\lambda}_0 = \dfrac{\psi(D_0^{-1}(Y_0 - C_0 X_0))}{|D_0| \psi(Y_0)}$ and for $l = 1, 2, \dots$ define

$$\overline{\lambda}_l = \frac{\phi(B_l^{-1}(X_l - A_l X_{l-1})) \psi(D_l^{-1}(Y_l - C_l X_l))}{|B_l||D_l| \phi(X_l) \psi(Y_l)}, \tag{5.4.3}$$

$$\overline{\Lambda}_k = \prod_{l=0}^{k} \overline{\lambda}_l. \tag{5.4.4}$$

Let $\{\mathcal{G}_k\}$ be the complete σ-field generated by $\{X_0, X_1, \dots, X_k, Y_0, Y_1, \dots, Y_k\}$ for $k \in \mathbb{N}$.

The process $\{\overline{\Lambda}_k\}$, $k \in \mathbb{N}$ is a $\overline{P}$-martingale with respect to the filtration $\{\mathcal{G}_k\}$.

Define P on $\{\Omega, \mathcal{F}\}$ by setting the restriction of the Radon–Nikodym derivative $\dfrac{dP}{d\overline{P}}$ to $\mathcal{G}_k$ equal to $\overline{\Lambda}_k$. It can be shown that on $\{\Omega, \mathcal{F}\}$ and under P, V_k and W_k are normally distributed with means 0 and respective covariance identity matrices $I_{m \times m}$ and $I_{d \times d}$, where

$$V_{k+1} \overset{\triangle}{=} B_{k+1}^{-1}(X_{k+1} - A_{k+1}X_k),$$

$$W_k \overset{\triangle}{=} D_k^{-1}(Y_k - C_k X_k).$$

Let $g : \mathbb{R}^m \to \mathbb{R}$ be a "test" function. Write $\alpha_k(.)$, $k \in \mathbb{N}$ for the density

$$\overline{E}[\overline{\Lambda}_k g(x_k) \mid \mathcal{Y}_k] = \int_{\mathbb{R}^m} g(x)\alpha_k(x)dx. \tag{5.4.5}$$

Then we have the following result:

Theorem 5.4.1

$$\alpha_{k+1}(x) = \frac{\psi(D_{k+1}^{-1}(Y_{k+1} - C_{k+1}x))}{|D_{k+1}||B_{k+1}|\psi(Y_{k+1})} \int \phi(B_{k+1}^{-1}(x - A_{k+1}z))\alpha_k(z)dz. \tag{5.4.6}$$

Proof The proof is similar to the scalar case and is left as an exercise. ∎

Remark 5.4.2 The linearity of (5.4.1) and (5.4.2) implies that (5.4.6) is proportional to a normal distribution with mean $\hat{X}_{k|k} = E[X_k \mid \mathcal{Y}_k]$ and error covariance matrix $\Sigma_{k|k} = E[(X_k - \hat{X}_{k|k})(X_k - \hat{X}_{k|k})' \mid \mathcal{Y}_k]$. Our purpose now is to give recursive estimates of $\hat{X}_{k|k}$ and $\Sigma_{k|k}$ using the recursion for $\alpha_k(x)$. □

Write

$$\Psi(x, Y_{k+1}) = \frac{\psi(D_{k+1}^{-1}(Y_{k+1} - C_{k+1}x))}{|D_{k+1}||B_{k+1}|\psi(Y_{k+1})}(2\pi)^{-m/2}|\Sigma_{k|k}|^{-1/2}.$$

Then

$$\alpha_{k+1}(x) = \Psi(x, Y_{k+1}) \int_{\mathbb{R}^m} \exp\left\{(-1/2)(B_{k+1}^{-1}(x - A_{k+1}z))'(B_{k+1}^{-1}(x - A_{k+1}z))\right.$$

$$\left. + (z - \hat{X}_{k|k})'\Sigma_{k|k}^{-1}(z - \hat{X}_{k|k})\right\} dz \tag{5.4.7}$$

$$= K(x) \int_{\mathbb{R}^m} \left\{\exp(-1/2)z'\overline{\Sigma}_{k|k}^{-1}z - 2\beta_{k+1}'z\right\} dz, \tag{5.4.8}$$

where $K(x)$ is independent of the variable z and

$$\overline{\Sigma}_{k|k}^{-1} = A_{k+1}'B_{k+1}^{-2}A_{k+1} + \Sigma_{k|k}^{-1},$$

$$\beta_{k+1}' = x'B_{k+1}^{-2}A_{k+1} + \hat{X}_{k|k}'\Sigma_{k|k}^{-1}.$$

The next step is to complete the "square" in the argument of the exponential in (5.4.8) in order to rewrite the integrand as a normal density which integrates out to 1.

Now

$$z'\overline{\Sigma}_{k|k}^{-1}z - 2\beta_{k+1}'z = (z - \overline{\Sigma}_{k|k}\beta_{k+1})'\overline{\Sigma}_{k|k}^{-1}(z - \overline{\Sigma}_{k|k}\beta_{k+1})$$

$$- \beta_{k+1}'\overline{\Sigma}_{k|k}\beta_{k+1}, \tag{5.4.9}$$

after substitution of (5.4.9) into (5.4.8) and integration we are left with only the x variable. Completing the "square" with respect to x,

$$\alpha_{k+1}(x) = K_1 \exp\left\{(-\frac{1}{2})(x - \hat{X}_{k+1|k+1})'\Sigma_{k+1|k+1}^{-1}(x - \hat{X}_{k+1|k+1})\right\},$$

where K_1 is a constant independent of x and

$$\Sigma_{k+1|k+1}^{-1} = B_{k+1}^{-2} - B_{k+1}^{-2}A_{k+1}\overline{\Sigma}_{k|k}A_{k+1}'B_{k+1}^{-2}$$
$$+ C_{k+1}'D_{k+1}^{-2}C_{k+1},$$

$$\Sigma_{k+1|k+1}^{-1}\hat{X}_{k+1|k+1} = B_{k+1}^{-2}A_{k+1}\overline{\Sigma}_{k|k}\Sigma_{k|k}^{-1}\hat{X}_{k|k} + C_{k+1}'D_{k+1}^{-2}Y_{k+1}.$$

The one-step ahead prediction version is

$$\hat{X}_{k+1|k} = E[X_{k+1} \mid \mathcal{Y}_k] = A_{k+1}\hat{X}_{k|k},$$

$$\Sigma_{k+1|k} = E[(X_{k+1} - \hat{X}_{k+1|k})(X_{k+1} - \hat{X}_{k+1|k})' \mid \mathcal{Y}_k]$$

$$= A_{k+1}\Sigma_{k|k}A_{k+1}' + B_{k+1}^{-2}.$$

Then, with $H_{k+1} \overset{\triangle}{=} \Sigma_{k+1|k}C_{k+1}'(C_{k+1}\Sigma_{k+1|k}C_{k+1}' + D_{k+1}^{-2})$ use of the Matrix Inversion Lemma 5.4.3 gives the Kalman filter equations in the form:

$$\hat{X}_{k+1|k+1} = A_{k+1}\hat{X}_{k+1|k} + H_{k+1}(Y_{k+1} - C_{k+1}\hat{X}_{k+1|k}),$$

$$\Sigma_{k+1|k+1} = \Sigma_{k+1|k} - \Sigma_{k+1|k}C_{k+1}'(C_{k+1}\Sigma_{k+1|k}C_{k+1}' + D_{k+1}^{-2}C_{k+1}\Sigma_{k+1|k}).$$

Lemma 5.4.3 *(Matrix Inversion Lemma) Assuming the required inverses exist, the Matrix Inversion Lemma states that:*

$$(A_{11} - A_{12}A_{22}^{-1}A_{21})^{-1} = A_{11}^{-1} + A_{11}^{-1}A_{12}(A_{22} - A_{21}A_{11}^{-1}A_{12})^{-1}A_{21}A_{11}^{-1}.$$

5.5 The EM algorithm

The EM algorithm ([3], [9]) is a widely used iterative numerical method for computing maximum likelihood parameter estimates of partially observed models such as linear Gaussian state space models. For such models, direct computation of the MLE is difficult. The EM algorithm has the appealing property that successive iterations yield parameter estimates with nondecreasing values of the likelihood function.

Suppose that we have observations $y_1, \ldots, y_K$ available, where K is a fixed positive integer. Let $\{P_\theta, \theta \in \Theta\}$ be a family of probability measures on $(\Omega, \mathcal{F})$, all absolutely continuous with respect to a fixed probability measure P_0. The log-likelihood function for computing an estimate of the parameter θ based on the information available in $\mathcal{Y}_K$ is

$$\mathcal{L}_K(\theta) = E_0\left[\log \frac{dP_\theta}{dP_0} \mid \mathcal{Y}_K\right],$$

and the maximum likelihood estimate (MLE) is defined by

$$\hat{\theta} \in \underset{\theta \in \Theta}{\operatorname{argmax}} \mathcal{L}_K(\theta).$$

Let $\hat{\theta}_0$ be the initial parameter estimate. The EM algorithm generates a sequence of parameter estimates $\{\theta_j\}$, $j \geq 1$, as follows.

Each iteration of the algorithm consists of two steps:

Step 1 (E-step). Set $\tilde{\theta} = \hat{\theta}_j$ and compute $Q(\theta, \tilde{\theta})$, where

$$Q(\theta, \tilde{\theta}) = E_{\tilde{\theta}}\left[\log \frac{dP_\theta}{dP_{\tilde{\theta}}} \mid \mathcal{Y}_K\right].$$

Step 2 (M-step). Find $\hat{\theta}_{j+1} \in \text{argmax}_{\theta \in \Theta}\, Q(\theta, \theta_j)$.

Using Jensen's inequality (2.3.3) it can be shown (see Theorem 1 in [9]) that the sequence of model estimates $\{\hat{\theta}_j, j \geq 1\}$ from the EM algorithm are such that the sequence of likelihoods $\{\mathcal{L}_K(\hat{\theta}_j)\}$, $j \geq 1$, is monotonically increasing with equality if and only if $\hat{\theta}_{j+1} = \hat{\theta}_j$.

Sufficient conditions for convergence of the EM algorithm are given in [37]. We briefly summarize them here:

Assume that

(i) The parameter space Θ is a subset of some finite dimensional Euclidean space $\mathbb{R}^r$.
(ii) $\{\theta \in \Theta : \mathcal{L}_K(\theta) \geq \mathcal{L}_K(\hat{\theta}_0)\}$ is compact for any $\mathcal{L}_K(\hat{\theta}_0) > -\infty$.
(iii) $\mathcal{L}_K$ is continuous in Θ and differentiable in the interior of Θ.
   (As a consequence of (i), (ii) and (iii), clearly $\mathcal{L}_K(\hat{\theta}_j)$ is bounded from above.)
(iv) The function $Q(\theta, \hat{\theta}_j)$ is continuous in both θ and $\hat{\theta}_j$.

Then by Theorem 2 in [37], the limit of the sequence EM estimates $\{\hat{\theta}_j\}$ has a stationary point $\bar{\theta}$ of $\mathcal{L}_K$. Also $\{\mathcal{L}_K(\hat{\theta}_j)\}$ converges monotonically to $\bar{\mathcal{L}}_t = \mathcal{L}_t(\bar{\theta})$ for some stationary point $\bar{\theta}$. To make sure that $\bar{\mathcal{L}}_t$ is a maximum value of the likelihood, it is necessary to try different initial values $\hat{\theta}_0$.

5.6 Discrete-time model parameter estimation

In all the existing literature on parameter estimation of linear Gaussian models via the EM algorithm, filtered estimates of the above quantities are computed via Kalman smoothing, which requires large memory numerical implementation. This problem is solved in [13] by providing *finite-dimensional filters for (the components of) such integral processes*. The authors further show that finite-dimensional filters exist for moments of all orders of the state process.

Assume that the state and observation processes are given by the vector dynamics

$$X_{k+1} = A_{k+1}X_k + B_{k+1}V_{k+1} \in \mathbb{R}^m, \tag{5.6.1}$$

$$Y_k = C_k X_k + D_k W_k \in \mathbb{R}^d. \tag{5.6.2}$$

A_k, C_k are matrices of appropriate dimensions, V_k and W_k are normally distributed with means 0 and respective covariance identity matrices $I_{m \times m}$ and $I_{d \times d}$. Assume that D_k is nonsingular, B_k is nonsingular and symmetric (for notational convenience). X_0

is a Gaussian random variable with zero mean and covariance matrix B_0^2 (of dimension $m \times m$).

The linear model given by (5.6.1) and (5.6.2) is determined by the matrices A, B, C and D which need to be known. These parameters are estimated using the expectation maximization (EM) algorithm.

Maximum likelihood estimation of the parameters via the EM algorithm requires computation of the filtered estimates of quantities such as
$T_k^{(0)} = \sum_{l=0}^{k} X_l \otimes X_l$, $T_k^{(1)} = \sum_{l=1}^{k} X_l \otimes X_{l-1}$, $T_k^{(2)} = \sum_{l=1}^{k} X_{l-1} \otimes X_{l-1}$, and $U_k = \sum_{l=0}^{k} X_l \otimes Y_l$.

Consider the time-invariant version of (5.6.1), (5.6.2):

$$X_{k+1} = AX_k + BV_{k+1} \in \mathbb{R}^m, \tag{5.6.3}$$

$$Y_k = CX_k + DW_k \in \mathbb{R}^d. \tag{5.6.4}$$

The aim is to compute ML estimates of the parameters $\theta = (A, B, C, D)$ given the observations $\mathcal{Y}_k = \sigma\{Y_s : s \le k\}$. This is done via the EM algorithm.

Notation

Let e_i, $e_j \in \mathbb{R}^m$ denote unit vectors with 1 in the i-th and j-th positions, respectively. For $i, j \in \{1, \ldots, m\}$,

$$T_k^{ij(0)} = \sum_{l=0}^{k} \langle X_l, e_i \rangle \langle X_l, e_j \rangle, \tag{5.6.5}$$

$$T_k^{ij(1)} = \sum_{l=0}^{k} \langle X_l, e_i \rangle \langle X_{l-1}, e_j \rangle, \tag{5.6.6}$$

here $\langle ., . \rangle$ denotes the scalar product.

Also let $f_j \in \mathbb{R}^d$ denote the unit vector with 1 in the j-th position. For $n \in \{1, \ldots, d\}$ write

$$U_k^{in} = \sum_{l=0}^{k} \langle X_l, e_i \rangle \langle Y_l, f_n \rangle. \tag{5.6.7}$$

Note that $T_k^{ij(0)}$, $T_k^{ij(1)}$ and U_k^{in} are merely the elements of the matrices $T_k^{(0)}$, $T_k^{(1)}$, $T_k^{(2)}$, and U_k respectively.

Now the expression for $\mathcal{Q}(\theta, \tilde{\theta})$ is derived.

To update the set of parameters from $\tilde{\theta}$ to θ, the following density is introduced:

$$\frac{dP_\theta}{dP_{\tilde{\theta}}} \bigg|_{\mathcal{G}_k} = \prod_{l=0}^{k} \gamma_l, \quad \text{where} \quad \gamma_0 = \frac{|\tilde{D}| \phi(D^{-1}(Y_0 - CX_0))}{|D| \phi(\tilde{D}^{-1}(Y_0 - \tilde{C}X_0))},$$

$$\gamma_l = \frac{|\tilde{D}| \phi(D^{-1}(Y_l - CX_l))}{|D| \phi(\tilde{D}^{-1}(Y_l - \tilde{C}X_l))} \frac{|\tilde{B}| \psi(B^{-1}(X_l - AX_{l-1}))}{|B| \psi(\tilde{B}^{-1}(X_l - \tilde{A}X_{l-1}))}.$$

Now

$$E_{\tilde{\theta}}\left[\log\frac{dP_{\theta}}{dP_{\tilde{\theta}}}\Big|_{\mathcal{G}_k}\Big|\mathcal{Y}_k\right] = -k\log|B| - (k+1)\log|D|$$

$$+ \frac{1}{2}E_{\tilde{\theta}}\left[\sum_{l=1}^{k}(X_l - AX_{l-1})'B^{-2}(X_l - AX_{l-1})\mid\mathcal{Y}_k\right]$$

$$- \frac{1}{2}E_{\tilde{\theta}}\left[\sum_{l=1}^{k}(Y_l - CX_l)'(DD')^{-1}(Y_l - CX_l)\mid\mathcal{Y}_k\right]$$

$$+ R(\tilde{\theta}) = \mathcal{Q}(\theta,\tilde{\theta}), \tag{5.6.8}$$

where $R(\tilde{\theta})$ does not involve θ.

To implement the M-step set the derivatives $\dfrac{\partial\mathcal{Q}}{\partial\theta} = 0$. This yields

$$A = E_{\tilde{\theta}}\left[\sum_{l=1}^{k}X_l\otimes X_{l-1}\mid\mathcal{Y}_k\right]\left(E_{\tilde{\theta}}\left[\sum_{l=1}^{k}X_{l-1}\otimes X_{l-1}\mid\mathcal{Y}_k\right]\right)^{-1}, \tag{5.6.9}$$

$$B^2 = \frac{1}{k}E_{\tilde{\theta}}\left[\sum_{l=1}^{k}(X_l - AX_{l-1})\otimes(X_l - AX_{l-1})\mid\mathcal{Y}_k\right], \tag{5.6.10}$$

$$C = E_{\tilde{\theta}}\left[\sum_{l=0}^{k}Y_l\otimes X_l\mid\mathcal{Y}_k\right]\left(E_{\tilde{\theta}}\left[\sum_{l=0}^{k}X_l\otimes X_l\mid\mathcal{Y}_k\right]\right)^{-1}, \tag{5.6.11}$$

$$DD' = \frac{1}{k+1}E_{\tilde{\theta}}\left[\sum_{l=0}^{k}(Y_l - CX_l)\otimes(Y_l - CX_l)\mid\mathcal{Y}_k\right]. \tag{5.6.12}$$

Next, finite-dimensional recursive filters for $T_k^{ij(0)}$, $T_k^{ij(1)}$ and U_k^{in} are derived. That is, these filters can be described in terms of a finite number of statistics.

5.7 Finite-dimensional filters

Initially, assume that all processes are defined on an "ideal" probability space $(\Omega, \mathcal{F}, \overline{P})$. Suppose that under $\overline{P}$:

1. $\{X_k\}$, $k \in \mathbb{N}$ is an i.i.d. $N(0, I_{m\times m})$ sequence with density function ψ,
2. $\{Y_k\}$, $k \in \mathbb{N}$ is an i.i.d. $N(0, I_{d\times d})$ sequence with density function ϕ.

Write $\overline{\lambda}_0 = \dfrac{\phi(D_0^{-1}(Y_0 - C_0 X_0))}{|D_0|\phi((Y_0))}$. For each $l = 1, 2, \ldots$ define

$$\overline{\lambda}_l = \frac{\phi(D_l^{-1}(Y_l - C_l X_l))}{|D_l|\phi((Y_l))}\frac{\psi(B_l^{-1}(X_l - A_l X_{l-1}))}{|B_l|\phi((X_l))}.$$

For each $k \geq 0$ set $\overline{\Lambda}_k = \prod_{l=0}^{k}\overline{\lambda}_l$.

Let $\mathcal{G}_k$ be the complete σ-field generated by $\{X_0, X_1, \ldots, X_k, Y_0, Y_1, \ldots, Y_k\}$ for $k \in \mathbb{N}$.

Define P on $\{\Omega, \mathcal{F}\}$ by setting the restriction of the Radon–Nikodym derivative $\dfrac{dP}{d\overline{P}}$ to $\mathcal{G}_k$ equal to $\overline{\Lambda}_k$. Then:

Lemma 5.7.1 *On* $\{\Omega, \mathcal{F}\}$ *and under* P, $\{V_k\}$, $k \in \mathbb{N}$, $\{W_k\}$, $k \in \mathbb{N}$ *are i.i.d.* $N(0, I_{d \times d})$, $N(0, I_{m \times m})$ *sequences respectively, where*

$$V_l \overset{\Delta}{=} D_l^{-1}(Y_l - C_l X_l),$$

$$W_l \overset{\Delta}{=} B_l^{-1}(X_l - A_l X_{l-1}).$$

Definition 5.7.2 *Define the measure-valued processes*

$$\alpha_k(x) = \overline{E}[\Lambda_k I(X_k \in dx) \mid \mathcal{Y}_k],$$

$$\beta_k^{ij(M)}(x) = \overline{E}[\Lambda_k T_k^{ij(M)} I(X_k \in dx) \mid \mathcal{Y}_k], \quad M = 0, 1, 2,$$

$$\delta_k^{in}(x) = \overline{E}[\Lambda_k U_k^{in} I(X_k \in dx) \mid \mathcal{Y}_k]. \tag{5.7.1}$$

Then for any "test" function $g : \mathbb{R}^m \to \mathbb{R}$, *write*

$$\overline{E}[\Lambda_k g(X_k) \mid \mathcal{Y}_k] = \int_{\mathbb{R}^m} \alpha_k(x) g(x) dx,$$

$$\overline{E}[\Lambda_k T_k^{ij(M)} g(X_k) \mid \mathcal{Y}_k] = \int_{\mathbb{R}^m} \beta_k^{ij(M)}(x) g(x) dx, \quad M = 0, 1, 2,$$

$$\overline{E}[\Lambda_k U_k^{in} g(X_k) \mid \mathcal{Y}_k] = \int_{\mathbb{R}^m} \delta_k^{in}(x) g(x) dx. \tag{5.7.2}$$

The following theorem ([13]) gives recursive expressions for the unnormalized densities $\alpha_k(x)$, $\beta_k^{ij(M)}(x)$, $M = 0, 1, 2$, and $\delta_k^{in}(x)$. The recursions are derived under the measure $\overline{P}$ where $\{X_l\}$ and $\{Y_l\}$, $l = 0, 1, \ldots$ are independent sequences of random variables.

Theorem 5.7.3 *For* $k = 0, 1, \ldots$ *the unnormalized densities* $\alpha_k(x)$, $\beta_k^{ij(M)}(x)$, $M = 0, 1, 2$, *and* $\delta_k^{in}(x)$ *defined by (5.7.1) are given by the following recursions.*

$$\alpha_k(x) = \Phi(x, Y_k) \int_{\mathbb{R}^m} \alpha_{k-1}(z) \psi(B_k^{-1}(x - A_k z)) dz, \tag{5.7.3}$$

$$\beta_k^{ij(0)}(x) = \Phi(x, Y_k) \bigg[\int_{\mathbb{R}^m} \beta_{k-1}^{ij(0)}(z) \psi(B_k^{-1}(x - A_k z)) dz$$

$$+ \langle x, e_i \rangle \langle x, e_j \rangle \int_{\mathbb{R}^m} \alpha_{k-1}(z) \psi(B_k^{-1}(x - A_k z)) dz \bigg], \tag{5.7.4}$$

$$\beta_k^{ij(1)}(x) = \Phi(x, Y_k) \bigg[\int_{\mathbb{R}^m} \beta_{k-1}^{ij(1)}(z) \psi(B_k^{-1}(x - A_k z)) dz$$

$$+ \langle x, e_i \rangle \int_{\mathbb{R}^m} \langle z, e_j \rangle \alpha_{k-1}(z) \psi(B_k^{-1}(x - A_k z)) dz \bigg], \tag{5.7.5}$$

$$\beta_k^{ij(2)}(x) = \Phi(x, Y_k) \bigg[\int_{\mathbb{R}^m} \beta_{k-1}^{ij(2)}(z) \psi(B_k^{-1}(x - A_k z)) dz$$

$$+ \int_{\mathbb{R}^m} \langle z, e_i \rangle \langle z, e_j \rangle \alpha_{k-1}(z) \psi(B_k^{-1}(x - A_k z)) dz \bigg], \tag{5.7.6}$$

$$\delta_k^{in}(x) = \Phi(x, Y_k)\left[\int_{\mathbb{R}^m} \delta_{k-1}^{in}(z)\psi(B_k^{-1}(x - A_k z))dz\right.$$

$$\left. + \langle x, e_i\rangle\langle Y_k, f_n\rangle \int_{\mathbb{R}^m} \alpha_{k-1}(z)\psi(B_k^{-1}(x - A_k z))dz\right], \tag{5.7.7}$$

where $\Phi(x, Y_k) = \dfrac{\phi(D_k^{-1}(Y_k - C_k x))}{|B_k||D_k|\phi((Y_k)}.$

Remarks 5.7.4

1. Using (5.7.3), (5.7.4) and (5.7.7) are written as

$$\beta_k^{ij(0)}(x) = \Phi(x, Y_k)\int_{\mathbb{R}^m} \beta_{k-1}^{ij(0)}(z)\psi(B_k^{-1}(x - A_k z))dz + \langle x, e_i\rangle\langle x, e_j\rangle\alpha_k(x), \tag{5.7.8}$$

$$\delta_k^{in}(x) = \Phi(x, Y_k)\int_{\mathbb{R}^m} \delta_{k-1}^{in}(z)\psi(B_k^{-1}(x - A_k z))dz + \langle x, e_i\rangle\langle Y_k, f_n\rangle\alpha_k(x). \tag{5.7.9}$$

2. The above theorem does not require V_l and W_l to be Gaussian. The recursions (5.7.3), (5.7.4) and (5.7.7) hold for arbitrary densities ψ and ϕ as long as ϕ is strictly positive.
3. *Initial conditions*: Note that at $k = 0$, the following holds for any Borel "test" function $g(x)$.

$$\overline{E[\overline{\Lambda}_0 g(x) \mid \mathcal{Y}_0]} = \overline{E}\left[\frac{\phi(D_0^{-1}(Y_0 - C_0 x))}{|D_0|\phi(Y_0)} g(x) \mid \mathcal{Y}_0\right]$$

$$= \frac{1}{|D_0|\phi(Y_0)}\int_{\mathbb{R}^m} \phi(D_0^{-1}(Y_0 - C_0 x))\psi(x)g(x)dx.$$

$$\tag{5.7.10}$$

Equating (5.7.1) and (5.7.10) yields

$$\alpha_0(x) = \frac{\phi(D_0^{-1}(Y_0 - C_0 x))}{|D_0|\phi(Y_0)}\psi(x). \tag{5.7.11}$$

Similarly the initial conditions for $\beta_k^{ij(M)}(x)$ $M = 0, 1, 2$ and $\delta_k^{in}(x)$ are

$$\beta_0^{ij(0)}(x) = \langle x, e_i\rangle\langle x, e_j\rangle\alpha_0(x), \quad \beta_0^{ij(1)}(x) = 0,$$

$$\beta_0^{ij(2)}(x) = 0, \quad \delta_0^{in}(x) = \langle x, e_i\rangle\langle Y_0, f_n\rangle\alpha_0(x).$$

$$\tag{5.7.12}$$

4. Theorems 5.2 and 5.3 in [13] derive finite-dimensional filters for $T_k^{ij(M)}$, $M = 0, 1, 2$, and U_k^{in} defined in (5.6.5), (5.6.6), and (5.6.7). In particular, the densities α_k, $\beta_k^{ij(M)}$, $M = 0, 1, 2$, and δ_k^{in} are characterized in terms of a finite number of statistics. $\qquad\square$

Recall from Section 5.4 that α_k is an unnormalized normal density with mean $\mu_k = E[X_k \mid \mathcal{Y}_k]$ and variance $R_k = E[(X_k - \mu_k)(X_k - \mu_k)' \mid \mathcal{Y}_k]$ given by the well-known Kalman filter equations.

$$\mu_k = R_k B_k^{-2} A_k \sigma_k^{-1} R_{k-1}^{-1}\mu_{k-1} + R_k' C_k'(D_k' D_k)^{-1}Y_k, \tag{5.7.13}$$

$$R_k = \left[(A_k R_{k-1} A_k' + B_k^2)^{-1} + C_k'(D_k' D_k)^{-1}C_k\right]^{-1}. \tag{5.7.14}$$

Here σ_k is a symmetric $m \times m$ matrix defined as

$$\sigma_k = A'_k B_k^{-2} A_k + R_{k-1}^{-1}. \tag{5.7.15}$$

Due to the presence of the quadratic term $\langle x, e_i \rangle \langle x, e_j \rangle$, the density $\beta_k^{(0)}$ in (5.7.8) is not Gaussian. However, as will be shown below (Theorem 5.2 in [13]) it is possible to express it as a quadratic expression in x multiplied by $\alpha_k(x)$ for all k. The important conclusion then is that by updating the coefficients of the quadratic expression, together with the Kalman filter above, gives finite-dimensional filters for computing $T_k^{ij(0)}$. A similar result also holds for $T_k^{ij(1)}$, $T_k^{ij(2)}$ and U_k^{in}.

Theorems 5.7.5 and 5.7.6 that follow derive finite-dimensional sufficient statistics for the densities $\beta_k^{ij(M)}$, $M = 0, 1, 2$, and δ_k^{in}. To simplify notation, define

$$\Sigma_k = B_k^{-2} A_k \sigma_k^{-1}, \quad S_k = \sigma_{k+1}^{-1} R_k^{-1} \mu_k. \tag{5.7.16}$$

Theorem 5.7.5 *At time k, the density $\beta_k^{ij(M)}(x)$ (initialized according to (5.7.12)) is completely defined by the five statistics $a_k^{ij(M)}$, $b_k^{ij(M)}$, $d_k^{ij(M)}$, R_k and μ_k as follows:*

$$\beta_k^{ij(M)}(x) = \left[a_k^{ij(M)} + b_k^{ij(M)'} x + x' d_k^{ij(M)} x \right] \alpha_k(x), \quad M = 0, 1, 2, \tag{5.7.17}$$

where $a_k^{ij(M)} \in \mathbb{R}$, $b_k^{ij(M)} \in \mathbb{R}^m$, and $d_k^{ij(M)} \in \mathbb{R}^{m \times m}$ is a symmetric matrix with elements $d_k(p, q)$, $p = 1, \ldots, m$, $q = 1, \ldots, m$.

Furthermore, $a_k^{ij(M)}$, $b_k^{ij(M)}$, $d_k^{ij(M)}$, $M = 0, 1, 2$, are given by the following recursions:

$$a_{k+1}^{ij(0)} = a_k^{ij(0)} + b_k^{ij(0)'} S_k + \text{Tr}\left[d_k^{ij(0)} \sigma_{k+1}^{-1} \right] + S_k' d_k^{ij(0)} S_k, \quad a_0^{ij(0)} = 0, \tag{5.7.18}$$

$$b_{k+1}^{ij(0)} = \Sigma_{k+1} \left(b_k^{ij(0)} + 2d_k^{ij(0)} S_k \right), \quad b_0^{ij(0)} = 0_{m \times 1}, \tag{5.7.19}$$

$$d_{k+1}^{ij(0)} = \Sigma_{k+1} d_k^{ij(0)} \Sigma_{k+1}' + \frac{1}{2}(e_i e_j' + e_j e_i'), \quad d_0^{ij(0)} = \frac{e_i e_j' + e_j e_i'}{2}, \tag{5.7.20}$$

$$a_{k+1}^{ij(1)} = a_k^{ij(1)} + b_k^{ij(1)'} S_k + \text{Tr}\left[d_k^{ij(1)} \sigma_{k+1}^{-1} \right] + S_k' d_k^{ij(1)} S_k, \quad a_0^{ij(1)} = 0, \tag{5.7.21}$$

$$b_{k+1}^{ij(1)} = \Sigma_{k+1} \left(b_k^{ij(1)} + 2d_k^{ij(1)} S_k \right) + e_i e_j' S_k, \quad b_0^{ij(1)} = 0_{m \times 1}, \tag{5.7.22}$$

$$d_{k+1}^{ij(1)} = \Sigma_{k+1} d_k^{ij(1)} \Sigma_{k+1}' + \frac{1}{2}(e_i e_j' \Sigma_{k+1}' + \Sigma_{k+1} e_j e_i'), \quad d_0^{ij(1)} = 0_{m \times m}, \tag{5.7.23}$$

$$a_{k+1}^{ij(2)} = a_k^{ij(2)} + b_k^{ij(2)'} S_k + \text{Tr}\left[d_k^{ij(2)} \sigma_{k+1}^{-1} \right] + S_k'(d_k^{ij(2)} + e_i e_j') S_k$$
$$+ \text{Tr}\left[e_i e_j' \sigma_{k+1}^{-1} \right], \quad a_0^{ij(2)} = 0, \tag{5.7.24}$$

$$b_{k+1}^{ij(2)} = \Sigma_{k+1} \left(b_k^{ij(2)} + (2d_k^{ij(2)} + 2e_i e_j') S_k \right), \quad b_0^{ij(2)} = 0_{m \times 1}, \tag{5.7.25}$$

$$d_{k+1}^{ij(2)} = \Sigma_{k+1} \left(d_k^{ij(2)} + \frac{e_i e_j' + e_j e_i'}{2} \right) \Sigma_{k+1}', \quad d_0^{ij(2)} = 0_{m \times m}, \tag{5.7.26}$$

where $\text{Tr}[.]$ denotes the trace of a matrix (which is the sum of the diagonal elements), σ_k is defined in (5.7.15) and μ_k, R_k are obtained from the Kalman filter (5.7.13) and (5.7.14).

Proof Only the proof of the theorem for $M = 0$ is given; the proofs for $M = 1, 2$ are similar and left as an exercise.

From (5.7.12), at time $k = 0$, $\beta_0^{ij(0)}(x)$ is of the form (5.7.17) with $a_0^{ij(0)} = 0$, $b_0^{ij(0)} = 0$, $d_0^{ij(0)} = (e_i e_j' + e_j e_i')/2$. Assume that (5.7.17) holds at time k. Then at time $k + 1$, using (5.7.17) and the recursion (5.7.8) it follows that

$$\beta_{k+1}^{ij(0)}(x) = \Phi(x, Y_k) \int_{\mathbb{R}^m} \left[a_k^{ij(0)} + b_k^{ij(0)'} z + z' d_k^{ij(0)} z \right] \alpha_k(z)$$

$$\times \psi(B_k^{-1}(x - A_k z)) dz + \langle x, e_i \rangle \langle x, e_j \rangle \alpha_{k+1}(x). \tag{5.7.27}$$

Denote the first term on the RHS of (5.7.27) as I_1.

$$I_1 = K(x) \int_{\mathbb{R}^m} \exp\left[-\frac{1}{2}\{(x - A_{k+1}z)' B_{k+1}^{-2}(x - A_{k+1}z) + (z - \mu_k)' R_k^{-1}(z - \mu_k)\} \right]$$

$$\times \left[a_k^{ij(0)} + b_k^{ij(0)'} z + z' d_k^{ij(0)} z \right] dz$$

$$= K_1(x) \int_{\mathbb{R}^m} \exp\left[-\frac{1}{2}(z'\sigma_{k+1}z - \xi_{k+1}'z) \right] \left[a_k^{ij(0)} + b_k^{ij(0)'} z + z' d_k^{ij(0)} z \right] dz, \tag{5.7.28}$$

where σ_k is defined in (5.7.15),

$$\xi_{k+1} = 2(x' B_{k+1}^{-2} A_{k+1} + \mu_k' R_k^{-1}),$$

$$K(x) = \Phi(x, Y)(2\pi)^{-m} |B_{k+1}|^{-1} |R_k^{-1/2} \bar{\alpha}_k,$$

$$\bar{\alpha}_k = \int_{\mathbb{R}^m} \alpha_k(z) dz,$$

$$K_1(x) = K(x) \exp\left[-\frac{1}{2}(x' B_{k+1}^{-2} x + \mu_k' R_k^{-1} \mu_k) \right]. \tag{5.7.29}$$

Completing the "square" in the exponential term in (5.7.28) yields

$$I_1 = K_1(x) \exp\left[-\frac{1}{2}\left(-\frac{\xi_{k+1}' \sigma_{k+1}^{-1} \xi_{k+1}}{4} \right) \right]$$

$$\times \int_{\mathbb{R}^m} \exp\left[-\frac{1}{2}\left(z - \frac{\sigma_{k+1}^{-1} \xi_{k+1}}{2} \right)' \sigma_{k+1} \left(z - \frac{\sigma_{k+1}^{-1} \xi_{k+1}}{2} \right) \right]$$

$$\times \left[a_k^{ij(0)} + b_k^{ij(0)'} z + z' d_k^{ij(0)} z \right] dz. \tag{5.7.30}$$

Now consider the integral in (5.7.30),

$$\int_{\mathbb{R}^m} \exp\left[-\frac{1}{2}\left(z - \frac{\sigma_{k+1}^{-1} \xi_{k+1}}{2} \right)' \sigma_{k+1} \left(z - \frac{\sigma_{k+1}^{-1} \xi_{k+1}}{2} \right) \right]$$

$$\times \left[a_k^{ij(0)} + b_k^{ij(0)'} z + z' d_k^{ij(0)} z \right] dz$$

$$= (2\pi)^{m/2} |\sigma_{k+1}|^{-1/2} \left[a_k^{ij(0)} + b_k^{ij(0)'} E[z] + E[z' d_k^{ij(0)} z] \right], \tag{5.7.31}$$

since the exponential term is an unnormalized Gaussian density in z with normalization constant $(2\pi)^{m/2}|\sigma_{k+1}|^{-1/2}$. So

$$E[z] = \frac{\sigma_{k+1}^{-1}\xi_{k+1}}{2}, \tag{5.7.32}$$

$$E[z'd_k^{ij(0)}z] = E\{(z - E[z])'d_k^{ij(0)}(z - E[z])\} + E[z']d_k^{ij(0)}E[z],$$

$$= \text{Tr}[d_k^{ij(0)}\sigma_{k+1}^{-1}] + \frac{1}{4}(\sigma_{k+1}^{-1}\xi_{k+1})'d_k^{ij(0)}(\sigma_{k+1}^{-1}\xi_{k+1}). \tag{5.7.33}$$

Therefore from (5.7.30), (5.7.31), (5.7.32), (5.7.33) and (5.7.27) it follows that

$$\beta_{k+1}^{ij(M)}(x) = \alpha_{k+1}(x)\Big[a_k^{ij(0)} + \frac{1}{2}b_k^{ij(0)'}\sigma_{k+1}^{-1}\xi_{k+1} + \text{Tr}[d_k^{ij(0)}\sigma_{k+1}^{-1}]$$

$$+ \frac{1}{4}\xi_{k+1}'\sigma_{k+1}^{-1}d_k^{ij(0)}\sigma_{k+1}^{-1}\xi_{k+1} + x'e_ie_j'x\Big]. \tag{5.7.34}$$

Substituting for ξ_{k+1} (which is affine in x) yields

$$\beta_{k+1}^{ij(0)}(x) = \Big[a_{k+1}^{ij(0)} + b_{k+1}^{ij(0)'}x + x'd_{k+1}^{ij(0)}x\Big]\alpha_{k+1}(x),$$

where $a_{k+1}^{ij(0)}$, $b_{k+1}^{ij(0)'}$ and $d_{k+1}^{ij(0)}$ are given by (5.7.18), (5.7.19) and (5.7.20). ∎

The proof of the following theorem (Theorem 5.3 in [13]) is very similar and hence omitted.

Theorem 5.7.6 *At time k, the density $\delta_k^{in}(x)$ (initialized according to (5.7.12)) is completely defined by the four statistics $\bar{a}_k^{in}$, $\bar{b}_k^{in}$, R_k and μ_k as follows:*

$$\delta_k^{in}(x) = \Big[\bar{a}_k^{in} + \bar{b}_k^{in'}x\Big]\alpha_k(x), \tag{5.7.35}$$

where $\bar{a}_k^{in} \in \mathbb{R}$, $\bar{b}_k^{in'} \in \mathbb{R}^m$, are given by the following recursions:

$$\bar{a}_{k+1}^{in} = \bar{a}_k^{in} + \bar{b}_k^{in'}S_k, \quad \bar{a}_0^{in} = 0, \tag{5.7.36}$$

$$\bar{b}_{k+1}^{in} = \Sigma_{k+1}\bar{b}_k^{in}, \quad \bar{b}_0^{in} = e_i\langle Y_{k+1}, f_n\rangle, \tag{5.7.37}$$

where Σ_k and S_k are defined in (5.7.16) and μ_k, R_k are obtained from the Kalman filter (5.7.13) and (5.7.14).

Having characterized the densities $\beta_k^{ij(M)}$, $M = 0, 1, 2$, and δ_k^{in} by their finite-dimensional sufficient statistics, finite-dimensional filters for $T_k^{ij(M)}$ and U_k^{in} are now derived.

Theorem 5.7.7 *(Theorem 5.4 in [13]) Finite-dimensional filters for $T_k^{ij(M)}$, $M = 0, 1, 2$, and U_k^{in} are given by*

$$E[T_k^{ij(M)} \mid \mathcal{Y}_k] = a_k^{ij(M)} + b_k^{ij(M)'}\mu_k + \text{Tr}\Big[d_k^{ij(M)}R_k\Big] + \mu_k'd_k^{ij(M)}\mu_k, \tag{5.7.38}$$

$$E[U_k^{in} \mid \mathcal{Y}_k] = \bar{a}_k^{in} + \bar{b}_k^{in'}\mu_k. \tag{5.7.39}$$

Proof Using the abstract Bayes' Theorem (4.1.1),

$$E[T_k^{ij(M)} \mid \mathcal{Y}_k] = \frac{\overline{E}[\Lambda_k T_k^{ij(M)} \mid \mathcal{Y}_k]}{\overline{E}[\Lambda_k \mid \mathcal{Y}_k]} = \frac{\int_{\mathbb{R}^m} \beta_k^{ij(M)} dx}{K}, \qquad (5.7.40)$$

where the constant $K = \int_{\mathbb{R}^m} \alpha_k(x) dx$. But since $\alpha_k(x)$ is an unnormalized density, from (5.7.17),

$$\int_{\mathbb{R}^m} \beta_k^{ij(M)} dx = K E\left[a_k^{ij(M)} + b_k^{ij(M)'} x + x' d_k^{ij(M)} x \right]$$

$$= K \left[a_k^{ij(M)} + b_k^{ij(M)'} \mu_k + \text{Tr}\left[d_k^{ij(M)} R_k \right] + \mu_k' d_k^{ij(M)} \mu_k \right]. \quad (5.7.41)$$

Substituting in (5.7.40) proves the theorem. The proof of (5.7.39) is left as an exercise. ∎

Remark 5.7.8 Theorem (5.7.7) gives finite-dimensional filters for the time sum of the states U_k^{in} and time sum of the square of the states T_k^{ij}. Theorem 6.2 in [13] shows that *finite-dimensional filters exist for the time sum of any arbitrary integral power of the states*. □

For notational simplicity, assume that the state and observation processes are scalar-valued i.e. $m = d = 1$ in (5.6.1) and (5.6.2).

Let $T_k = \sum_{i=0}^{k} X_i^p$ and define the unnormalized density

$$\beta_k(x) = \overline{E}[\Lambda_k T_k I(X_k \in dx) \mid \mathcal{Y}_k].$$

The first step is to obtain a recursion for $\beta_k(x)$. It can be easily shown that

$$\beta_k(x) = \Phi(x, Y_k) \int_{\mathbb{R}} \beta_{k-1}(z) \psi(B_k^{-1}(x - A_k z)) dz + x^p \alpha_k(x),$$

$$(5.7.42)$$

where $\Phi(x, Y_k) = \dfrac{\phi(D_k^{-1}(Y_k - C_k x))}{B_k D_k \phi(Y_k)}$

Next, $\beta_k(x)$ is characterized in terms of finite sufficient statistics, as shown in Theorem 5.7.3. Also for $p = 1$ and 2, Theorems 5.7.6 and 5.7.5 give finite-dimensional sufficient statistics. Theorem 6.2 in [13] shows that β_k can be characterized in terms of finite-dimensional statistics for any $p \in \mathbb{N}$.

Theorem 5.7.9 *([13] Theorem 6.2) At time k, the density $\beta_k(x)$ in (5.7.42) is completely defined by $p + 3$ statistics $a_k(0), a_k(1), \ldots, a_k(p), R_k$ and μ_k as follows:*

$$\beta_k(x) = \left(\sum_{i=0}^{p} a_k(i) x^i \right) \alpha_k(x), \qquad (5.7.43)$$

where

$$a_{k+1}(n) = \sum_{i=n}^{p} \sum_{j=n}^{i} a_k(i)\eta_{ij} \binom{j}{n} (R_k^{-1}\mu_k)^{j-n}(A_{k+1}B_{k+1}^{-2})^n, \ 0 \le n < p,$$

$$a_{k+1}(p) = 1 + a_k(p)\eta_{pp}(A_{k+1}B_{k+1}^{-2})^p, \tag{5.7.44}$$

and

$$\eta_{ij} = \begin{cases} \binom{i}{j} 1.3...(i-j-1)\sigma_{k+1}^{-(j+1)} & \text{if } i-j \text{ is even, } i > j, \\ 0 & \text{if } i-j \text{ is odd, } i > j, \\ \sigma_{k+1}^{-j} & \text{if } i = j. \end{cases} \tag{5.7.45}$$

Proof   At time $k = 0$, $\beta_0(x) = x^p \alpha_0(x)$ and so satisfies (5.7.43).

Assume that (5.7.43) holds at time k. Then at time $k + 1$, using similar arguments to Theorem 5.7.5, it follows that

$$\beta_{k+1}(x) = \Phi(x, Y_k) \int_{\mathbb{R}} \psi(B_{k+1}^{-1}(x - A_{k+1}z)) \left(\sum_{i=0}^{p} a_k(i)z^i\right) \alpha_k(z)dz + x^p \alpha_k(x). \tag{5.7.46}$$

Denote the RHS as I_1.

$$I_1 = K_1(x)\exp\left\{-\frac{1}{2}\left(-\frac{\delta'_{k+1}\sigma_{k+1}^{-1}\delta_{k+1}}{4}\right)\right\}$$

$$\times \int_{\mathbb{R}} \left(\sum_{i=0}^{p} a_k(i)z^i\right) \exp\left\{-\frac{1}{2}\left(z - \frac{\sigma_{k+1}^{-1}\delta_{k+1}}{2}\right)^2 \sigma_{k+1}\right\} dz. \tag{5.7.47}$$

The integral in (5.7.47) is

$$(2\pi)^{1/2}|\sigma_{k+1}|^{-1/2} E\left[\sum_{i=0}^{p} a_k(i)z^i\right]$$

$$= (2\pi)^{1/2}|\sigma_{k+1}|^{-1/2} \sum_{i=0}^{p} a_k(i) \sum_{j=0}^{i} \binom{i}{j} E\left[(z - E[z])^{i-j}\right] (E[z])^j. \tag{5.7.48}$$

Recall from (5.7.32) that $E[z]$ is affine in x:

$$E[z] = \sigma_{k+1}^{-1}[R_k^{-1}\mu_k + A_{k+1}B_{k+1}^{-2}x]. \tag{5.7.49}$$

Also $E[(z - E[z])^2]$ is independent of x. Indeed ([31], p. 111),

$$E\left[(z - E[z])^{i-j}\right] = \begin{cases} 0 & \text{if } i-j \text{ is even, } i > j, \\ 1.3...(i-j-1)\sigma_{k+1}^{-1} & \text{if } i-j \text{ is even, } i > j, \\ 1 & \text{if } i = j. \end{cases} \tag{5.7.50}$$

Thus

$$\beta_{k+1}(x) = \alpha_{k+1}(x) \left(\sum_{i=0}^{p} \sum_{j=0}^{p} \sum_{n=0}^{j} a_k(i) \eta_{ij} \binom{j}{n} (R_k^{-1} \mu_k)^{j-n} (A_{k+1} B_{k+1}^{-2})^n x^n + x^p \right)$$

$$= \alpha_{k+1}(x) \left(\sum_{n=0}^{p} \sum_{i=n}^{p} \sum_{j=n}^{i} a_k(i) \eta_{ij} \binom{j}{n} (R_k^{-1} \mu_k)^{j-n} (A_{k+1} B_{k+1}^{-2})^n x^n + x^p \right).$$

$$\tag{5.7.51}$$

Equation (5.7.51) is of the form (5.7.43) with $a_{k+1}(i)$, $i = 0, \dots, p$ given by (5.7.44). ∎

Remark 5.7.10 The filters derived in Theorems 5.7.3, 5.7.5 and 5.7.6 have one major problem: they require B_k to be invertible. In practice, B_k is often not invertible. A simple transformation that expresses the filters in the terms of the inverse of the predicted Kalman covariance matrix is used. This inverse exists even if B_k is singular as long as a certain uniform controllability condition holds. Both the uniform controllability condition and the transformation used are well-known in Kalman filter literature (Chapter 7 in [20]). □

First define the Kalman predicted state estimate $\mu_{k|k-1} \triangleq E[X_k \mid \mathcal{Y}_{k-1}]$ and the predicted state covariance $R_{k|k-1} \triangleq E[(X_k - \mu_{k|k-1})(X_k - \mu_{k|k-1})' \mid \mathcal{Y}_{k-1}]$. It is left as an exercise to show that

$$R_{k|k-1} = B_k^2 + A_k R_{k-1} A_k'. \tag{5.7.52}$$

The first step is to provide a sufficient condition for $R_{k|k-1}$ to be non-singular.

Definition 5.7.11 ([20] Chapter 7) The state space model (5.6.1), (5.6.2) is said to be uniformly completely controllable if there exist a positive integer N_1 and positive constants α, β such that

$$\alpha I \leq C(k, k - N_1) \leq \beta I \quad \text{for all } k \geq N_1. \tag{5.7.53}$$

Here

$$C(k, k - N_1) \triangleq \sum_{l=k-N_1}^{k} \phi(k, l+1) B_l B_l' \phi(k, l+1)'. \tag{5.7.54}$$

$$\phi(k_2, k_1) = \begin{cases} A_{k_2} A_{k_2-1} \dots A_{k_1+1} & \text{if } k_2 > k_1, \\ I & \text{if } k_2 = k_1. \end{cases} \tag{5.7.55}$$

Lemma 5.7.12 If the state space model (5.6.1), (5.6.2) is uniformly completely controllable and $R_0 \geq 0$ then R_k and $R_{k|k-1}$ are positive definite matrices (and hence nonsingular) for all $k \geq N_1$.

Proof   See [20], p. 238, Lemma 7.3. ∎

The following lemma will be used in the sequel.

Lemma 5.7.13 *(Lemma 7.3 in [13]) Assume $R_{k|k-1}^{-1}$ exists. Then with σ_k and Σ_k defined in (5.7.15) and (5.7.16), respectively,*

$$\sigma_k^{-1} = R_{k-1} - R_{k-1} A_k' R_{k|k-1}^{-1} A_k R_{k-1}, \tag{5.7.56}$$

$$\Sigma_{k+1} = R_{k+1|k}^{-1} A_{k+1} R_k. \tag{5.7.57}$$

Furthermore, the Kalman filter (5.7.13), (5.7.14) can be expressed in "standard" form

$$\mu_k = A_k \mu_{k-1} + R_{k|k-1} C_k' [C_k R_{k|k-1} C_k' + D_k D_k']^{-1} (Y_k - C_k A_k \mu_{k-1}),$$

$$R_k = R_{k|k-1} - R_{k|k-1} C_k' [C_k R_{k|k-1} C_k' + D_k D_k']^{-1} C_k R_{k|k-1},$$

$$R_{k|k-1} = B_k^2 + A_k R_{k-1} A_k'. \tag{5.7.58}$$

Proof Straightforward use of the Matrix Inversion Lemma 5.4.3 on (5.7.15) yields

$$\sigma_k^{-1} = R_{k-1} - R_{k-1} A_k' (B_k^2 + A_k R_{k-1} A_k')^{-1} A_k R_{k-1}. \tag{5.7.59}$$

Substituting (5.7.52) in (5.7.59) proves (5.7.56).

To prove (5.7.57), first note that

$$\Sigma_{k+1} \stackrel{\triangle}{=} B_{k+1}^{-2} A_{k+1} \sigma_{k+1}^{-1}$$

$$= B_{k+1}^{-2} A_{k+1} R_k - B_{k+1}^{-2} A_{k+1} R_k A_{k+1}' R_{k+1|k}^{-1} A_{k+1} R_k$$

$$= B_{k+1}^{-2} A_{k+1} R_k - B_{k+1}^{-2} (R_{k+1|k} - B_{k+1}^{-2}) R_{k+1|k}^{-1} A_{k+1} R_k,$$

because of (5.7.52). So

$$\Sigma_{k+1} = B_{k+1}^{-2} A_{k+1} R_k - B_{k+1}^{-2} A_{k+1} R_k + R_{k+1|k}^{-1} A_{k+1} R_k$$

$$= R_{k+1|k}^{-1} A_{k+1} R_k.$$

To prove (5.7.58), consider the Kalman filter equations (5.7.13) and (5.7.14). Using Lemma 5.7.12 to the first term on the RHS of (5.7.13) yields the "standard" Kalman filter equations. ∎

Now the filters derived earlier in this section are expressed in terms of $R_{k|k-1}$ instead of B_k. As shown below (Theorem 7.4 in [13]), the advantage of doing so is that B_k no longer needs to be invertible, as long as the uniformly controllability condition in Definition 5.7.11 holds.

Theorem 5.7.14 *Consider the linear dynamical system (5.6.1) and (5.6.2) with B_k not necessarily invertible. Assume that the system is uniformly completely controllable, i.e. (5.7.53) holds. Then at time k, with σ_k^{-1} given by (5.7.56) and Σ_k defined in (5.7.57), the following model holds.*

1. *The density $\alpha_k(x)$ (defined in (5.7.1)) is an unnormalized Gaussian density with mean $\mu \in \mathbb{R}^m$ and covariance matrix $R_k \in \mathbb{R}^{m \times m}$. These are recursively computed via the standard Kalman filter equations (5.7.58).*

2. The density $\beta_k^{ij(M)}(x)$ (initialized according to (5.7.12)) is completely defined by the five statistics $a_k^{ij(M)}$, $b_k^{ij(M)}$, $d_k^{ij(M)}$, R_k and μ_k as follows:

$$\beta_k^{ij(M)}(x) = \left[a_k^{ij(M)} + b_k^{ij(M)'} x + x' d_k^{ij(M)} x \right] \alpha_k(x), \quad M = 0, 1, 2,$$

where $a_k^{ij(M)} \in \mathbb{R}$, $b_k^{ij(M)} \in \mathbb{R}^m$, and $d_k^{ij(M)} \in \mathbb{R}^{m \times m}$ is a symmetric matrix with elements $d_k(p, q)$, $p = 1, \ldots, m$, $q = 1, \ldots, m$. These statistics are recursively computed by Equations (5.7.18) to (5.7.26).

3. The density $\delta_k^{in}(x)$ (initialized according to (5.7.12)) is completely defined by the four statistics $\bar{a}_k^{in}$, $\bar{b}_k^{in}$, R_k and μ_k as follows:

$$\delta_k^{in}(x) = \left[\bar{a}_k^{in} + \bar{b}_k^{in'} x \right] \alpha_k(x), \tag{5.7.60}$$

where $\bar{a}_k^{in} \in \mathbb{R}$, $\bar{b}_k^{in'} \in \mathbb{R}^m$ are given by the recursions (5.7.36) and (5.7.37).

Finally, finite-dimensional filters for $T_k^{ij(M)}$ and U_k^{in} in terms of the above statistics are given by (5.7.38) and (5.7.39).

Proof It only remains to show that subject to the uniform complete controllability condition (5.7.53), the filtering equations (5.7.18)–(5.7.26) and (5.7.36), (5.7.37) in Theorem 5.7.14 hold even if the matrices B_{k+1} are singular. The proof of this is as follows. If B_{k+1} is singular,

1. Add $\epsilon \times N(0, 1)$ noise to each component of X_{k+1}. This is done by replacing B_{k+1} in (5.6.1) with the nonsingular matrix $B_{k+1}^\epsilon = B_{k+1} + \epsilon I_m$, where $\epsilon \in \mathbb{R}$. Denote the resulting state process X_{k+1}^ϵ.
2. Define $R_{k+1|k}^\epsilon$ as in (5.7.52) with B_{k+1} replaced by B_{k+1}^ϵ. Express the filters in term of $R_{k+1|k}^\epsilon$ as Theorem 5.7.14.
3. As $\epsilon \to 0$, $R_{k+1|k}^\epsilon \to R_{k+1|k}$.
4. Then using the bounded conditional convergence theorem (p. 214, [4]), the conditional estimates of X_k^ϵ, $X_k^\epsilon X_k^{\epsilon'}$, $T_k^{ij(0)}(x^\epsilon)$, and $U_k^{in}(x^\epsilon)$ converge to the conditional estimates of X_k, $X_k X_k'$, $T_k^{ij(0)}(x)$, and $U_k^{in}(x)$, respectively. ∎

5.8 Continuous-time vector dynamics

Consider the classical linear Gaussian model for the signal and observation processes. That is, the signal $\{x_t\}$, $t \geq 0$, is described by the equation

$$dx_t = A_t x_t dt + B_t dw_t, \quad x_0 \in \mathbb{R}^m, \tag{5.8.1}$$

and the observation process $\{y_t\}$, $t \geq 0$, is described by the equation

$$dy_t = C_t x_t dt + D_t dv_t, \quad y_0 = 0 \in \mathbb{R}^n. \tag{5.8.2}$$

Here w and v are independent r-dimensional and n-dimensional Brownian motions, respectively, defined on a probability space $(\Omega, \mathcal{F}, \overline{P})$ with complete filtrations $\mathcal{F}_t = \sigma\{x_s, y_s : s \leq t\}$, and $\mathcal{Y}_t = \sigma\{y_s : s \leq t\}$, $t \geq 0$. Further, w and v are independent of x_0. We assume that x_0 is random variable with normal density

$$p_0(x) = (2\pi)^{-n/2} |P_0|^{-1/2} \exp(-1/2) \left\{ (x - \hat{x}_0)' P_0^{-1} (x - \hat{x}_0) \right\}.$$

The matrix functions $A_t \in \mathbb{R}^{m \times m}$, $B_t \in \mathbb{R}^{m \times r}$, $C_t \in \mathbb{R}^{n \times m}$, and $D_t \in \mathbb{R}^{n \times n}$ are measurable functions of t. We assume D_t is a positive definite matrix.

We model the above dynamics by supposing that initially we have an "ideal" probability space $(\Omega, \mathcal{F}, P)$ such that under P,

1. w is an r-dimensional Brownian motion and $\{x_t\}$ is defined by (5.8.1),
2. y is an n-dimensional Brownian motion, independent of w and x_0, and having quadratic variation $\langle y \rangle_t = D_t > 0$; i.e., D_t is a positive definite matrix.

Write $\nabla = \left(\dfrac{\partial}{\partial x_1}, \ldots, \dfrac{\partial}{\partial x_n} \right)'$.

For any function $g : \mathbb{R}^m \to \mathbb{R}$, write

$$\nabla^2 = \begin{bmatrix} \dfrac{\partial^2 g}{\partial x_1^2} & \cdots & \dfrac{\partial^2 g}{\partial x_1 \partial x_m} \\ \vdots & \vdots & \vdots \\ \dfrac{\partial^2 g}{\partial x_m \partial x_1} & \cdots & \dfrac{\partial^2 g}{\partial x_m^2} \end{bmatrix}.$$

For a vector field $g(x) = (g_1(x), g_2(x), \ldots, g_m(x))'$ defined on $\mathbb{R}^m$, define

$$\mathrm{div}(g) = \frac{\partial g_1}{\partial x_1} + \frac{\partial g_2}{\partial x_2} + \cdots + \frac{\partial g_m}{\partial x_m}.$$

Define

$$\Lambda_t = \exp \left\{ \int_0^t (C_s x_s)' D_s^{-1} (D_s^{-1})' dy_s - \frac{1}{2} \int_0^t x_s' C_s' (D_s^{-1})' D_s^{-1} C_s x_s \, ds \right\},$$

which is also given by

$$\Lambda_t = 1 + \int_0^t \Lambda_s x_s' C_s' (D_s^{-1})' D_s^{-1} dy_s. \tag{5.8.3}$$

To see this apply the Itô rule to the function $\log \Lambda_t$.

Then Λ_t is an $\mathcal{F}_t$-martingale and $E[\Lambda_t] = 1$.

A new probability measure $\overline{P}$ can be defined by setting $\left. \dfrac{d\overline{P}}{dP} \right|_{\mathcal{F}_t} = \Lambda_t$.

Define the process v_t by the formula

$$dv_t = D_t^{-1}(y_t - C_t x_t dt), \qquad v_0 = 0.$$

Then Girsanov's theorem 4.3.3 implies that $\{v_t\}$ is a standard n-dimensional Brownian motion process under $\overline{P}$. Therefore under $\overline{P}$,

$$dy_t = C_t x_t dt + D_t dv_t.$$

Note that under $\overline{P}$, the process $\{x_t\}$ still satisfies (5.8.1). Consequently, under $\overline{P}$ the processes $\{x_t\}$ and $\{y_t\}$ satisfy the real world dynamics (5.8.1) and (5.8.2). However, P is a more convenient measure with which to work.

For any "test" function $\phi : \mathbb{R}^m \to \mathbb{R}$ which is in C^2 and has compact support, write

$$\sigma(\phi)_t = E[\Lambda_t \phi(x_t) \mid \mathcal{Y}_t].$$

In the case when the measure defined by $\sigma(.)_t$ has a density $q(x, t)$ we have

$$\sigma(\phi)_t = \int_{\mathbb{R}^m} \phi(x) q(x, t) dx.$$

Using the vector Itô rule (Theorem 3.6.9) we establish

$$\phi(x_t) = \phi(x_0) + \int_0^t (\nabla \phi(x_s))' A_s x_s ds + \int_0^t (\nabla \phi(x_s))' B_s dw_s$$

$$+ \frac{1}{2} \int_0^t \text{Tr}[\nabla^2 \phi(x_s)' B_s B_s'] ds. \qquad (5.8.4)$$

In view of (5.8.3) and (5.8.4) and using the Itô product rule (Example 3.7.15),

$$\Lambda_t \phi(x_t) = \phi(x_0) + \int_0^t \Lambda_s (\nabla \phi(x_s))' A_s x_s ds + \int_0^t \Lambda_s (\nabla \phi(x_s))' B_s dw_s$$

$$+ \frac{1}{2} \int_0^t \Lambda_s \text{Tr}[\nabla^2 \phi(x_s)' B_s B_s'] ds$$

$$+ \int_0^t \Lambda_s \phi(x_s) x_s' C_s' (D_s^{-1})' D_s^{-1} dy_s. \qquad (5.8.5)$$

Conditioning both sides of (5.8.5) on $\mathcal{Y}_t$ and using the fact that B_t and y_t are independent and that y_t has independent increments under P (it is Wiener) (see [15] Lemma 3.2 p. 261), we have

Theorem 5.8.1 *Suppose $\phi \in C^2$ is a real-valued function with compact support. Then*

$$\sigma(\phi)_t = \sigma(\phi)_0 + \int_0^t \sigma((\nabla \phi(x_s))' A_s x_s) ds$$

$$+ \frac{1}{2} \int_0^t \sigma(\text{Tr}[\nabla^2 \phi(x_s)' B_s B_s']) ds$$

$$+ \int_0^t \sigma(\phi(x_s) x_s') C_s' (D_s^{-1})' D_s^{-1} dy_s. \qquad (5.8.6)$$

If $\sigma(.)_t$ has a density $q(x, t)$, we integrate by parts each term of

$$\int_{\mathbb{R}^m} q(x, t) \phi(x) dx = \int_{\mathbb{R}^m} q(x, 0) \phi(x) dx$$

$$+ \int_0^t \int_{\mathbb{R}^m} q(x, t)(\nabla \phi(x))' A_s x) dx ds$$

$$+ \frac{1}{2} \int_0^t \int_{\mathbb{R}^m} q(x, t) \text{Tr}[\nabla^2 \phi(x)' B_s B_s'] dx ds$$

$$+ \int_0^t \int_{\mathbb{R}^m} q(x, t) \phi(x) x' C_s' (D_s^{-1})' D_s^{-1} dx dy_s.$$

For instance, if $m = 2$,

$$\int_0^t \int_{\mathbb{R}^m} q(x,t)(\nabla\phi(x))' A_s x)dx ds = \int_0^t \int_{\mathbb{R}^m} q(x,t)\left(\frac{\partial\phi}{\partial x_1}, \frac{\partial\phi}{\partial x_2}\right)$$

$$\times (a_{11}x_1 + a_{12}x_2, a_{21}x_1 + a_{22}x_2)' dx ds$$

$$= \int_0^t \int_{\mathbb{R}^m} q(x,t)(a_{11}x_1 + a_{12}x_2)\frac{\partial\phi}{\partial x_1}dx ds$$

$$+ \int_0^t \int_{\mathbb{R}^m} q(x,t)(a_{21}x_1 + a_{22}x_2)\frac{\partial\phi}{\partial x_2}dx ds$$

$$= -\int_0^t \int_{\mathbb{R}^m} \phi(x)\frac{\partial q(x,t)(a_{11}x_1 + a_{12}x_2)}{\partial x_1}dx ds$$

$$- \int_0^t \int_{\mathbb{R}^m} \phi(x)\frac{\partial q(x,t)(a_{21}x_1 + a_{22}x_2)}{\partial x_2}dx ds$$

$$= -\int_0^t \int_{\mathbb{R}^m} \phi(x)\left[\frac{\partial q(x,t)(a_{11}x_1 + a_{12}x_2)}{\partial x_1}\right.$$

$$\left. + \frac{\partial q(x,t)(a_{21}x_1 + a_{22}x_2)}{\partial x_2}\right]dx ds$$

$$= -\int_0^t \int_{\mathbb{R}^m} \phi(x)\text{div}(A_s x q(x,s))dx ds.$$

Similarly,

$$\int_0^t \int_{\mathbb{R}^m} q(x,s)\text{Tr}[\nabla^2\phi(x)' B_s B_s']dx ds = \int_0^t \int_{\mathbb{R}^m} \phi(x)\text{Tr}[\nabla^2 q(x,s) B_s B_s']dx ds,$$

which holds for all "test" functions ϕ; hence

Lemma 5.8.2

$$q(x,t) = q(x,0) - \int_0^t \text{div}(A_s x q(x,s))ds + \frac{1}{2}\int_0^t \text{Tr}[\nabla^2 q(x,s)B_s B_s']ds$$

$$+ \int_0^t q(x,s)x'C_s'(D_s^{-1})'D_s^{-1}dy_s, \tag{5.8.7}$$

with $q(x,0) = p_0(x)$, the density of x_0.

Remark 5.8.3 Equation (5.8.7) is a stochastic partial differential equation for the unnormalized conditional density of x_t given $\mathcal{Y}_t$. In general, the solution of this equation is a conditional density function, evolving stochastically in time. For the linear, Gaussian dynamics (5.8.1) and (5.8.2), however, $q(x,t)$ has a simple form. □

Theorem 5.8.4 *The solution of (5.8.7) is*

$$q(x,t) = (2\pi)^{-m/2}|\Sigma_t|^{-1/2}v_t \exp(-1/2)\left\{(x - m_t)'\Sigma_t^{-1}(x - m_t)\right\}. \tag{5.8.8}$$

Here $m_t = \overline{E}[x_t \mid \mathcal{Y}_t]$, $m_0 = \hat{x}_0$, $\Sigma_t = \overline{E}[(x_t - m_t)'(x_t - m_t) \mid \mathcal{Y}_t]$, $\Sigma_0 = P_0$, and v_t is a normalizing factor.

It is well known that m_t and Σ_t are given by the Kalman filter equations

$$dm_t = A_t m_t dt + \Sigma_t C_t'(D_t^{-1})' D_t^{-1}(dy_t - C_t m_t dt)$$

$$= A_t m_t dt + \Sigma_t C_t'(D_t^{-1})' dv_t,$$

$$(\text{since under } \overline{P}, \ dv_t = D_t^{-1}(dy_t - C_t m_t dt), \tag{5.8.9}$$

$$\frac{d\Sigma_t}{dt} = \Sigma_t A_t' + A_t \Sigma_t + B_t B_t' - \Sigma_t C_t'(D_t^{-1})' D_t^{-1} C_t \Sigma_t. \tag{5.8.10}$$

Note that Σ_t is deterministic and can be computed off-line. Also

$$v_t = \exp\left\{ \int_0^t m_s' C_s'(D_s^{-1})' D_s^{-1} dy_s - \frac{1}{2}\int_0^t m_s' C_s'(D_s^{-1})' D_s^{-1} C_s m_s ds \right\}. \tag{5.8.11}$$

Proof ([2]) We have to show that for any "test" function $\phi(.)$

$$\sigma(\phi)_t = \sigma(\phi)_0 + \int_0^t \sigma((\nabla\phi(x_s))' A_s x_s) ds + \frac{1}{2}\int_0^t \sigma(\text{Tr}[\nabla^2\phi(x_s)' B_s B_s']) ds$$

$$+ \int_0^t \sigma(\phi(x_s)x_s') C_s'(D_s^{-1})' D_s^{-1} dy_s$$

$$= \int_{\mathbb{R}^m} \phi(x) q(x, t) dx, \tag{5.8.12}$$

where $q(x, t)$ is given by (5.8.8).

Let $\xi = \Sigma_t^{-1/2}(x - m_t)$, $x = m_t + \Sigma_t^{1/2}\xi$ and $dx = |\Sigma_t|^{1/2} d\xi$. Hence the dx integral on the right hand side of (5.8.12) is equal to

$$\int_{\mathbb{R}^m} (2\pi)^{-m/2}\phi(m_t + \Sigma_t^{1/2}\xi) v_t e^{-|\xi|^2/2} d\xi.$$

Now:

$$d\left[\int_{\mathbb{R}^m} (2\pi)^{-m/2}\phi(m_t + \Sigma_t^{1/2}\xi) v_t e^{-|\xi|^2/2} d\xi \right]$$

$$= \int_{\mathbb{R}^m} (2\pi)^{-m/2} d[\phi(m_t + \Sigma_t^{1/2}\xi) v_t] e^{-|\xi|^2/2} d\xi,$$

and using the product rule,

$$d(\phi v_t) = \phi dv_t + v_t d\phi + d\langle \phi, v \rangle_t$$

$$d\phi(m_t + \Sigma_t^{1/2}\xi) = \frac{\partial\phi'}{\partial x}(dm_t + d\Sigma_t^{1/2}\xi) + \frac{1}{2}\text{Tr}[\frac{\partial^2\phi}{\partial x^2} d\langle m, m\rangle_t].$$

From (5.8.9), (5.8.10) and (5.8.11),

$$dv_t = v_t m_s' C_s'(D_s^{-1})' D_s^{-1} dy_s,$$

$$d\Sigma_t = (\Sigma_t A_t' + A_t \Sigma_t + B_t B_t' - \Sigma_t C_t'(D_t^{-1})' D_t^{-1} C_t \Sigma_t) dt,$$

$$dm_t = (A_t m_t - \Sigma_t C_t'(D_t^{-1})' D_t^{-1} C_t m_t) dt + \Sigma_t C_t'(D_t^{-1})' D_t^{-1} dy_t,$$

$$d\langle m, m\rangle_t = \Sigma_t C_t'(D_t^{-1})' D_t^{-1} C_t \Sigma_t dt,$$

$$d\langle \phi, v\rangle_t = v_t \frac{\partial\phi'}{\partial x} \Sigma_t C_t'(D_t^{-1})' D_t^{-1} C_t m_t dt.$$

Therefore

$$
d(\phi v_t) = v_t \left\{ \left(\frac{\partial \phi'}{\partial x} A_t m_t + \frac{\partial \phi'}{\partial x} d\Sigma_t^{1/2} \xi + \frac{1}{2} \text{Tr}[\frac{\partial^2 \phi}{\partial x^2} \Sigma_t C_t'(D_t^{-1})' D_t^{-1} C_t \Sigma_t] \right) dt \right.
$$
$$
\left. + [\phi m_s' C_s'(D_s^{-1})' D_s^{-1} + \frac{\partial \phi'}{\partial x} \Sigma_t C_t'(D_t^{-1})' D_t^{-1}] dy_s \right\},
$$

and

$$
d\sigma(\phi)_t = (2\pi)^{-m/2} v_t \int_{\mathbb{R}^m} \left\{ \left(\frac{\partial \phi'}{\partial x} A_t m_t + \frac{\partial \phi'}{\partial x} d\Sigma_t^{1/2} \xi \right. \right.
$$
$$
+ \frac{1}{2} \text{Tr}[\frac{\partial^2 \phi}{\partial x^2} \Sigma_t C_t'(D_t^{-1})' D_t^{-1} C_t \Sigma_t] \Big) dt
$$
$$
\left. + [\phi m_s' C_s'(D_s^{-1})' D_s^{-1} + \frac{\partial \phi'}{\partial x} \Sigma_t C_t'(D_t^{-1})' D_t^{-1}] dy_s \right\} e^{-|\xi|^2/2} d\xi.
$$

However, using integration by parts,

$$
\int_{\mathbb{R}^m} \frac{\partial \phi'}{\partial x} d\Sigma_t^{1/2} \xi e^{-|\xi|^2/2} d\xi = \frac{1}{2} \int_{\mathbb{R}^m} \text{Tr}[\frac{\partial^2 \phi}{\partial x^2} \frac{d\Sigma_t}{dt}] e^{-|\xi|^2/2} d\xi,
$$
$$
\int_{\mathbb{R}^m} \frac{\partial \phi'}{\partial x} \Sigma_t C_t'(D_t^{-1})' D_t^{-1} e^{-|\xi|^2/2} d\xi = \int_{\mathbb{R}^m} \phi \xi' \Sigma_t^{1/2} C_t'(D_t^{-1})' D_t^{-1} e^{-|\xi|^2/2} d\xi,
$$

$\xi' = x' \Sigma_t^{-1/2} - m_t' \Sigma_t^{-1/2}$. It follows that

$$
d\sigma(\phi)_t = (2\pi)^{-m/2} v_t \int_{\mathbb{R}^m} \left\{ \left(\frac{\partial \phi'}{\partial x} A_t m_t \right. \right.
$$
$$
\left. + \frac{1}{2} \text{Tr}\left[\frac{\partial^2 \phi}{\partial x^2} (\Sigma_t A_t' + A_t \Sigma_t)\right] + \frac{1}{2} \text{Tr}\left[\frac{\partial^2 \phi}{\partial x^2} B_t B_t'\right] \right) dt
$$
$$
\left. + \phi x' C_t'(D_t^{-1})' D_t^{-1} dy_s \right\} e^{-|\xi|^2/2} d\xi.
$$

Using integration by parts again,

$$
\int_{\mathbb{R}^m} \text{Tr}\left[\frac{\partial^2 \phi}{\partial x^2} \Sigma_t A_t'\right] e^{-|\xi|^2/2} d\xi = \int_{\mathbb{R}^m} \text{Tr}\left[A_t \Sigma_t \frac{\partial^2 \phi}{\partial x^2}\right] e^{-|\xi|^2/2} d\xi
$$
$$
= \int_{\mathbb{R}^m} \frac{\partial \phi'}{\partial x} A_t \Sigma_t^{1/2} \xi e^{-|\xi|^2/2} d\xi
$$
$$
= \int_{\mathbb{R}^m} \frac{\partial \phi'}{\partial x} A_t x e^{-|\xi|^2/2} d\xi - \int_{\mathbb{R}^m} \frac{\partial \phi'}{\partial x} A_t m_t e^{-|\xi|^2/2} d\xi,
$$
$$
\int_{\mathbb{R}^m} \text{Tr}\left[\frac{\partial^2 \phi}{\partial x^2} A_t \Sigma_t\right] e^{-|\xi|^2/2} d\xi = \int_{\mathbb{R}^m} \frac{\partial \phi'}{\partial x} A_t \Sigma_t^{1/2} \xi e^{-|\xi|^2/2} d\xi
$$
$$
= \int_{\mathbb{R}^m} \frac{\partial \phi'}{\partial x} A_t x e^{-|\xi|^2/2} d\xi - \int_{\mathbb{R}^m} \frac{\partial \phi'}{\partial x} A_t m_t e^{-|\xi|^2/2} d\xi.
$$

Hence

$$
d\sigma(\phi)_t = (2\pi)^{-m/2} v_t \int_{\mathbb{R}^m} \left\{ \left(\frac{\partial \phi}{\partial x}' A_t x + \frac{1}{2} \mathrm{Tr} \left[\frac{\partial^2 \phi}{\partial x^2} B_t B_t' \right] \right) dt \right.
$$

$$
\left. + \phi x' C_t' (D_t^{-1})' D_t^{-1} dy_s \right\} e^{-|\xi|^2/2} d\xi
$$

$$
= \sigma((\nabla \phi(x))' A_t x) ds + \frac{1}{2} \sigma(\mathrm{Tr}[\nabla^2 \phi(x)' B_t B_t') dt
$$

$$
+ \sigma(\phi(x) x') C_t' (D_t^{-1})' D_t^{-1} dy_t,
$$

which is the desired result.   ∎

5.9 Continuous-time model parameters estimation

The linear model given by (5.8.1) and (5.8.2) is determined by the matrices A, B, C and D which need to be known. These parameters are estimated using the expectation maximization (EM) algorithm which we describe here.

Maximum likelihood estimation of the parameters via the EM algorithm requires computation of the filtered estimates of quantities such as

$$
\int_0^t x_s \otimes dx_s, \quad \int_0^t x_s \otimes dy_s, \quad \int_0^t x_s \otimes x_s ds.
$$

Remark 5.9.1 In all the existing literature on parameter estimation of linear Gaussian models via the EM algorithm, filtered estimates of the above quantities are computed via Kalman smoothing, which requires large memory numerical implementation. This problem is solved in [14] by providing *finite-dimensional filters for (the components of) such integral processes*. It is further shown that finite-dimensional filters exist for integrals and stochastic integrals of moments of all orders of the state process etc.   □

Consider the time-invariant version of (5.8.1), (5.8.2):

$$
dx_t = Ax_t dt + Bdw_t, \quad x_0 \in \mathbb{R}^m,
$$

$$
dy_t = Cx_t dt + Ddv_t, \quad y_0 = 0 \in \mathbb{R}^n.
$$

The aim is to compute ML estimates of the parameters $\theta = (A, C)$ given the observations $\mathcal{Y}_t = \sigma\{y_s : s \le t\}$ and assuming B, D are known. This is done via the EM algorithm.

Remark 5.9.2 Unlike the discrete-time case, in continuous time, it is not possible to obtain ML estimates of the variance terms B and D because measures corresponding to Wiener processes with different variances are not absolutely continuous (see Chapter 6.1 in [15]). Estimates for B and D are given in terms of the quadratic variations of the state and observation processes.   □

Notation

Let $e_i, e_j \in \mathbb{R}^m$ denote unit vectors with 1 in the i-th and j-th positions, respectively. Write

$$T_t^{ij} = \int_0^t \langle x_s, e_i \rangle \langle e_j, x_s \rangle ds = \int_0^t x_s'(e_i e_j')x_s ds, \tag{5.9.1}$$

$$L_t^{ij} = \int_0^t \langle x_s, e_i \rangle \langle e_j, dx_s \rangle = \int_0^t x_s'(e_i e_j')dx_s; \tag{5.9.2}$$

here $\langle ., . \rangle$ denotes the scalar product.

Also let $f_j \in \mathbb{R}^n$ denote the unit vector with 1 in the j-th position. Write

$$U_t^{ij} = \int_0^t \langle x_s, e_i \rangle \langle f_j, dy_s \rangle = \int_0^t x_s'(e_i f_j')dy_s. \tag{5.9.3}$$

Now the expression for $Q(\theta, \tilde{\theta})$ is derived.

To update the estimates from $\tilde{A}$ to A, introduce the density

$$\left. \frac{dP_\theta}{dP_{\tilde{\theta}}} \right|_{\mathcal{G}_t} = \exp\left\{ \int_0^t x_s'(A' - \tilde{A}')(BB')^{\#}dx_s \right.$$
$$\left. - \frac{1}{2} \int_0^t x_s'(A' - \tilde{A}')(BB')^{\#}(A' - \tilde{A}')x_s ds \right\},$$

where # denotes the pseudo inverse. Then

$$E\left[\log \frac{dP(A)}{dP(\tilde{A})} \mid \mathcal{Y}_t \right] = E\left[\int_0^t x_s' A'(BB')^{\#}dx_s \right.$$
$$\left. - \frac{1}{2} \int_0^t x_s' A(BB')^{\#} Ax_s ds \mid \mathcal{Y}_t \right] + R(\tilde{A}), \tag{5.9.4}$$

where $R(\tilde{A})$ does not involve A.

Similarly, to update the estimates from $\tilde{C}$ to C, introduce the density

$$\left. \frac{dP(C)}{dP(\tilde{C})} \right|_{\mathcal{G}_t} = \exp\left\{ \int_0^t x_s'(C' - \tilde{C}')(DD')^{-1}dx_s \right.$$
$$\left. - \frac{1}{2} \int_0^t x_s'(C' - \tilde{C}')(DD')^{-1}(C' + \tilde{C})x_s ds \right\},$$

Consequently,

$$E\left[\log \frac{dP(C)}{dP(\tilde{C})} \mid \mathcal{Y}_t \right] = E\left[\int_0^t x_s'C'(DD')^{-1}dy_s \right.$$
$$\left. - \frac{1}{2} \int_0^t x_s'C'(DD')^{-1}Cx_s ds \mid \mathcal{Y}_t \right] + S(\tilde{C}), \tag{5.9.5}$$

where $R(\tilde{C})$ does not involve C.

Adding (5.9.4) and (5.9.5) yields

$$
\mathcal{Q}(\theta, \tilde{\theta}) = E\left[\int_0^t x_s' A'(BB')^{\#}\mathrm{d}x_s - \frac{1}{2}\int_0^t x_s' A(BB')^{\#} Ax_s \mathrm{d}s \mid \mathcal{Y}_t\right]
$$
$$
+ E\left[\int_0^t x_s' C'(DD')^{-1}\mathrm{d}y_s - \frac{1}{2}\int_0^t x_s' C'(DD')^{-1} Cx_s \mathrm{d}s \mid \mathcal{Y}_t\right]
$$
$$
+ E[R(\tilde{\theta}) \mid \mathcal{Y}_t].
$$

To implement the M-step set the derivatives $\dfrac{\partial \mathcal{Q}}{\partial \theta} = 0$. This yields

$$
A = E\left[\int_0^t \mathrm{d}x_s \otimes x_s \mid \mathcal{Y}_t\right]\left(E\left[\int_0^t x_s \otimes x_s \mathrm{d}s \mid \mathcal{Y}_t\right]\right)^{-1}
$$
$$
= \hat{L}_t' \hat{T}_t^{-1},
$$
$$
C = E\left[\int_0^t \mathrm{d}y_s \otimes x_s \mid \mathcal{Y}_t\right]\left(E\left[\int_0^t x_s \otimes x_s \mathrm{d}s \mid \mathcal{Y}_t\right]\right)^{-1}
$$
$$
= \hat{J}_t' \hat{H}_t^{-1},
$$

where $\hat{T}_t$ and $\hat{L}_t \in \mathbb{R}^{m \times m}$ denote matrices with elements $\hat{T}_t^{ij} \triangleq E[T_t^{ij} \mid \mathcal{Y}_t]$ and $\hat{L}_t^{ij} \triangleq E[L_t^{ij} \mid \mathcal{Y}_t]$, $i, j \in \{1, \ldots, m\}$. Also $\hat{U}_t \in \mathbb{R}^{m \times n}$ denotes the matrix with elements $\hat{U}_t^{ij} \triangleq E[U_t^{ij} \mid \mathcal{Y}_t]$, $i \in \{1, \ldots, m\}$, $j \in \{1, \ldots, n\}$.

Remark 5.9.3 The terms $\hat{T}_t^{ij}$, $\hat{L}_t^{ij}$ and $\hat{U}_t^{ij}$ are computed in terms of finite-dimensional filters in Theorems 3.2, 3.8 and 3.5 in [14], thus obtaining a filter-based EM algorithm. □

Definition 5.9.4 For any "test" function $g : \mathbb{R}^m \to \mathbb{R}$, define a measure-valued process $E[\Lambda_t T_t^{ij} g(x_t) \mid \mathcal{Y}_t]$. This has a density $\beta_t^{ij}(x)$ so that

$$
E[\Lambda_t T_t^{ij} g(x_t) \mid \mathcal{Y}_t] = \int_{\mathbb{R}^m} \beta_t^{ij}(x) g(x_t)\mathrm{d}x.
$$

The existence of the density $\beta_t^{ij}(x)$ follows from the existence and uniqueness of solutions of stochastic partial differential equations. This is established in, for instance, Section 4.2 of [2].

The following theorem (Theorem 3.2 in [14]) shows the surprising result that we can describe the measure $\beta_t^{ij}(x)$ exactly as a quadratic in x multiplying the $q(x, t)$ of (5.8.6).

Theorem 5.9.5 At time t, the density $\beta_t^{ij}(x)$ is completely described by the five statistics $\bar{a}_t^{ij}$, $\bar{b}_t^{ij}$, $\bar{c}_t^{ij}$, Σ_t, and m_t as follows:

$$
\beta_t^{ij}(x) = (\bar{a}_t^{ij} + x'\bar{b}_t^{ij} + x'\bar{c}_t^{ij}x)q(x, t). \tag{5.9.6}
$$

Here $\bar{a}_t^{ij} \in \mathbb{R}$, $\bar{b}_t^{ij} \in \mathbb{R}^m$, and $\bar{c}_t^{ij} \in L_s(\mathbb{R}^m, \mathbb{R}^m)$, the space of symmetric $m \times m$ matrices.

Further,

$$\frac{d\bar{a}_t^{ij}}{dt} = \mathrm{Tr}\left(\bar{c}_t^{ij} B_t B_t'\right) + \bar{b}_t^{ij'} B_t B_t' \Sigma_t^{-1} m_t, \quad \bar{a}_0^{ij} = 0 \in \mathbb{R}, \tag{5.9.7}$$

$$\frac{d\bar{b}_t^{ij}}{dt} = -\left(A_t' + \Sigma_t^{-1} B_t B_t'\right) \bar{b}_t^{ij} + 2\bar{c}_t^{ij} B_t B_t' \Sigma_t^{-1} m_t, \quad \bar{b}_0^{ij} = 0 \in \mathbb{R}^m, \tag{5.9.8}$$

$$\frac{d\bar{c}_t^{ij}}{dt} = -\left(A_t' + \Sigma_t^{-1} B_t B_t'\right) \bar{c}_t^{ij} - \bar{c}_t^{ij} \left(A_t + B_t B_t' \Sigma_t^{-1}\right)$$

$$+ \frac{1}{2}(e_j e_i' + e_i e_j'), \quad \bar{c}_0^{ij} = 0 \in L_s(\mathbb{R}^m, \mathbb{R}^m). \tag{5.9.9}$$

Proof Apply the Itô product rule to $dT_t^{ij} = x_t'(e_i e_j') x_t dt$ and

$$d\Lambda_t \phi(x_t) = \Lambda_t (\nabla \phi(x_t))' A_t x_t dt + \Lambda_t (\nabla \phi(x_t))' B_t dw_t$$

$$+ \frac{1}{2} \Lambda_t \mathrm{Tr}[\nabla^2 \phi(x_t)' B_t B_t'] dt$$

$$+ \Lambda_t \phi(x_t) x_t' C_t' (D_t^{-1})' D_t^{-1} dy_t, \tag{5.9.10}$$

to get

$$\Lambda_t \phi(x_t) T_t^{ij} = \int_0^t \Lambda_s T_s^{ij} (\nabla \phi(x_s))' A_s x_s ds$$

$$+ \int_0^t \Lambda_s T_s^{ij} (\nabla \phi(x_s))' B_s dw_s$$

$$+ \frac{1}{2} \int_0^t \Lambda_s T_s^{ij} \mathrm{Tr}[\nabla^2 \phi(x_s)' B_s B_s'] ds$$

$$+ \int_0^t \Lambda_s T_s^{ij} \phi(x_s) x_s' C_s' (D_s^{-1})' D_s^{-1} dy_s$$

$$+ \int_0^t \Lambda_s \phi(x_s) x_s' (e_i e_j') x_s ds. \tag{5.9.11}$$

Conditioning both sides of (5.9.11) on $\mathcal{Y}_t$ under the "ideal" world probability measure P and using Lemma 3.2 p. 261 of Hajek and Wong [15] gives

$$E[\Lambda_t \phi(x_t) T_t^{ij} \mid \mathcal{Y}_t] = \int_0^t E[\Lambda_s T_s^{ij} (\nabla \phi(x_s))' A_s x_s \mid \mathcal{Y}_s] ds$$

$$+ \frac{1}{2} \int_0^t E[\Lambda_s T_s^{ij} \mathrm{Tr}[\nabla^2 \phi(x_s)' B_s B_s'] \mid \mathcal{Y}_s] ds$$

$$+ \int_0^t E[\Lambda_s T_s^{ij} \phi(x_s) x_s' C_s' (D_s^{-1})' D_s^{-1} \mid \mathcal{Y}_s] dy_s$$

$$+ \int_0^t E[\Lambda_s \phi(x_s) x_s' (e_i e_j') x_s \mid \mathcal{Y}_s] ds.$$

In terms of the densities $\beta_t^{ij}(x)$ and $q(x, t)$, integrate by parts each term of

$$\int_{\mathbb{R}^m} \beta_t^{ij}(x)\phi(x)dx = \int_0^t \int_{\mathbb{R}^m} \beta_t^{ij}(x)(\nabla\phi(x))'A_sx)dxds$$

$$+ \frac{1}{2}\int_0^t \int_{\mathbb{R}^m} \beta_t^{ij}(x)\text{Tr}[\nabla^2\phi(x)'B_s B_s']dxds$$

$$+ \int_0^t \int_{\mathbb{R}^m} \beta_t^{ij}(x)\phi(x)x'C_s'(D_s^{-1})'D_s^{-1}dxdy_s$$

$$+ \int_0^t q(x, t)\phi(x_s)x_s'(e_i e_j')x_s ds.$$

that is, $\beta_t^{ij}(x)$ must satisfy the stochastic partial differential equation

$$\beta_t^{ij}(x) = - \int_0^t \text{div}(A_s x\beta_s^{ij}(x))ds$$

$$+ \frac{1}{2}\int_0^t \text{Tr}[\nabla^2\beta_s^{ij}(x)B_s B_s']ds$$

$$+ \int_0^t \beta_s^{ij}(x)x'C_s'(D_s^{-1})'D_s^{-1}dy_s$$

$$+ \int_0^t q(x, s)x_s'(e_i e_j')x_s ds. \tag{5.9.12}$$

We look for a solution of (5.9.12) of the form

$$\bar{\beta}_s^{ij}(x) = (\bar{a}_s^{ij} + x'\bar{b}_s^{ij} + x'\bar{c}_s^{ij} x)q(x, s). \tag{5.9.13}$$

As noted just after Definition 5.9.4, if such solution exists, it is unique.

To simplify notation drop superscripts i, j on $\bar{a}$, $\bar{b}$ and $\bar{c}$,

$$\text{div}(\bar{\beta}_s(x)A_sx) = \text{div}((\bar{a}_s + x'\bar{b}_s + x'\bar{c}_sx)q(x, s)A_sx)$$

$$= (\bar{b}_s + 2\bar{c}_s)'A_sxq(x, s) + (\bar{a}_s + \bar{b}_sx + x'\bar{c}_sx)\text{div}(A_sxq(x, s))$$

$$\nabla\bar{\beta}_s(x) = \nabla((\bar{a}_s + x'\bar{b}_s + x'\bar{c}_sx)q(x, s))$$

$$= (\bar{b}_s + 2\bar{c}_sx)q(x, s) + (\bar{a}_s + x'\bar{b}_s + x'\bar{c}_sx)\nabla q(x, s)$$

$$\nabla^2\bar{\beta}_s(x) = 2\bar{c}_sxq(x, s) + 2(\bar{b}_s + 2\bar{c}_sx)(\nabla q(x, s))'$$

$$+ (\bar{a}_s + x'\bar{b}_s + x'\bar{c}_sx)\nabla^2 q(x, s)$$

$$\text{Tr}[\nabla^2\beta_s^{ij}(x)B_s B_s'] = 2q(x, s)\text{Tr}(\bar{c}_s B_s B_s') + 2(\bar{b}_s + 2\bar{c}_sx)B_s B_s'\nabla q(x, s)$$

$$+ (\bar{a}_s + x'\bar{b}_s + x'\bar{c}_sx)\text{Tr}[\nabla^2 q(x, s)B_s B_s'].$$

Now from (5.8.8),

$$\nabla q(x, s) = -\Sigma_s^{-1}(x - m_s)q(x, s).$$

Consequently, substitution of $\bar{\beta}_t(x)$, given by (5.9.13) in the differential form of the right hand side of (5.9.12), yields

$$
\begin{aligned}
&- ((\bar{b}_s + 2\bar{c}_s)x)' A_s x q(x, s) ds - (\bar{a}_s + \bar{b}_s x + x'\bar{c}_s x) \mathrm{div}(A_s x q(x, s)) ds \\
&+ q(x, s) \mathrm{Tr}(\bar{c}_s B_s B_s') ds + (\bar{b}_s + 2\bar{c}_s x) B_s B_s' \Sigma_s^{-1} (x - m_s) q(x, s) ds \\
&+ \frac{1}{2} (\bar{a}_s + x'\bar{b}_s + x'\bar{c}_s x) \mathrm{Tr}[\nabla^2 q(x, s) B_s B_s'] ds \\
&+ (\bar{a}_s + x'\bar{b}_s + x'\bar{c}_s x) q(x, s) x' C_s' (D_s^{-1})' D_s^{-1} dy_s \\
&+ q(x, s) x_s'(e_i e_j') x_s ds.
\end{aligned}
\tag{5.9.14}
$$

Also,

$$
d\bar{\beta}s(x) = (d\bar{a}_s + d\bar{b}_s' x + x'd\bar{c}_s x) q(x, s) + (\bar{a}_s^{ij} + x'\bar{b}_s + x'\bar{c}_s x) dq(x, s).
\tag{5.9.15}
$$

Consequently, $\bar{\beta}_t(x)$, given by (5.9.13), is a solution of (5.9.12) if (5.9.14) equals (5.9.15). However, $q(x, s)$ solves the equation (5.8.7), so

$$
\begin{aligned}
dq(x, s) = &- \mathrm{div}(A_s x q(x, s)) ds \\
&+ \frac{1}{2} \mathrm{Tr}[\nabla^2 q(x, s) B_s B_s'] ds \\
&+ q(x, s) x' C_s' (D_s^{-1})' D_s^{-1} dy_s.
\end{aligned}
$$

Therefore, substituting the above expression for $dq(x, s)$ into (5.9.15) yields

$$
\begin{aligned}
d\bar{\beta}s(x) = &(d\bar{a}_s + d\bar{b}_s' x + x'd\bar{c}_s x) q(x, s) \\
&- (\bar{a}_s^{ij} + x'\bar{b}_s + x'\bar{c}_s x) \mathrm{div}(A_s x q(x, s)) ds \\
&+ \frac{1}{2} (\bar{a}_s^{ij} + x'\bar{b}_s + x'\bar{c}_s x) \mathrm{Tr}[\nabla^2 q(x, s) B_s B_s'] ds \\
&+ (\bar{a}_s^{ij} + x'\bar{b}_s + x'\bar{c}_s x) x' C_s' (D_s^{-1})' D_s^{-1} q(x, s) dy_s.
\end{aligned}
\tag{5.9.16}
$$

Finally, equating the coefficients of x, x' and the constants in (5.9.14) and (5.9.16), it is seen the result holds if (5.9.7), (5.9.8) and (5.9.9) hold. ∎

A solution of the ordinary differential equations (5.9.8) and (5.9.9) is now obtained. Write G_t for the matrix solution of

$$
\frac{dG_t}{dt} = -(A_t' + \Sigma_t^{-1} B_t B_t') G_t, \quad G_0 = I_{m \times m}.
\tag{5.9.17}
$$

Note that G_t is deterministic and can be calculated off-line. Also, as an exponential matrix, G_t has an inverse G_t^{-1}.

Lemma 5.9.6 *The explicit solutions of (5.9.8) and (5.9.9) are*

$$
\bar{b}_t^{ij} = 2G_t \left(\int_0^t G_s^{-1} \bar{c}_s^{ij} B_s B_s' \Sigma_s^{-1} m_s ds \right),
$$

$$
\bar{c}_t^{ij} = \frac{1}{2} G_t \left(\int_0^t G_s^{-1} (e_j e_i' + e_i e_j')(G_s')^{-1} ds \right) G_t'.
$$

Proof   The above equations follow using variation of constants.   ∎

Remark 5.9.7 We proceed similarly with the processes U_t^{ij} and L_t^{ij} leaving the details as exercises.   □

Definition 5.9.8 *For any "test" function* $g : \mathbb{R}^m \to \mathbb{R}$, *define a measure-valued process* $E[\Lambda_t U_t^{ij} g(x_t) \mid \mathcal{Y}_t]$. *This has a density* $\gamma_t^{ij}(x)$ *so that*

$$E[\Lambda_t U_t^{ij} g(x_t) \mid \mathcal{Y}_t] = \int_{\mathbb{R}^m} \gamma_t^{ij}(x) g(x_t) \mathrm{d}x.$$

The existence of the density $\gamma_t^{ij}(x)$ follows from the existence and uniqueness of solutions of stochastic partial differential equations (see Section 4.2 of [2]).

The following theorem (Theorem 3.5 in [14]) shows that we can describe the measure $\gamma_t^{ij}(x)$ exactly as a quadratic in x multiplying the $q(x, t)$ of (5.8.7).

Theorem 5.9.9 *At time t, the density* $\gamma_t^{ij}(x)$ *is completely described by the five statistics* $\breve{a}_t^{ij}$, $\breve{b}_t^{ij}$, $\breve{c}_t^{ij}$, Σ_t, *and* m_t *as follows:*

$$\gamma_t^{ij}(x) = (\breve{a}_t^{ij} + x'\breve{b}_t^{ij} + x'\breve{c}_t^{ij}x)q(x, t).$$

Here $\breve{a}_t^{ij} \in \mathbb{R}$, $\breve{b}_t^{ij} \in \mathbb{R}^m$, *and* $\breve{c}_t^{ij} \in L_s(\mathbb{R}^m, \mathbb{R}^m)$, *the space of symmetric* $m \times m$ *matrices. Further,*

$$\frac{\mathrm{d}\breve{a}_t^{ij}}{\mathrm{d}t} = \mathrm{Tr}\left(\breve{c}_t^{ij} B_t B_t'\right) + \breve{b}_t^{ij'} B_t B_t' \Sigma_t^{-1} m_t, \quad \breve{a}_0^{ij} = 0 \in \mathbb{R}, \tag{5.9.18}$$

$$\mathrm{d}\breve{b}_t^{ij} = [-\left(A_t' + \Sigma_t^{-1} B_t B_t'\right)\breve{b}_t^{ij} + 2\breve{c}_t^{ij} B_t B_t' \Sigma_t^{-1} m_t]\mathrm{d}t + \mathrm{d}y_t' f_j e_i,$$

$$\breve{b}_0^{ij} = 0 \in \mathbb{R}^m, \tag{5.9.19}$$

$$\frac{\mathrm{d}\breve{c}_t^{ij}}{\mathrm{d}t} = -\left(A_t' + \Sigma_t^{-1} B_t B_t'\right)\breve{c}_t^{ij} - \breve{c}_t^{ij}\left(A_t + B_t B_t' \Sigma_t^{-1}\right)$$

$$+ \frac{1}{2}(e_i f_j' C_t + C_t' f_j e_i'), \quad \breve{c}_0^{ij} = 0 \in L_s(\mathbb{R}^m, \mathbb{R}^m). \tag{5.9.20}$$

Proof   Apply the Itô product rule to $\mathrm{d}U_t^{ij} = x_t'(e_i f_j')\mathrm{d}y_s$ and

$$\mathrm{d}\Lambda_t \phi(x_t) = \Lambda_t(\nabla\phi(x_t))' A_t x_t \mathrm{d}t + \Lambda_t(\nabla\phi(x_t))' B_t \mathrm{d}w_t$$

$$+ \frac{1}{2}\Lambda_t \mathrm{Tr}[\nabla^2\phi(x_t)' B_t B_t']\mathrm{d}t$$

$$+ \Lambda_t \phi(x_t) x_t' C_t' (D_t^{-1})' D_t^{-1} \mathrm{d}y_t.$$

and condition both sides on $\mathcal{Y}_t$ under the "ideal" world probability measure P using Lemma 3.2, p. 261 of [15].

In terms of the densities $\gamma_t^{ij}(x)$ and $q(x, t)$, integration by parts yields

$$\gamma_t^{ij}(x) = -\int_0^t \text{div}(A_s x \gamma_s^{ij}(x))ds + \frac{1}{2}\int_0^t \text{Tr}[\nabla^2 \gamma_s^{ij}(x)B_s B_s']ds$$

$$+ \int_0^t \gamma_s^{ij}(x)x'C_s'(D_s^{-1})'D_s^{-1}dy_s + \int_0^t q(x, s)x_s'(e_i f_j')dy_s$$

$$+ \int_0^t q(x, s)(x'C_s' f_j e_i' x)ds. \tag{5.9.21}$$

We look for a solution of (5.9.21) of the form

$$\breve\gamma_s^{ij}(x) = (\breve a_s^{ij} + x'\breve b_s^{ij} + x'\breve c_s^{ij}x)q(x, s). \tag{5.9.22}$$

As noted just after Definition 5.9.4, if such solution exists, it is unique.

Recall that $q(x, s)$ solves the equation (5.8.7):

$$dq(x, s) = -\text{div}(A_s x q(x, s))ds$$

$$+ \frac{1}{2}\text{Tr}[\nabla^2 q(x, s)B_s B_s']ds$$

$$+ q(x, s)x'C_s'(D_s^{-1})'D_s^{-1}dy_s.$$

So (5.9.22) is a solution of (5.9.21) if (5.9.18), (5.9.19) and (5.9.20) hold.　∎

Now the ordinary differential equations (5.9.19) and (5.9.20) are solved explicitly. Note that $f_j' dy_t = dy_t' f_j = dy_t^j$, where y_t^j denotes the j-th component of y_t.

Lemma 5.9.10 *The explicit solutions of (5.9.19) and (5.9.20) are*

$$\breve b_t^{ij} = 2G_t \left[\int_0^t G_s^{-1}\breve c_s^{ij} B_s B_s' \Sigma_s^{-1} m_s ds + \int_0^t G_s e_i dy_s\right],$$

$$\breve c_t^{ij} = \frac{1}{2}G_t \left[\int_0^t G_s^{-1}(e_i f_j' C_s + C_s' f_j e_i')(G_s')^{-1}ds\right]G_t'.$$

Definition 5.9.11 *For any "test" function $g : \mathbb{R}^m \to \mathbb{R}$, define a measure-valued process $E[\Lambda_t U_t^{ij} g(x_t) \mid \mathcal{Y}_t]$. This has a density $\lambda_t^{ij}(x)$, so that*

$$E[\Lambda_t U_t^{ij} g(x_t) \mid \mathcal{Y}_t] = \int_{\mathbb{R}^m} \lambda_t^{ij}(x)g(x_t)dx.$$

The existence of the density $\lambda_t^{ij}(x)$ follows from the existence and uniqueness of solutions of stochastic partial differential equations (see Section 4.2 of [2]).

The following theorem (Theorem 3.8 in [14]) shows that one can describe the measure $\lambda_t^{ij}(x)$ exactly as a quadratic in x multiplying the $q(x, t)$ of (5.8.7).

Theorem 5.9.12 *At time t, the density $\lambda_t^{ij}(x)$ is completely described by the five statistics $\tilde a_t^{ij}$, $\tilde b_t^{ij}$, $\tilde c_t^{ij}$, Σ_t, and m_t as follows:*

$$\lambda_t^{ij}(x) = (\tilde a_t^{ij} + x'\tilde b_t^{ij} + x'\tilde c_t^{ij}x)q(x, t).$$

Here $\tilde{a}_t^{ij} \in \mathbb{R}$, $\tilde{b}_t^{ij} \in \mathbb{R}^m$, and $\tilde{c}_t^{ij} \in L_s(\mathbb{R}^m, \mathbb{R}^m)$, the space of symmetric $m \times m$ matrices. Further,

$$\frac{d\tilde{a}_t^{ij}}{dt} = \mathrm{Tr}\left(\tilde{c}_t^{ij} B_t B_t'\right) + \tilde{b}_t^{ij'} B_t B_t' \Sigma_t^{-1} m_t - \mathrm{Tr}\left(B_t B_t' e_i e_j'\right),$$

$$\tilde{a}_0^{ij} = 0 \in \mathbb{R}, \tag{5.9.23}$$

$$\frac{d\tilde{b}_t^{ij}}{dt} = -\left(A_t' + \Sigma_t^{-1} B_t B_t'\right) \tilde{b}_t^{ij} + 2\tilde{c}_t^{ij} B_t B_t' \Sigma_t^{-1} m_t - (e_j e_i') B_t B_t' \Sigma_t^{-1} m_t,$$

$$\tilde{b}_0^{ij} = 0 \in \mathbb{R}^m, \tag{5.9.24}$$

$$\frac{d\tilde{c}_t^{ij}}{dt} = -\left(A_t' + \Sigma_t^{-1} B_t B_t'\right) \tilde{c}_t^{ij} - \tilde{c}_t^{ij}\left(A_t + B_t B_t' \Sigma_t^{-1}\right)$$

$$+ \frac{1}{2}(e_i e_j'(A_t + B_t B_t' \Sigma_t^{-1}) + \left(A_t' + \Sigma_t^{-1} B_t B_t'\right) e_j e_i'),$$

$$\tilde{c}_0^{ij} = 0 \in L_s(\mathbb{R}^m, \mathbb{R}^m). \tag{5.9.25}$$

Proof Apply the Itô product rule to $\Lambda_t U_t^{ij} \phi(x_t)$ and condition on $\mathcal{Y}_t$ under the "ideal" world probability measure P using Lemma 3.2, p. 261 of [15].

In terms of the densities $\lambda_t^{ij}(x)$ and $q(x, t)$, integrate by parts to get

$$\lambda_t^{ij}(x) = -\int_0^t \mathrm{div}(A_s x \lambda_s^{ij}(x)) ds + \frac{1}{2}\int_0^t \mathrm{Tr}[\nabla^2 \lambda_s^{ij}(x) B_s B_s'] ds$$

$$+ \int_0^t \lambda_s^{ij}(x) x' C_s'(D_s^{-1})' D_s^{-1} dy_s + \int_0^t q(x, s) x'(e_i e_j') A_s x ds$$

$$- \int_0^t \mathrm{div}(x'(e_i e_j') B_t B_t' q(x, s)) ds. \tag{5.9.26}$$

We look for a solution of (5.9.26) of the form

$$\bar{\gamma}_s^{ij}(x) = (\tilde{a}_s^{ij} + x' \tilde{b}_s^{ij} + x' \tilde{c}_s^{ij} x) q(x, s). \tag{5.9.27}$$

Recall that $q(x, s)$ solves the equation (5.8.7). So (5.9.27) is a solution of (5.9.26) if (5.9.23), (5.9.24) and (5.9.25) hold. ∎

Now the ordinary differential equations (5.9.24) and (5.9.25) are solved explicitly.

Lemma 5.9.13 *The explicit solutions of (5.9.24) and (5.9.25) are*

$$\tilde{b}_t^{ij} = G_t\left[\int_0^t G_s^{-1}(2\tilde{c}_s^{ij} - e_j e_i')\Sigma_s^{-1} B_s B_s' m_s ds\right],$$

$$\tilde{c}_t^{ij} = \frac{1}{2}G_t\left[\int_0^t [G_s^{-1}(e_i e_j'(A_t + B_t B_t' \Sigma_t^{-1})\right.$$

$$\left. + \left(A_t' + \Sigma_t^{-1} B_t B_t'\right) e_j e_i')(G_s')^{-1}] ds\right] G_t'. \tag{5.9.28}$$

Remark 5.9.14 Note that from the definition of G_t, (5.9.17), that the integrand in (5.9.28) includes only half of the four terms in the derivative of $G_t^{-1}(e_i e_j' + e_j e_i')(G_t')^{-1}$), and so the integral cannot be evaluated in closed form.   □

Theorem 5.9.15 *(Theorem 3.10 in [14]) Finite-dimensional filters for T_t^{ij}, U_t^{ij}, and L_t^{ij}, defined in (5.9.1), (5.9.3), and (5.9.2), are given by*

$$E[T_t^{ij} \mid \mathcal{Y}_t] = \bar{a}_t^{ij} + m_t' \bar{a}_t^{ij} + \sum_{p,q=1}^{m} \bar{c}_t^{ij}(p,q)\Sigma_t(p,q) + m_t' \bar{c}_t^{ij} m_t,$$

$$E[U_t^{ij} \mid \mathcal{Y}_t] = \breve{a}_t^{ij} + m_t' \breve{a}_t^{ij} + \sum_{p,q=1}^{m} \breve{c}_t^{ij}(p,q)\Sigma_t(p,q) + m_t' \breve{c}_t^{ij} m_t, \quad (5.9.29)$$

$$E[L_t^{ij} \mid \mathcal{Y}_t] = \tilde{a}_t^{ij} + m_t' \tilde{a}_t^{ij} + \sum_{p,q=1}^{m} \tilde{c}_t^{ij}(p,q)\Sigma_t(p,q) + m_t' \tilde{c}_t^{ij} m_t. \quad (5.9.30)$$

Proof   Recall from (5.8.8) that $q(x,t)$ is an unnormalized Gaussian density with mean m_t and variance Σ_t. Therefore,

$$\int_{\mathbb{R}^m} q(x,t)\mathrm{d}x = v_t.$$

Note that for $u \in \mathbb{R}^m$,

$$\int_{\mathbb{R}^m} u'x q(x,t)\mathrm{d}x = (u'm_t)v_t.$$

Also, for any matrix

$$M \in L(\mathbb{R}^m, \mathbb{R}^m) \text{ with entries } M(p,q), 1 \le p,q \le m,$$

$$\int_{\mathbb{R}^m} x'Mx q(x,t)\mathrm{d}x = \int_{\mathbb{R}^m} (x - m_t)'M(x - m_t)q(x,t)\mathrm{d}x$$

$$+ m_t'Mm_t \int_{\mathbb{R}^m} q(x,t)\mathrm{d}x$$

$$= \left(\sum_{p,q=1}^{m} M(p,q)\Sigma_t(p,q) + m_t'Mm_t \right) v_t.$$

Now from Bayes' Theorem (4.1.1), we have

$$\overline{E}[T_t^{ij} \mid \mathcal{Y}_t] = \frac{E[\Lambda_t T_t^{ij} \mid \mathcal{Y}_t]}{E[\Lambda_t \mid \mathcal{Y}_t]} = \frac{\int_{\mathbb{R}^m} \beta_t^{ij}(x)\mathrm{d}x}{\int_{\mathbb{R}^m} q(x,t)\mathrm{d}x}$$

$$= \bar{a}_t^{ij} + m_t' \bar{a}_t^{ij} + \sum_{p,q=1}^{m} \bar{c}_t^{ij}(p,q)\Sigma_t(p,q) + m_t' \bar{c}_t^{ij} m_t,$$

by (5.9.6) and because the factors v_t cancel. The proof of equations (5.9.29) and (5.9.30) are similar.   ∎

Estimation of B and D

First consider the tensor product of x_t with itself:

$$x_t \otimes x_t = x_0 \otimes x_0 + \int_0^t x_s \otimes dx_s + \int_0^t dx_s \otimes x_s + \int_0^t BB'ds. \qquad (5.9.31)$$

Conditioning both sides of (5.9.31) on $\mathcal{Y}_t$, we have

$$\Sigma_t = \overline{E}[x_0 \otimes x_0 \mid \mathcal{Y}_t] + \overline{E}\left[\int_0^t x_s \otimes dx_s + \int_0^t dx_s \otimes x_s \mid \mathcal{Y}_t\right] + \int_0^t BB'ds. \qquad (5.9.32)$$

$\overline{E}[x_0 \otimes x_0 \mid \mathcal{Y}_t]$ in (5.9.32) is the smoothed second moment and is given in terms of finite-dimensional statistics; see Theorem 12.11, section 12.4 in [25]. The components of the conditioned stochastic integral in (5.9.32) are given by the filtered estimates of L_t^{ij}. Consequently, we have a procedure for estimating the matrix BB'.

Similarly, consider the tensor product of y_t with itself:

$$y_t \otimes y_t = \int_0^t y_s \otimes dy_s + \int_0^t dy_s \otimes y_s + DD't.$$

This expression simply amounts to evaluating DD' in terms of the quadratic variation of y.

5.10 Direct parameter estimation

In the previous sections maximum likelihood arguments were used to estimate recursively the parameters for the linear model (5.8.1), (5.8.2). Here a direct approach to the estimation problem as well as rates of convergence are discussed.

The following theorem ([12]) is a continuous-time version of Kronecker's Lemma (see for example [26] or [29] for the discrete-time case). This result is applied to discuss rates of convergence of the estimates.

Suppose $(\Omega, \mathcal{F}, \mathcal{F}_t, P)$, $t \geq 0$ is a stochastic basis and M is a continuous locally square integrable martingale. Further, u_t is a positive nondecreasing predictable process such that

$$u_t > c > 0 \quad \text{a.s.}$$

Write $z_t \overset{\Delta}{=} \int_{t_0}^t u_r^{-1} dM_r$, for $0 \leq t_0 \leq t$.

Theorem 5.10.1 *Suppose*

$$\lim_{t \to \infty} z_t(\omega) = \xi(\omega) < \infty \quad a.s.$$

Then $\lim_{t \to \infty} \left(\dfrac{1}{u_t}\right)(M_t - M_{t_0})$ *exists a.s.*
If $\lim_{t \to \infty} u_t(\omega) = +\infty$, *this limit is* 0.

Proof   For any s, $t_0 < s < t$, because u is nondecreasing,

$$M_t - M_s = \int_s^t u_r dz_r = \int_s^t u_r d(z_r - z_s)$$

$$= u_t(z_t - z_s) - \int_s^t (z_r - z_s) du_r \quad \text{a.s.}$$

Consequently,

$$|M_t - M_s| \le 2u_t \sup_{r \ge s} |z_r - z_s|. \tag{5.10.1}$$

Suppose that $\lim_{t \to \infty} u_t(\omega) = u(\omega) < \infty$. Then

$|M_t - M_s| \le 2u \sup_{r \ge s} |z_r - z_s|$.

From the hypothesis that $\lim_{t \to \infty} z_t(\omega) = \xi(\omega) < \infty$ a.s., for any $\epsilon > 0$ there is an s'_ϵ such that, if $r \ge s \ge s'_\epsilon$, $|z_r - z_s| < \epsilon/2u$. Consequently, if $r \ge s \ge s'_\epsilon$,

$$|M_t - M_s| \le \epsilon.$$

That is, $M_t(\omega)$ satisfies a Cauchy condition and converges to a limit $\mu(\omega)$. Then $\lim_{t \to \infty} \left(\dfrac{1}{u_t}\right)(M_t - M_{t_0}) = \dfrac{1}{u}(\mu - M_{t_0})$.

Suppose now that $\lim_{t \to \infty} u_t(\omega) = +\infty$. Given $\epsilon > 0$, again using the Cauchy condition for z, there is an s_ϵ such that, if $r \ge s \ge s_\epsilon \vee t_0$,

$$|z_r - z_s| < \frac{\epsilon}{3}.$$

Consequently, $\sup_{r \ge s_\epsilon \vee t_0} |z_r - z_s| \le \epsilon/3$. From (5.10.1), if $t \ge s_\epsilon \vee t_0$,

$$\frac{1}{u_t}|M_t - M_{s_\epsilon \vee t_0}| \le \frac{2\epsilon}{3}.$$

Suppose $t_0 \le s_\epsilon \vee t_0 < t_0$. Now $\lim_{t \to \infty} u_t(\omega) = +\infty$, so there is a t_ϵ such that, $t > t_\epsilon$,

$$u_t \ge \frac{3|M_{s_\epsilon \vee t_0} - M_{t_0}|}{\epsilon}.$$

That is $\dfrac{1}{u_t}|M_t - M_{s_\epsilon \vee t_0}| \le \dfrac{\epsilon}{3}$. Now

$$\frac{1}{u_t}|M_t - M_{t_0}| \le \frac{1}{u_t}|M_{s_\epsilon \vee t_0} - M_{t_0}| + \frac{1}{u_t}|M_t - M_{s_\epsilon \vee t_0}|.$$

So if $t > s_\epsilon \vee t_\epsilon \vee t_0$,

$$\frac{1}{u_t}|M_t - M_{t_0}| \le \epsilon,$$

and the result is proved. ∎

The signal coefficient

From (5.8.1),

$$\int_0^t dx_s \otimes x_s = A \int_0^t x_s \otimes x_s ds + \int_0^t dw_s \otimes x_s,$$

which we rewrite

$$L_t = AT_t + M_t,$$

and

$$E[L_t \mid \mathcal{Y}_t] = AE[T_t \mid \mathcal{Y}_t] + E[M_t \mid \mathcal{Y}_t].$$

An estimate for A is, therefore,

$$\hat{A}_t = \hat{L}_t \hat{H}_t^{-1},$$

and the error $\hat{A}_t - A = \hat{M}_t \hat{H}_t^{-1}$.

Now as a special case of Theorem 5.10.1 we investigate the convergence of this error to zero.

Consider a function $\rho(t)$, $t \geq 0$, which is positive nondecreasing and such that $\lim_{t \to \infty} \int_0^t \rho^{-1}(s) ds = \lambda < \infty$. Note from Theorem 5.10.1 this last condition implies that $\lim_{t \to \infty} t \rho(t)^{-1} = 0$. An example of such a function is

$$\rho(t) = \max(1, t(\log t)(\log \log t)^{\alpha}), \quad \alpha > 1.$$

Clearly any function which grows faster than t^{α}, $\alpha > 1$, at infinity satisfies the condition. The strongest results are those for functions which have the slowest growth at infinity.

Consider the (matrix) martingale M_t. M is locally square integrable; $\langle M \rangle$ will denote the predictable nonnegative process such that $M_t M_t' - \langle M \rangle_t$ is a local martingale.

In fact $\langle M \rangle_t = BB' \int_0^t x_s' x_s ds$ and $\mathrm{Tr}\langle M \rangle_t = \mathrm{Tr}(BB') \int_0^t x_s' x_s ds$. Consider the martingale

$$R_t \overset{\Delta}{=} \int_0^t \rho(\mathrm{Tr}\langle M \rangle_s)^{-1/2} dM_s.$$

Lemma 5.10.2 R_t *is a square integrable martingale, so* $\lim_{t \to \infty} R_t = \xi(\omega) < \infty$ *exists a.s.*

Proof

$$E[\mathrm{Tr}(R_t R_t')] = E\left[\int_0^t \rho(\mathrm{Tr}\langle M \rangle_s)^{-1} d(\mathrm{Tr}\langle M \rangle_s)\right].$$

Now $\int_0^t \rho(\mathrm{Tr}\langle M \rangle_s)^{-1} d(\mathrm{Tr}\langle M \rangle_s) < \lambda$ a.s. So

$$\lim_{t \to \infty} E[\mathrm{Tr}(R_t R_t')] \leq \lambda < \infty.$$

and R_t is a square integrable martingale for $0 \leq t \leq \infty$.   ■

Corollary 5.10.3 *From Theorem 5.10.1, if ρ is continuous,*

$$\lim_{t \to \infty} \rho(\mathrm{Tr}\langle M \rangle_t)^{-1/2} M_t$$

exists.

If $\lim_{t \to \infty} \mathrm{Tr}\langle M \rangle_t = +\infty$, this limit is zero. ($\mathrm{Tr}\langle M \rangle_t$ is an increasing process so $\lim_{t \to \infty} \mathrm{Tr}\langle M \rangle_t$ exists and is either finite or $+\infty$).

Corollary 5.10.4

$$\lim_{t \to \infty} \rho \left(\int_0^t x_s' x_s ds\right)^{-(1/2)} M_t$$

exists a.s.

Proof   Note that, apart from the positive constant $B^* = \text{Tr}(BB')$, $\text{Tr}\langle M\rangle_t$ is $\displaystyle\int_0^t x_s' x_s \, ds$.

Therefore as ρ is nondecreasing,

$$\rho((B^* + 1)^{-1} \text{Tr}\langle M\rangle_t) = \rho\left((B^* + 1)^{-1} B^* \int_0^t x_s' x_s \, ds\right)$$

$$\leq \rho\left(\int_0^t x_s' x_s \, ds\right),$$

so

$$\rho\left(\int_0^t x_s' x_s \, ds\right)^{-1} \leq \rho\left((B^* + 1)^{-1} \text{Tr}\langle M\rangle_t\right)^{-1}.$$

With

$$\bar{R}_t \stackrel{\triangle}{=} \int_0^t \rho\left(\int_0^u x_s' x_s \, ds\right)^{-1/2} dM_u,$$

we have

$$E[\text{Tr}(\bar{R}_t \bar{R}_t')] \leq (B^* + 1)\lambda < \infty.$$

Therefore, $\lim_{t\to\infty} \bar{R}_t$ exists and is finite a.s., so from Theorem 5.10.1

$$\lim_{t\to\infty} \rho\left(\int_0^t x_s' x_s \, ds\right)^{-1/2} M_t$$

exists a.s.   ∎

Corollary 5.10.5 *Suppose x satisfies the stability property*

$$L = \sup_t \frac{1}{t} \int_0^t x_s' x_s \, ds < \infty,$$

and

$$\lim_{t\to\infty} \rho(t) M_t = \infty.$$

Then

$$\lim_{t\to\infty} \rho(t)^{-1/2} M_t = 0 \quad \text{a.s.}$$

Proof

$$\rho((B^* + 1)^{-1}(L + 1)^{-1} \text{Tr}\langle M\rangle_t)$$

$$= \rho\left((B^* + 1)^{-1}(L + 1)^{-1} B^* \int_0^t x_s' x_s \, ds\right)$$

$$= \rho\left((B^* + 1)^{-1}(L + 1)^{-1} B^* t \frac{1}{t} \int_0^t x_s' x_s \, ds\right)$$

$$\leq \rho\left((B^* + 1)^{-1}(L + 1)^{-1} B^* t L\right) \leq \rho(t).$$

Therefore

$$\rho(t)^{-1} \le \rho((B^* + 1)^{-1}(L + 1)^{-1}\mathrm{Tr}\langle M\rangle_t).$$

With

$$\tilde{R}_t \overset{\triangle}{=} \int_0^t \rho(s)^{-1/2}\mathrm{d}M_s,$$

we have

$$E[\mathrm{Tr}\tilde{R}_t \tilde{R}_t'] = E\left[\int_0^t \rho(s)^{-1}\mathrm{d}(\mathrm{Tr}\langle M\rangle_s)\right]$$
$$\le (B^* + 1)(L + 1)\lambda < \infty.$$

Therefore, $\lim_{t\to\infty} \tilde{R}_t$ exists and is finite a.s. Thus from Theorem 5.10.1

$$\lim_{t\to\infty} \rho(t)^{-1/2} M_t = 0 \quad \text{a.s.}$$

■

Theorem 5.10.6 *Suppose x satisfies the stability property of Corollary 5.10.5 and $\lim_{t\to\infty} \rho(t) = \infty$. Further, suppose x satisfies the excitation condition*

$$\rho(t)^{-1}\hat{H}_t > K > 0,$$

where $T_t = \int_0^t x_s x_s' \mathrm{d}s$ and $\hat{H}_t = E[T_t \mid \mathcal{Y}_t]$. Then

$$\lim_{t\to\infty} \hat{M}_t \hat{H}_t^{-1} = 0 \quad \text{a.s.}$$

with convergence at a rate $\rho(t)^{1/2}$.
   Then

$$\lim_{t\to\infty} \rho(t)^{-1/2} = 0 \quad \text{a.s.}$$

Proof The stability property states that $\sup_t \dfrac{1}{t}\displaystyle\int_0^t x_s'x_s\mathrm{d}s \le L$ a.s. Therefore

$$\sup_t \frac{1}{t}E\left[\int_0^t x_s'x_s\mathrm{d}s\right] \le L < \infty,$$

and because $\lim_{t\to\infty} t\rho(t)^{-1} = 0$,

$$\sup_t \frac{1}{\rho(t)}E[\mathrm{Tr}\langle M\rangle_t] < \infty,$$

and the set of random variables $\{\rho(t)^{-1/2} M_t\}$ is bounded in L^2. We can, therefore, condition the convergence of Corollary 5.10.5 and deduce

$$\lim_{t\to\infty} \rho(t)^{-1/2}\hat{M}_t = 0 \quad \text{a.s.}$$

Now

$$\hat{M}_t \hat{H}_t^{-1} = \rho(t)^{-1/2} \hat{M}_t (\rho(t)^{-1/2} \hat{H}_t)^{-1}$$
$$< \rho(t)^{-1/2} \hat{M}_t \rho(t)^{-1/2} K^{-1}.$$

Therefore, $\lim_{t\to\infty} \rho(t)^{1/2} \hat{M}_t \hat{H}_t^{-1} = 0$ a.s. and the result follows. ∎

The observation coefficient

From (5.8.2),

$$\int_0^t dy_s \otimes x_s = C \int_0^t x_s \otimes x_s ds + \int_0^t dv_s \otimes x_s,$$

which we rewrite

$$U_t = CT_t + N_t,$$

and

$$E[U_t \mid \mathcal{Y}_t] = C E[T_t \mid \mathcal{Y}_t] + E[N_t \mid \mathcal{Y}_t].$$

An estimate for C is, therefore,

$$\hat{A}_t = \hat{J}_t \hat{H}_t^{-1},$$

and the error $\hat{C}_t - A = \hat{N}_t \hat{H}_t^{-1}$.

Similar discussions allow us to conclude that, under the stability and excitation conditions, the error $\hat{N}_t \hat{H}_t^{-1}$ converges to zero almost surely at a rate $\rho(t)^{1/2}$.

5.11 Continuous-time nonlinear filtering

Suppose $(\Omega, \mathcal{F}, P)$ is a probability space with a complete filtration $\{\mathcal{F}_t\}, t \geq 0$, on which are given two independent $\mathcal{F}_t$-Brownian motion processes B_t and y_t with quadratic variations $Q(.)$ and $R(.)$ respectively. Let x_0 be a real valued random variable with distribution $\pi_0(.)$.

Consider the Borel functions

$g : \mathbb{R} \times [0, \infty) \to \mathbb{R}, s : \mathbb{R} \times [0, \infty) \to \mathbb{R}$, where

$$|g(x_1, t) - g(x_2, t)| \leq k|x_1 - x_2|,$$
$$|s(x_1, t) - s(x_2, t)| \leq k|x_1 - x_2|.$$

Write $\mathcal{Y}_t = \sigma\{y_s : s \leq t\}$ for the complete filtration generated by the observation process y.

Remark 5.11.1 The stochastic differential equation

$$dx_t = g(x_t)dt + s(x_t)dB_t,$$

with initial state x_0, has a strong solution. □

Consider the Borel function $h : \mathbb{R} \times [0, \infty) \to \mathbb{R}$, where we suppose

$$|h(x, t)| \le k(1 + |x|).$$

Define

$$\Lambda_t = \exp\left\{\int_0^t h(x_s)R_s^{-1}\mathrm{d}y_s - \frac{1}{2}\int_0^t h^2(x_s)R_s^{-1}\mathrm{d}s\right\},$$

which is also given by

$$\Lambda_t = 1 + \int_0^t \Lambda_s h(x_s)R_s^{-1}\mathrm{d}y_s. \tag{5.11.1}$$

To see this apply the Itô rule to the function $\log \Lambda_t$.

Then Λ_t is an $\mathcal{F}_t$-martingale and $E[\Lambda_t] = 1$.

A new probability measure $\overline{P}$ can be defined by setting

$$\left.\frac{\mathrm{d}\overline{P}}{\mathrm{d}P}\right|_{\mathcal{F}_t} = \Lambda_t.$$

Define the process b_t by the formula $b_t = y_t - \int_0^t h(x_s)\mathrm{d}s$. Then $\{b_t\}$ is a Wiener process under $\overline{P}$ with quadratic variation $R(.)$. Therefore under $\overline{P}$,

$$y_t = \int_0^t h(x_s)\mathrm{d}s + b_t.$$

For any real valued function ϕ for which the expectation is defined, write

$$\sigma(\phi)_t = E[\Lambda_t\phi(x_t) \mid \mathcal{Y}_t]. \tag{5.11.2}$$

In the case when the measure defined by $\sigma(.)_t$ has a density $q(x, t)$, we have

$$\sigma(\phi)_t = \int_{\mathbb{R}} \phi(x)q(x, t)\mathrm{d}x.$$

Using the Itô rule, we establish

$$\phi(x_t) = \phi(x_0) + \int_0^t \frac{\partial\phi(x_s)}{\partial x}s(x_s)\mathrm{d}B_s$$

$$+ \int_0^t \left(\frac{1}{2}\frac{\partial^2\phi(x_s)}{\partial x^2}s^2(x_s) + g(x_s)\frac{\partial\phi(x_s)}{\partial x}\right)\mathrm{d}s. \tag{5.11.3}$$

In view of (5.11.1) and (5.11.3) and using the Itô product rule (Example 3.7.15),

$$\Lambda_t\phi(x_t) = \phi(x_0) + \int_0^t \Lambda_s\mathrm{d}\phi(x_s) + \int_0^t \phi(x_s)\mathrm{d}\Lambda_s + [\Lambda, \phi]_t$$

$$= \phi(x_0) + \int_0^t \Lambda_s\frac{\partial\phi(x_s)}{\partial x}s(x_s)\mathrm{d}B_s$$

$$+ \int_0^t \Lambda_s\left(\frac{1}{2}s^2(x_s)\frac{\partial^2\phi(x_s)}{\partial x^2} + g(x_s)\frac{\partial\phi(x_s)}{\partial x}\right)\mathrm{d}s$$

$$+ \int_0^t \Lambda_s h(x_s)R_s^{-1}\phi(x_s)\mathrm{d}y_s. \tag{5.11.4}$$

Conditioning both sides of (5.11.4) on $\mathcal{Y}_t$ and using the fact that B_t and y_t are independent and that y_t has independent increments under P (it is Wiener) (see [15] Lemma 3.2 of Chapter 7), we obtain a stochastic differential equation for (5.11.2).

Theorem 5.11.2 *Suppose $\phi \in C^2$ is a real valued function with compact support. Then*

$$\sigma(\phi)_t = \sigma(\phi)_0 + \int_0^t \sigma(A\phi)_s ds + \int_0^t \sigma(h(x_s)R_s^{-1}\phi(x_s)) dy_s, \qquad (5.11.5)$$

where $A\phi(x) = \dfrac{1}{2}s^2(x_s)\dfrac{\partial^2\phi(x_s)}{\partial x^2} + g(x_s)\dfrac{\partial\phi(x_s)}{\partial x}.$

If $\sigma(.)_t$ has a density $q(x, t)$, we integrate by parts each term of (5.11.5) using the fact that $\phi \in C^2$ has compact support:

$$\int_{\mathbb{R}} \phi(x)q(x, t)dx = \int_{\mathbb{R}} \phi(x)q_0(t)dx + \frac{1}{2}\int_0^t \int_{\mathbb{R}} q(x, s)s^2(x)\frac{\partial^2\phi(x)}{\partial x^2}$$

$$+ \int_0^t \int_{\mathbb{R}} q(x, s)g(x)\frac{\partial\phi(x)}{\partial x}dxds + \int_0^t \int_{\mathbb{R}} q(x, s)h(x)R_s^{-1}\phi(x)dxdy_s,$$

or

$$\int_{\mathbb{R}} \phi(x)q(x, t)dx = \int_{\mathbb{R}} \phi(x)q_0(x)dx + \frac{1}{2}\int_0^t \int_{\mathbb{R}} \phi(x)\frac{\partial^2(q(x, s)s^2(x))}{\partial x^2}$$

$$- \int_0^t \int_{\mathbb{R}} \phi(x_s)\frac{\partial(q(x, s)g(x))}{\partial x}dxds$$

$$+ \int_0^t \int_{\mathbb{R}} q(x, s)h(x)R_s^{-1}\phi(x)dxdy_s,$$

for all "test" functions ϕ, hence

Corollary 5.11.3 *q satisfies the linear stochastic differential equation*

$$q(x, t) = q_0(x) + \int_0^t (A^*q)(x, s)ds + \int_0^t q(x, s)h(x, s)R_s^{-1}dy_s.$$

Here $(A^*q)(x, t) = \dfrac{1}{2}\dfrac{\partial^2 s^2(x_t)q(x, t)}{\partial x^2} - \dfrac{\partial g(x_t)q(x, t)}{\partial x}$ *and $q_0(x)$ is the density such that*
$\pi_0(dx) = q_0(x)dx.$

The correlated case

Here we consider nonlinear dynamics with correlated noises.

Suppose $(\Omega, \mathcal{F}, P)$ is a probability space with a complete filtration $\{\mathcal{F}_t\}, t \geq 0$, on which are given two $\mathcal{F}_t$-Brownian motion processes $B_t \in \mathbb{R}^d$ and $W_t \in \mathbb{R}^m$ such that

$$\langle B^i, W^j \rangle_t = \int_0^t \rho_s^{ij} ds, \quad 1 \leq i, \quad 1 \leq j \leq m.$$

$x_0 \in \mathbb{R}^d$ has distribution $\pi_0(.)$ and is independent of B_t and W_t.

Consider the Borel functions

$g : \mathbb{R}^d \times [0, \infty) \to \mathbb{R}^d$, $s : \mathbb{R}^d \times [0, \infty) \to \mathcal{L}(\mathbb{R}^d, \mathbb{R}^d)$, $h : \mathbb{R}^d \times [0, \infty) \to \mathbb{R}^m$ and the continuous and nonsingular matrix $\alpha : [0, \infty) \to \mathcal{L}(\mathbb{R}^m, \mathbb{R}^m)$. We assume here that

$$|g(x_1, t) - g(x_2, t)| \le k|x_1 - x_2|,$$
$$||s(x_1, t) - s(x_2, t)|| \le k|x_1 - x_2|,$$
$$|h(x, t)| \le k(1 + |x|),$$
$$||\alpha(y)|| \ge \delta > 0 \quad \text{for some} \quad \delta \text{ and}$$
$$||\alpha(y_t^1) - \alpha(y_t^2)|| \le k|y_t^1 - y_t^2|.$$
$$dx_t = g(x_t)dt + s(x_t)dB_t,$$
$$dy_t = h(x_t)dt + \alpha(y_t)dW_t.$$

Write $\mathcal{Y}_t = \sigma\{y_s : s \le t\}$ for the complete filtration generated by the observation process y.

Define

$$\Lambda_t^{-1} = \exp\left\{-\int_0^t (\alpha(y_s)^{-1}h(x_s))'dW_s - \frac{1}{2}\int_0^t |\alpha(y_s)^{-1}h(x_s)|^2 ds\right\}$$

$$= \exp\left\{-\int_0^t (\alpha(y_s)^{-1}h(x_s))'\alpha(y_s)^{-1}dy_s + \frac{1}{2}\int_0^t |\alpha(y_s)^{-1}h(x_s)|^2 ds\right\}.$$

Consequently,

$$\Lambda_t = \exp\left\{\int_0^t (\alpha(y_s)^{-1}h(x_s))'\alpha(y_s)^{-1}dy_s - \frac{1}{2}\int_0^t |\alpha(y_s)^{-1}h(x_s)|^2 ds\right\}.$$

By Girsanov's Theorem, a new probability measure $\overline{P}$ can be defined by setting $\left.\dfrac{d\overline{P}}{dP}\right|_{\mathcal{F}_t} = \Lambda_t^{-1}$, and under $\overline{P}$ the processes $V_t \in \mathbb{R}^d$ and $\bar{y}_t \in \mathbb{R}^m$ are standard Brownian motions, where

$$dV_t^i = dB_t^i + \langle \rho^i, \alpha^{-1}h\rangle dt, \quad \rho^i \in \mathbb{R}^d,$$

and

$$d\bar{y}_t = \alpha^{-1}dy = dW_t + \alpha^{-1}hdt.$$

Furthermore, under $\overline{P}$, $\langle V^i, y^j\rangle_t = \int_0^t \rho_s^{ij} ds$

For any real valued function ϕ for which the expectation is defined write

$$\sigma(\phi)_t = E[\Lambda_t\phi(x_t) \mid \mathcal{Y}_t].$$

Theorem 5.11.4 *Suppose $\phi \in C^2(\mathbb{R}^d)$ is a real valued function with compact support. Then*

$$\sigma(\phi)_t = \sigma(\phi)_0 + \int_0^t \sigma(A\phi)_s ds$$

$$+ \int_0^t \{\sigma(\nabla\phi.s.\rho) + \alpha^{-1}(y_s)\sigma(\phi h)\}'\alpha^{-1}(y_s)dy_s,$$

where $A\phi(x) = \dfrac{1}{2}\displaystyle\sum_{i,j=1}^d (ss')^{ij}(x_s)\dfrac{\partial^2\phi(x_s)}{\partial x^i \partial x^j} + \sum_{i=1}^d g^i(x_s)\dfrac{\partial\phi(x_s)}{\partial x^i}.$

Proof   The proof is left as an exercise.   ∎

5.12 Problems

1. Assume that the state and observation processes of a system are given by the vector dynamics (5.4.1) and (5.4.2). For $m, k \in \mathbb{N}$, $m < k$, write the unnormalized conditional density such that

$$\overline{E}[\overline{\Lambda}_k I(X_m \in dx) \mid \mathcal{Y}_k] = \gamma_{m,k}(x)dx.$$

Using the change of measure techniques described in Section 5.3, show that

$$\gamma_{m,k}(x) = \alpha_m(x)\beta_{m,k}(x),$$

where $\alpha_m(x)$ is given recursively by (5.3.6). Show that

$$\beta_{m,k}(x) = \overline{E}[\overline{\Lambda}_{m+1,k} \mid X_m = x, \mathcal{Y}_k]$$

$$= \frac{1}{\phi(y_{m+1})} \int_{\mathbb{R}^m} \phi_{m+1}(Y_{m+1} - C_{m+1}z)$$

$$\times \psi_{m+1}(z - A_{m+1}x)\beta_{m+1,k}(z)dz. \tag{5.12.1}$$

2. Show that the density $\beta_{m,k}(x)$ (5.12.1) is Gaussian and derive backward recursions for its conditional mean and covariance matrix ([10] page 101).

3. Assume that the state and observation processes are given by the vector dynamics

$$X_{k+1} = A_{k+1}X_k + V_{k+1} + W_{k+1} \in \mathbb{R}^m,$$

$$Y_k = C_k X_k + W_k \in \mathbb{R}^d.$$

A_k, C_k are matrices of appropriate dimensions, V_k and W_k are normally distributed with means 0 and respective covariance matrices Q_k and R_k, assumed nonsingular. Using measure change techniques derive recursions for the conditional mean and covariance matrix of the state X given the observations Y.

4. Let $m = n = 1$ in (5.8.1) and (5.8.2). The notation in Section 5.8 and Section 5.9 is used here.

Let Γ_t be the process defined as

$$\Gamma_t = \int_0^t x_s^p ds, \quad p = 1, 2, \dots.$$

Write

$$\overline{E}[\Lambda_t I_{(\Gamma_t \in dx)} \mid \mathcal{Y}_t] = \mu_t(x)dx.$$

Show that at time t, the density $\mu_t(x)$ is completely described by the $p + 3$ statistics $s_t(0), s_t(1), \dots, s_t(p), \Sigma_t$, and m_t as follows:

$$\mu_t(x) = \left[\sum_{i=1}^p s_t(i)\right]q(x, t),$$

where $s_0(i) = 0, i = 1, \ldots, p$, and

$$\frac{ds_t(p)}{dt} = -p(A_t + \Sigma_t^{-1} B_t^2) s_t(p) + 1,$$

$$\frac{ds_t(p-1)}{dt} = -(p-1)(A_t + \Sigma_t^{-1} B_t^2) s_t(p-1) + p s_t(p) \Sigma_t^{-1} B_t^2 m_t,$$

$$\frac{ds_t(i)}{dt} = -i(A_t + \Sigma_t^{-1} B_t^2) s_t(i) + \frac{1}{2}(i+1)(i+2) s_t(i+2)$$

$$+ (i+1) s_t(i+1) \Sigma_t^{-1} B_t^2, \quad i = 1, \ldots, p-2,$$

$$\frac{ds_t(0)}{dt} = B_t^2 s_t(2) + \Sigma_t^{-1} s_t(1) m_t.$$

5. Give a detailed proof of Lemma 5.7.1.
6. Prove (5.7.5), (5.7.6), (5.7.7) and (5.7.3).
7. Finish the proof of Theorem 5.7.5.
8. Give the proof of Theorem 5.7.6.
9. Prove (5.7.39).
10. Establish (5.7.52).
11. Give the proof of Theorem 5.11.4.

Financial applications

6.1 Volatility estimation

Suppose a price S evolves in discrete time, $k = 0, 1, \ldots$, with dynamics

$$S_{k+1} = S_k e^{\mu - \frac{\sigma_{k+1}^2}{2} + \sigma_{k+1} b_{k+1}}.$$

Here $\{b_k\}$ is a sequence of i.i.d. normal random variables with mean 0 and variance 1 ($N(0, 1)$) and σ_{k+1} represents the volatility of the price change between times k and $k + 1$.

$$E[S_{k+1} \mid S_k] = S_k e^{\mu}.$$

The price sequence $S_0, S_1, \ldots$ is observed as are the logarithmic increments

$$y_{k+1} = \log \frac{S_{k+1}}{S_k} = \mu - \frac{\sigma_{k+1}^2}{2} + \sigma_{k+1} b_{k+1}.$$

Let us suppose that $\log \sigma_k$ has dynamics

$$\log \sigma_{k+1} = a + b \log \sigma_k + \theta w_{k+1}.$$

Here again $\{w_k\}$ is a sequence of i.i.d. $N(0, 1)$ random variables. Writing $x_k = \log \sigma_k$, so that $\sigma_k = e^{x_k}$, we see

$$x_{k+1} = a + b x_k + \theta w_{k+1},$$

$$y_k = \mu - \frac{e^{2x_k}}{2} + e^{2x_k} b_k.$$

Now assume that under the reference probability measure $\overline{P}$ both $\{x_k\}$ and $\{y_k\}$ are sequences of i.i.d. $N(0, 1)$ random variables. Write

$$\mathcal{G}_k = \sigma\{x_0, \ldots, x_k, y_0, \ldots, y_{k-1}\},$$

and denoting by $\phi(.)$ the $N(0, 1)$ probability density function

$$\lambda_k \triangleq \frac{\phi(\theta^{-1}(x_k - a - b x_{k-1}))}{\theta \phi(x_k)} \frac{\phi(e^{-x_k}(y_k - \mu + \frac{1}{2}e^{2x_k}))}{e^{x_k} \phi(y_k)}, \tag{6.1.1}$$

for $k = 1, 2, \ldots$.

Set

$$\lambda_0 = \frac{\phi(e^{-x_0}(y_0 - \mu + \frac{1}{2}e^{2x_0}))}{e^{x_0}\phi(y_0)},$$

$$\Lambda_n = \prod_{k=0}^{n} \lambda_k.$$

Define a new probability measure P (the "real world" probability), by setting $\left.\dfrac{\mathrm{d}P}{\mathrm{d}\overline{P}}\right|_{\mathcal{G}_n} = \Lambda_n$.

We can then show that under P, $\{w_k\}$, $\{b_k\}$, $k = 0, 1, \ldots$ are sequences of i.i.d. $N(0, 1)$ random variables, where

$$w_k \overset{\triangle}{=} \theta^{-1}(x_k - a - bx_{k-1}),$$

$$b_k \overset{\triangle}{=} e^{-x_k}(y_k - \mu + \frac{1}{2}e^{2x_k}).$$

From Bayes' Theorem 4.1.1, for any Borel measurable function f,

$$E[f(x_k) \mid \mathcal{Y}_k] = \frac{\overline{E}[\Lambda_k f(x_k) \mid \mathcal{Y}_k]}{\overline{E}[\Lambda_k \mid \mathcal{Y}_k]}.$$

The numerator defines a measure; suppose it has a density $q_k(.)$ so that

$$\overline{E}[\Lambda_k f(x_k) \mid \mathcal{Y}_k] = \int_{-\infty}^{\infty} f(z)q_k(z)\mathrm{d}z, \tag{6.1.2}$$

and we have the recursion

Theorem 6.1.1

$$q_k(z) = \Phi(z, y) \int_{-\infty}^{\infty} \phi(\theta^{-1}(z - a - bx))q_{k-1}(x)\mathrm{d}x.$$

Here $\Phi(z, y) = \dfrac{e^{-z}\phi(e^{-z}(y_k - \mu + \frac{1}{2}e^{2z}))}{\theta\phi(y_k)}.$

This gives the formula for updating the unnormalized conditional density of $x_k = \log \sigma_k$ given $\mathcal{Y}_k$.

Putting $f(x) \equiv 1$ in (6.1.2) we see

$$\overline{E}[\Lambda_k \mid \mathcal{Y}_k] = \int_{-\infty}^{\infty} q_k(z)\mathrm{d}z, \tag{6.1.3}$$

so that the normalized conditional density of $x_k = \log \sigma_k$ given $\mathcal{Y}_k$ is

$$p_k(z) = \frac{q_k(z)}{\displaystyle\int_{-\infty}^{\infty} q_k(x)\mathrm{d}x}.$$

Furthermore, taking $f(x_k) = x_k$ we see

$$E[x_k \mid \mathcal{Y}_k] = \frac{\displaystyle\int_{-\infty}^{\infty} zq_k(z)dz}{\displaystyle\int_{-\infty}^{\infty} q_k(z)dz}.$$

This is the optimal estimate of the logarithm of the volatility given the observations of the price.

Calibration

Suppose H, F, G are integrable functions. Consider

$$S_n \triangleq \sum_{k=1}^{n} H(y_k)F(x_k)G(x_{k-1}). \tag{6.1.4}$$

We wish to estimate $E[S_k \mid \mathcal{Y}_k]$.

Consider an associate measure and suppose there is a density $L_k(z)$ such that

$$\overline{E}[\Lambda_k S_k f(x_k) \mid \mathcal{Y}_k] = \int_{-\infty}^{\infty} f(z)L_k(z)dz, \tag{6.1.5}$$

for any integrable function f. We can derive the following formula for updating L_k.

Theorem 6.1.2

$$L_k(z) = \Phi(z, y)\Big[\int_{-\infty}^{\infty} \phi(\theta^{-1}(z - a - bx))L_{k-1}(x)dx$$

$$+ H(y_k)F(z)\int_{-\infty}^{\infty} \phi(\theta^{-1}(z - a - bx))G(x)q_{k-1}(x)dx\Big],$$

where $\Phi(z, y) = \dfrac{e^{-z}\phi(e^{-z}(y_k - \mu + \frac{1}{2}e^{2z}))}{\theta\phi(y_k)}.$

Proof Using (6.1.1) and (6.1.4),

$$\overline{E}[\Lambda_k S_k f(x_k) \mid \mathcal{Y}_k] = \int_{-\infty}^{\infty} f(z)L_k(z)dz$$

$$= \overline{E}[\Lambda_{k-1} S_{k-1} f(x_k)\frac{\phi(\theta^{-1}(x_k - a - bx_{k-1}))}{\theta\phi(x_k)}$$

$$\times \frac{\phi(e^{-x_k}(y_k - \mu + \frac{1}{2}e^{2x_k}))}{e^{x_k}\phi(y_k)} \mid \mathcal{Y}_k]$$

$$+ \overline{E}[\Lambda_{k-1} f(x_k)H(y_k)F(y_k)G(x_{k-1})\frac{\phi(\theta^{-1}(x_k - a - bx_{k-1}))}{\theta\phi(x_k)}$$

$$\times \frac{\phi(e^{-x_k}(y_k - \mu + \frac{1}{2}e^{2x_k}))}{e^{x_k}\phi(y_k)} \mid \mathcal{Y}_k]$$

$$= \int_{-\infty}^{\infty} \int_{-\infty}^{\infty} \phi(\theta^{-1}(z - a - bx)) f(z) \Phi(z, y) L_{k-1}(x) dx dz$$

$$+ \int_{-\infty}^{\infty} \int_{-\infty}^{\infty} \phi(\theta^{-1}(z - a - bx)) \Phi(z, y)$$

$$\times H(y_k) F(z) G(x) f(z) q_{k-1}(x) dx dz.$$

This equality holds for all integrable f and the result follows. ∎

Corollary 6.1.3 *Taking $f(z) \equiv 1$ in (6.1.5) we see*

$$\overline{E}[\Lambda_k S_k \mid \mathcal{Y}_k] = \int_{-\infty}^{\infty} L_k(z) dz. \tag{6.1.6}$$

Further, from Bayes' Theorem 4.1.1,

$$E[S_k \mid \mathcal{Y}_k] = \frac{\overline{E}[\Lambda_k S_k \mid \mathcal{Y}_k]}{\overline{E}[\Lambda_k \mid \mathcal{Y}_k]} = \frac{\int_{-\infty}^{\infty} L_k(z) dz}{\int_{-\infty}^{\infty} q_k(z) dz}.$$

Special cases

1. For $s_k^1 = \sum_{i=1}^{k} x_i$ a measure γ_k^1 is defined by

$$\overline{E}[\Lambda_k S_k^1 f(x_k) \mid \mathcal{Y}_k] = \int_{-\infty}^{\infty} f(z) \gamma_k^1(z) dz. \tag{6.1.7}$$

This is updated by the formula

$$\gamma_k^1(z) = \Phi(z, y) \left(\int_{-\infty}^{\infty} \phi(\theta^{-1}(z - a - bx)) \gamma_{k-1}^1(x) dx \right.$$

$$\left. + z \int_{-\infty}^{\infty} \phi(\theta^{-1}(z - a - bx)) G(x) q_{k-1}(x) dx \right).$$

Then

$$E[S_k^1 \mid \mathcal{Y}_k] = \frac{\int_{-\infty}^{\infty} \gamma_k^1(z) dz}{\int_{-\infty}^{\infty} q_k(z) dz}.$$

2. For $s_k^2 = \sum_{i=1}^{k} x_{i-1}$ the corresponding measure γ_k^2 is updated by

$$\gamma_k^2(z) = \Phi(z, y) \left(\int_{-\infty}^{\infty} \phi(\theta^{-1}(z - a - bx)) \gamma_{k-1}^2(x) dx \right.$$

$$\left. + \int_{-\infty}^{\infty} x \phi(\theta^{-1}(z - a - bx)) G(x) q_{k-1}(x) dx \right).$$

3. For $J_k = \sum_{i=1}^{k} x_i x_{i-1}$ the corresponding measure β_k^1 is updated by

$$\beta_k^1(z) = \Phi(z, y)\Big(\int_{-\infty}^{\infty} \phi(\theta^{-1}(z - a - bx))\beta_{k-1}^1(x)dx$$

$$+ z\int_{-\infty}^{\infty} x\phi(\theta^{-1}(z - a - bx))G(x)q_{k-1}(x)dx\Big).$$

4. Similar formulae, which are all special cases of the expression for $L_k(z)$, are obtained for updating the measures:

$$\beta_k^2(z) \text{ associated with } \sum_{i=1}^{k} x_{i-1}^2,$$

$$\beta_k^3(z) \text{ associated with } \sum_{i=1}^{k} x_i^2,$$

$$v_k^1(z) \text{ associated with } \sum_{i=1}^{k} y_i e^{-2x_i},$$

$$v_k^2(z) \text{ associated with } \sum_{i=1}^{k} e^{-2x_i}.$$

In all cases the conditional expectation of the sum, given the observations, is obtained by normalizing the integral of the associated measure. For example,

$$E\left[\sum_{i=1}^{k} y_i e^{-2x_i} \mid \mathcal{Y}_k\right] = \frac{\int_{-\infty}^{\infty} v_k^1(z)dz}{\int_{-\infty}^{\infty} q_k(z)dz}.$$

6.2 Parameter estimation

Estimates of the sums above can be used to apply to the EM algorithm. Parameters in our model can be re-estimated recursively and, further, one parameter at a time can be updated.

For example, suppose after some iteration a parameter set (a, b, θ, μ) is obtained and we wish to re-estimate the parameter b, given the observations $y_1, y_2, \ldots, y_k$.

Consider a change of measure which replaces parameter b in our model by $\hat{b}$. This is given by a Radon–Nikodym derivative

$$\Lambda_k^{\hat{b}} \triangleq \prod_{i=1}^{k} \frac{\phi(\theta^{-1}(x_i - a - \hat{b}x_{i-1}))}{\phi(\theta^{-1}(x_i - a - bx_{i-1}))},$$

and setting $\dfrac{dP^{\hat{b}}}{dP^b}\Big|_{\mathcal{G}_k} = \Lambda_k^{\hat{b}}$.

The maximizing step determines the conditional expectation of $\log \Lambda_k^{\hat{b}}$ given the observations. That is, consider

$$E[\log \Lambda_k^{\hat{b}} \mid \mathcal{Y}_k] = E\left[-\frac{1}{2}\sum_{i=1}^{k}(\theta^{-1}(x_i - a - \hat{b}x_{i-1})^2) + R(b) \mid \mathcal{Y}_k\right],$$

where $R(b)$ does not involve b.

The first order condition gives the maximum value of $\hat{b}$ as:

$$\hat{b}_k = \frac{E[\sum_{i=1}^k x_i x_{i-1} - a \sum_{i=1}^k x_{i-1} \mid \mathcal{Y}_k]}{E[\sum_{i=1}^k x_{i-1}^2 \mid \mathcal{Y}_k]}$$

$$= \frac{\int_{-\infty}^{\infty} (\beta_k^1(z) - a\gamma_k^2(z))dz}{\int_{-\infty}^{\infty} \beta_k^2(z)dz}.$$

Similar arguments gives estimates

$$\hat{a}_k = \frac{1}{k}E\left[\sum_{i=1}^k x_i - b \sum_{i=1}^k x_{i-1} \mid \mathcal{Y}_k \right]$$

$$= \frac{\int_{-\infty}^{\infty} (\gamma_k^1(z) - b\gamma_k^2(z))dz}{k \int_{-\infty}^{\infty} q_k(z)dz}.$$

$$(\hat{\theta}_k)^2 = \frac{1}{2k}E\left[\sum_{i=1}^k (x_i - a - bx_{i-1})^2 \mid \mathcal{Y}_k \right]$$

$$= \frac{\int_{-\infty}^{\infty} F(z)dz}{2k \int_{-\infty}^{\infty} q_k(z)dz} - \frac{a}{2},$$

$$\hat{\mu}_k = \frac{k}{2} + E\left[\sum_{i=1}^k y_i e^{2x_i} \mid \mathcal{Y}_k \right]$$

$$= \frac{\frac{t}{2} \int_{-\infty}^{\infty} q_k(z)dz + \int_{-\infty}^{\infty} v_k^1(z)dz}{\int_{-\infty}^{\infty} v_k^1(z)dz}.$$

Here $F(z) = \beta_k^3(z) + b\beta_k^2(z) - 2b\beta_k^1(z) - 2a\gamma_k^1(z) + 2ab\gamma_k^2(z).$

6.3 Filtering a price process

Suppose in discrete time a price S has the form

$$S_{k+1} = S_k e^{Y_{k+1}},$$

where $Y_{k+1} = c_k + \sigma_k b_{k+1}$. Here $\{b_\ell\}$ is a sequence of i.i.d. normal random variables with mean 0 and variance 1 ($N(0, 1)$).

Suppose (c_k, σ_k) takes values in a finite set $B = \{(c_i, \sigma_i) : 1 \le i \le N\}$. Write

$$c = (c_1, c_2, \ldots, c_N)', \qquad \sigma = (\sigma_1, \sigma_2, \ldots, \sigma_N)',$$

and suppose that (c_k, σ_k) evolves as a Markov chain with state space B. We can identify B with

$$S = \{e_1, e_2, \ldots, e_N\},$$

where, as before, $e_i = (0, \ldots, 1, \ldots, 0)' \in \mathbb{R}^N$. Suppose $\phi : B \to S$ gives this bijection, so that for each i, $1 \le i \le N$,

$$\phi(c_i, \sigma_i) = e_i.$$

Write $X_k = \phi((c_k, \sigma_k))$ (where k now denotes the time parameter). Then

$$c_k = \langle c, X_k \rangle, \text{ and } \sigma_k = \langle \sigma, X_k \rangle.$$

We suppose X is a Markov chain on $(\Omega, \mathcal{F}, P)$ with state space S and transition matrix A.

The state space S could be quite small and X could represent the state of the economy as "good", "bad", or "average".

Of course X is not observed directly. Instead we observe logarithmic increments of the price process:

$$Y_{k+1} = \log \frac{S_{k+1}}{S_k} = c_k + \sigma_k b_{k+1} = \langle c, X_k \rangle + \langle \sigma, X_k \rangle b_{k+1}.$$

The $N(0, 1)$ random variable b models a purely random noise in the dynamics. The Markov chain X also models some random behavior, but hopefully random behavior with some structure.

The results of the previous sections can now be applied. For any price process $\{S_k\}$, $k = 1, 2, \ldots$, the steps are

1. calculate the sequence of logarithmic increments

$$Y_{k+1} = \log \frac{S_{k+1}}{S_k} = c_k + \sigma_k b_{k+1} = \langle c, X_k \rangle + \langle \sigma, X_k \rangle b_{k+1},$$

2. choose "appropriate" prior values for $\{(c_i, \sigma_i) : 1 \le i \le N\}$ and for the transition probabilities

$$a_{ji} = P(X_{k+1} = e_j \mid X_k = e_i) \ge 0,$$

with $\sum_{j=1}^{N} a_{ji} = 1$,

3. after n values of Y have been observed, calculate new estimates for c, σ and the a,

4. use these values, iteratively, to re-estimate the c, σ and the a. The EM algorithm implies the estimates improve monotonically, in the sense that the expected log-likelihood increases with each re-estimation. Consequently, the model is 'self-tuning'. This step is repeated until some stopping criterion is satisfied.

6.4 Parameter estimation for a modified Kalman filter

This application considers a slightly modified linear Gaussian model.

Consider the following model for the spot price of oil S:

$$dS_t = (\mu - \delta_t)S_t dt + \sigma_1 S_t dz_1(t). \tag{6.4.1}$$

Here z_1 is a standard Brownian motion and δ_t represents the "convenience yield". (This models the value of holding amounts of the commodity.) In fact it is supposed that δ follows similar stochastic dynamics of the form

$$d\delta_t = \kappa(\alpha - \delta_t)dt + \sigma_2 dz_2(t). \tag{6.4.2}$$

Here z_2 is a second standard Brownian motion with $\langle z_1(t), z_2(t)\rangle = \rho t$.

It is convenient to consider the logarithm of the stock price,

$$X_t = \log_e S_t.$$

Then X satisfies

$$dX_t = \kappa(\mu - \delta_t - \frac{1}{2}\sigma_1^2)dt + \sigma_1 dz_1(t). \tag{6.4.3}$$

If r is the risk-free interest rate (taken to be constant here) and λ is the market price of convenience yield risk (also assumed constant), S and δ follow similar processes under an equivalent martingale measure.

However, it is equations (6.4.2) and (6.4.3) which we discretize to give dynamics for the state vector $(X_t, \delta_t)'$ as:

$$(X_t, \delta_t) = c_t + Q_t(X_{t-1}, \delta_{t-1})' + \eta_t. \tag{6.4.4}$$

Here $c_t = ((\mu - \frac{1}{2}\sigma_1^2)\Delta t, \kappa\alpha\Delta t)' \in \mathbb{R}^2$ and

$$Q_t = \begin{pmatrix} 1 & -\Delta t \\ 0 & 1 - \kappa\Delta t \end{pmatrix}, \text{ a } 2 \times 2 \text{ matrix.}$$

The future price for oil for delivery at time $T \geq 0$ is given by:

$$F(S, \delta, T) = S \exp\left[-\delta\frac{(1 - e^{-\kappa T})}{\kappa} + A(T)\right],$$

where

$$A(T) = \left(r - \alpha + \frac{1}{2}\frac{\sigma_2^2}{\kappa^2} - \frac{\sigma_1\sigma_2\rho}{\kappa}\right)T + \frac{1}{4}\sigma_2^2\frac{(1 - e^{-\kappa T})}{\kappa^3}$$

$$+ \left(\alpha\kappa + \sigma_1\sigma_2\rho - \frac{\sigma_2^2}{\kappa}\right)\frac{(1 - e^{-\kappa T})}{\kappa^2}.$$

Here S is the spot price today, $T = 0$ and δ is the value of the convenience yield today, $T = 0$.

Consequently,

$$\log_e F(S, \delta, T) = \log_e S - \delta\frac{(1 - e^{-\kappa T})}{\kappa} + A(T). \tag{6.4.5}$$

It is these future prices, for various dates T, which are given in the market. That is, for different dates $T_1, T_2, \ldots, T_N$ we have observations

$$y_t^1 = \log_e F(S, \delta, T^1),$$

$$\vdots$$

$$y_t^N = \log_e F(S, \delta, T^N).$$

It is supposed these observations give the right hand side of (6.4.5) plus some "noise" term $\varepsilon_t \in \mathbb{R}^N$, where $\varepsilon_t = (\varepsilon_t^1, \ldots, \varepsilon_t^N)$ is a sequence for $t = 0, 1, \ldots$ of independent Gaussian random variables with $E[\varepsilon_t] = 0 \in \mathbb{R}^N$ and Var $\varepsilon_t = E[\varepsilon_t \varepsilon_t'] = H \in \mathbb{R}^{N \times N}$. The observation equation (6.4.5), plus ε_t noise on the right side, therefore has the form:

$$y_t = d_t + Z_t(X_t, \delta_t)' + \varepsilon_t, \tag{6.4.6}$$

for $t = 1, 2, \ldots, T$, where

$$y_t = \begin{pmatrix} y_t^1 \\ \vdots \\ y_t^N \end{pmatrix} = \begin{pmatrix} \log_e F(S, \delta, T^1) \\ \vdots \\ \log_e F(S, \delta, T^N) \end{pmatrix}$$

are the future prices at time t for delivery at times $t + T^1, \ldots, t + T^N$.

$$d_t = \begin{pmatrix} A(T_1) \\ \vdots \\ A(T_N) \end{pmatrix} \in \mathbb{R}^N, \quad Z_t = \begin{pmatrix} 1, & -\kappa^{-1}(1 - e^{\kappa T^1}) \\ 1, & -\kappa^{-1}(1 - e^{\kappa T^2}) \\ & \vdots \\ 1, & -\kappa^{-1}(1 - e^{\kappa T^N}) \end{pmatrix}.$$

The model, in summary, has dynamics (6.4.4) for the "signal" (X_t, δ_t),

$$(X_t, \delta_t) = c + Q(X_{t-1}, \delta_{t-1})' + \eta_t, \tag{6.4.7}$$

and dynamics (6.4.6) for the observations,

$$y_t = (y_t^1, \ldots, y_t^N),$$
$$y_t = d_t + Z_t(X_t, \delta_t)' + \varepsilon_t. \tag{6.4.8}$$

Note that, in spite of Schwartz's notation, c, Q, d and Z do not depend on t. They do include Δt, the time increment of fixed size.

Equations (6.4.7) and (6.4.8) are of the form where the classical Kalman filter can be applied. This considers linear dynamics for the signal $X_t = (X_t^1, \ldots, X_t^m) \in \mathbb{R}^m$, $t = 0, 1, \ldots$,

$$X_{t+1} = \bar{A} + AX_t + Bw_{t+1}, \quad A \in \mathbb{R}^{m \times m}, \tag{6.4.9}$$

and observations

$$y_t = \bar{C} + CX_t + Dv_t, \quad t = 0, 1, \ldots. \tag{6.4.10}$$

Note that, because of the inclusion of the terms $\bar{A}$ and $\bar{C}$, this model is slightly different from that considered previously.

One observes y_t, $t = 0, 1, \ldots, T, \ldots$, and wishes to make the best estimate of X_t. This is the quantity

$$\hat{X}_t = E[X_t \mid y_0, y_1, \ldots, y_t].$$

In fact $\hat{X}_t$ is also a Gaussian random variable with conditional mean

$$\mu_t = \hat{X}_t = E[X_t \mid y_0, y_1, \ldots, y_t],$$

and variance

$$R_t = E[(X_t - \mu_t)(X_t - \mu_t)' \mid y_0, y_1, \ldots, y_t].$$

In fact, the formulae are better written in terms of the one-step predictions:

$$\mu_{k|k-1} = E[x_k \mid y_0, y_1, \ldots, y_{k-1}]$$
$$= \bar{A} + A\mu_{k-1},$$

and

$$R_{k|k-1} = E[(X_k - \mu_{k|k-1})(X_k - \mu_{k|k-1})' \mid y_0, y_1, \ldots, y_{k-1}].$$

Then $R_{k|k-1} = B^2 + A R_{k-1} A'$.

Kalman filter

The (modified) Kalman filter then gives recursive updates:

$$\mu_{k+1} = \bar{A} + A\mu_k + R_{k+1|k} C'(C R_{k+1|k} C' + DD')^{-1}$$
$$\times (y_{k+1} - \bar{C} - C\bar{A} - CA\mu_k),$$
$$R_{k+1} = R_{k+1|k} - R_{k+1|k} C'(C R_{k+1|k} C' + DD')^{-1} C R_{k+1|k}.$$

As stated, $\mu_k = \hat{X}_k = E[X_t \mid y_0, y_1, \ldots, y_k]$ is the conditional mean, or best estimate, of X_k given $y_0, y_1, \ldots, y_k$. Similarly, $R_k = E[(X_k - \mu_k)(X_k - \mu_k)' \mid y_0, y_1, \ldots, y_k]$.

Parameter estimation

However, to implement the Kalman filter knowledge of the parameters $\bar{A}, A, B, \bar{C}, C, D$ is required.

Our algorithms, when modified for these "affine" dynamics, provide optimal ways of estimating these parameters.

In fact, consider the following recursions for $a_k^{ij(M)} \in \mathbb{R}$, $b_k^{ij(M)} \in \mathbb{R}^m$, $d_k^{ij(M)} \in \mathbb{R}^{m \times m}$ (a symmetric matrix with elements $d_k(p, q)$, $p = 1, \ldots, m$, $q = 1, \ldots, m$), $\bar{a}_k^{in}, \bar{b}_k^{in}, u_k^i, v_k^i$,

$\bar{u}_k^i, \bar{v}_k^i. \ M = 0, 1, 2, \ 1 \leq i, j \leq m, \ 1 \leq n \leq d, \ \sigma_k^{-1} = R_{k-1} - R_{k-1} A' R_{k|k-1}^{-1} A R_{k-1}.$

$$a_k^{ij(M)} = a_k^{ij(M)} + b_k^{ij(M)'} \sigma_{k+1}^{-1} R_k^{-1} \mu_k + \text{Tr}\left[d_k^{ij(M)} \sigma_{k+1}^{-1} \right]$$

$$+ \mu_k' R_k^{-1} \sigma_{k+1}^{-1} d_k^{ij(M)} \sigma_{k+1}^{-1} R_k^{-1} \mu_k - b_k^{ij(M)'} R_k A' R_{k+1|k}^{-1} \bar{A}$$

$$+ \bar{A}' R_{k+1|k}^{-1} A R_k d_k^{ij(M)} R_k A' R_{k+1|k}^{-1} \bar{A}$$

$$- 2 A' R_{k+1|k}^{-1} A R_k d_k^{ij(M)} \sigma_{k+1}^{-1} R_k^{-1} \mu_k,$$

$$a_0^{ij(M)} = 0 \in \mathbb{R},$$

$$b_{k+1}^{ij(0)} = R_{k+1|k}^{-1} A R_k \left(b_k^{ij(0)} + 2 d_k^{ij(0)} \sigma_{k+1}^{-1} R_k^{-1} \mu_k - 2 d_k^{ij(0)} R_k A' R_{k+1|k}^{-1} \bar{A} \right),$$

$$b_0^{ij(0)} = 0 \in \mathbb{R}^m,$$

$$d_{k+1}^{ij(0)} = R_{k+1|k}^{-1} A R_k d_k^{ij(0)} R_k A' R_{k+1|k}^{-1} + \frac{1}{2}(e_i e_j' + e_j e_i'),$$

$$d_0^{ij(0)} = \frac{e_i e_j' + e_j e_i'}{2} \in \mathbb{R}^{m \times m},$$

$$b_{k+1}^{ij(1)} = R_{k+1|k}^{-1} A R_k \left(b_k^{ij(1)} + 2 d_k^{ij(1)} \sigma_{k+1}^{-1} R_k^{-1} \mu_k - 2 d_k^{ij(1)} R_k A' R_{k+1|k}^{-1} \bar{A} \right)$$

$$+ e_i e_j' \sigma_{k+1}^{-1} R_k^{-1} \mu_k - e_i e_j' R_k A' R_{k+1|k}^{-1} \bar{A},$$

$$b_0^{ij(1)} = 0 \in \mathbb{R}^m,$$

$$d_{k+1}^{ij(1)} = R_{k+1|k}^{-1} A R_k d_k^{ij(1)} R_k A' R_{k+1|k}^{-1} + \frac{1}{2}(e_i e_j' R_k A' R_{k+1|k}^{-1} + R_{k+1|k}^{-1} A R_k e_j e_i'),$$

$$d_0^{ij(1)} = 0 \in \mathbb{R}^{m \times m},$$

$$a_{k+1}^{ij(2)} = a_k^{ij(2)} + b_k^{ij(2)'} \sigma_{k+1}^{-1} R_k^{-1} \mu_k + \text{Tr}\left[d_k^{ij(2)} \sigma_{k+1}^{-1} \right] + \mu_k' R_k^{-1} d_k^{ij(2)} \sigma_{k+1}^{-1} R_k^{-1} \mu_k$$

$$+ \text{Tr}\left[e_i e_j' \sigma_{k+1}^{-1} \right] + \mu_k' R_k^{-1} \sigma_{k+1}^{-1}(e_i e_j') \sigma_{k+1}^{-1} R_k^{-1} \mu_k$$

$$+ \bar{A}' R_{k+1|k}^{-1} A R_k (e_i e_j') R_k A' R_{k+1|k}^{-1} \bar{A}$$

$$- b_k^{ij(2)'} R_k A' R_{k+1|k}^{-1} \bar{A} + \bar{A}' R_{k+1|k}^{-1} A R_k d_k^{ij(2)} R_k A' R_{k+1|k}^{-1} \bar{A}$$

$$- 2 \bar{A}' R_{k+1|k}^{-1} A R_k d_k^{ij(2)} \sigma_{k+1}^{-1} R_k^{-1} \mu_k$$

$$- \bar{A}' R_{k+1|k}^{-1} A R_k (e_i e_j' + e_j e_i') \sigma_{k+1}^{-1} R_k^{-1} \mu_k,$$

$$a_0^{ij(2)} = 0 \in \mathbb{R},$$

$$b_{k+1}^{ij(2)} = R_{k+1|k}^{-1} A R_k \left(b_k^{ij(2)} + 2 d_k^{ij(2)} \sigma_{k+1}^{-1} R_k^{-1} \mu_k \right.$$

$$\left. - 2 d_k^{ij(2)} R_k A' R_{k+1|k}^{-1} \bar{A} \right)$$

$$+ R_{k+1|k}^{-1} A R_k (e_i e_j' + e_j e_i') \sigma_{k+1}^{-1} R_k^{-1} \mu_k$$

$$- R_{k+1|k}^{-1} A R_k (e_i e_j' + e_j e_i') R_k A' R_{k+1|k}^{-1} \bar{A},$$

$$b_0^{ij(2)} = 0 \in \mathbb{R}^m,$$

$$d_{k+1}^{ij(2)} = R_{k+1|k}^{-1} A R_k \left(d_k^{ij(2)} + \frac{e_i e_j' + e_j e_i'}{2} \right) R_k A' R_{k+1|k}^{-1},$$

$$d_0^{ij(2)} = 0 \in \mathbb{R}^{m \times m},$$

$$\bar{a}_{k+1}^{in} = \bar{a}_k^{in} + \bar{b}_k^{in'} \sigma_{k+1}^{-1} R_k^{-1} \mu_k - \bar{b}_k^{in'} R_k A' R_{k+1|k}^{-1} \bar{A},$$

$$\bar{a}_0^{in} = 0,$$

$$\bar{b}_{k+1}^{in} = R_{k+1|k}^{-1} A R_k \bar{b}_k^{in} + e_i e_j' y_{k+1},$$

$$\bar{b}_0^{in} = e_i \langle y_0, e_n \rangle,$$

$$u_{k+1}^i = u_k^i + v_k^{i'} \sigma_{k+1}^{-1} R_k^{-1} \mu_k - v_k^{i'} R_k A' R_{k+1|k}^{-1} \bar{A},$$

$$u_0^i = 0,$$

$$v_{k+1}^i = R_{k+1|k}^{-1} A R_k v_k^i + e_i, \quad v_0^i = e_i \in \mathbb{R}^m,$$

$$\bar{u}_{k+1}^i = \bar{u}_k^i + (\bar{v}_k^{i'} + e_i') \sigma_{k+1}^{-1} R_k^{-1} \mu_k$$
$$\quad - (\bar{v}_k^{i'} + e_i') R_k A' R_{k+1|k}^{-1} \bar{A}, \quad \bar{u}_0^i = 0 \in \mathbb{R},$$

$$\bar{v}_{k+1}^i = R_{k+1|k}^{-1} A R_k (\bar{v}_k^i + e_i), \quad \bar{v}_0^i = 0 \in \mathbb{R}^m.$$

where Tr[.] denotes the trace of a matrix (which is the sum of the diagonal elements). Write

$$H_k^{(0)} = \sum_{\ell=0}^k x_\ell x_\ell', \quad H_k^{(1)} = \sum_{\ell=0}^k x_\ell x_\ell',$$

$$H_k^{(2)} = \sum_{\ell=1}^k x_{\ell-1} x_{\ell-1}', \quad J_k = \sum_{\ell=0}^k x_\ell y_\ell',$$

$$L_k = \sum_{\ell=0}^k x_\ell, \quad \bar{L}_k = \sum_{\ell=1}^k x_{\ell-1},$$

$$\hat{H}_k^{(0)} = E[H_k^{(0)} \mid \mathcal{Y}_k], \quad \hat{H}_k^{(1)} = E[H_k^{(1)} \mid \mathcal{Y}_k],$$

etc.

Then for $M = 0, 1, 2$:

$$E[H_k^{ij(M)} \mid \mathcal{Y}_k] = a_k^{ij(M)} + b_k^{ij(M)'} \mu_k + \text{Tr}[d_k^{ij(M)} R_k] + \mu_k' d_k^{ij(M)} \mu_k,$$

$$E[J_k^{in} \mid \mathcal{Y}_k] = \bar{a}_k^{in(M)} + \bar{b}_k^{in'} \mu_k,$$

$$E[L_k^i \mid \mathcal{Y}_k] = u_k^i + v_k^{i'} \mu_k,$$

$$E[\bar{L}_k^i \mid \mathcal{Y}_k] = \bar{u}_k^i + \bar{v}_k^{i'} \mu_k.$$

These equations give recursive finite dimensional filters for estimating the matrices and vectors $H_k^{(M)}$, $M = 0, 1, 2$, J_k, L_k and $\bar{L}_k$ given the observations $y_0, y_1, \ldots, y_k$.

The revised estimates for the parameters A, B, C, D, $\bar{A}$, $\bar{C}$ are then (given $y_0, y_1, \ldots, y_k$):

$$\bar{A}_k = \frac{1}{k}(\hat{L}_k - A\hat{L}'_k),$$

$$\bar{A}_k = \frac{1}{k+1}\Big(\sum_{\ell=0}^{k} y_\ell - C\hat{L}_k\Big),$$

$$A_k = (\hat{H}_k^{(1)} - \bar{A}\hat{L}'_k)(\hat{H}_k^{(2)})^{-1},$$

$$C_k = (\hat{J}'_k - \bar{C}\hat{L}'_k)(\hat{H}_k^{(0)})^{-1},$$

$$B^2 = \frac{1}{k}\{\hat{H}_k^{(0)} - (A\hat{H}_k^{(1)'} + \hat{H}_k^{(1)}A') + A\hat{H}_k^{(2)}A'$$

$$- (\bar{A}\hat{L}'_k + \hat{L}_k\bar{A}') + (\bar{A}\hat{L}'_kA' + A\hat{L}_k\bar{A}') + k\bar{A}\bar{A}'\},$$

$$(DD') = \frac{1}{k+1}\Big\{\sum_{\ell=0}^{k} y_\ell y'_\ell - (\hat{J}'_kC' + C\hat{J}_k) + C\hat{H}_k^{(0)}C' - \bar{C}\Big(\sum_{\ell=0}^{k} y'_\ell\Big)$$

$$- \Big(\sum_{\ell=0}^{k} y_\ell\Big)\bar{C}' + (\bar{C}\hat{L}'_kC' + C\hat{L}_k\bar{C}') + (k+1)\bar{C}\bar{C}'\Big\}.$$

Given observations $y_0, y_1, \ldots, y_k$, the parameters are initialized and the above algorithms run to re-estimate the parameters one at a time. With the same $y_0, y_1, \ldots, y_k$ this process is iterated until some stopping rule is satisfied.

6.5 Estimating the implicit interest rate of a risky asset

In this section a risky asset is considered whose price at time t is supposed described by an equation of the form

$$dS_t = S_t(\rho_t dt + \sigma dB_t), \quad t \geq 0. \tag{6.5.1}$$

Here the drift coefficient ρ_t is the underlying interest rate of the risky asset, B is a standard Brownian motion and the integrals are taken to be Itô integrals.

This model is used frequently, and often the coefficients ρ and σ are supposed to be constant. Various forms and methods of estimating the volatility of diffusion coefficient σ can be found in the literature; see for example [33]. We shall suppose σ is constant and determined by one of these techniques.

We shall suppose that the implicit interest rate ρ_t behaves like a Markov chain with state space $\{r_1, \ldots, r_N\}$. r will denote the (column) vector $(r_1, \ldots, r_N)'$. Suppose $S_0 = S$. Then from (6.5.1),

$$S_t = S \exp\left\{\int_0^t (\rho_u - \frac{1}{2}\sigma^2)du + \sigma B_t\right\}.$$

Write $Y_t = \ln S_t - \ln S$. Then

$$Y_t = \int_0^t (\rho_u - \frac{1}{2}\sigma^2)du + \sigma B_t.$$

Now $\rho_t - \frac{1}{2}\sigma^2$ takes values in the set $\{r_1 - \frac{1}{2}\sigma^2, \ldots, r_N - \frac{1}{2}\sigma^2\}$. Write $g_i = r_i - \frac{1}{2}\sigma^2$ and g for the (column) vector $(g_1, \ldots, g_N)'$. Without loss of generality, we shall consider a Markov chain on $S = \{e_1, \ldots, e_N\}$ (see Example 2.6.17). Here, for $0 \le i \le N$, $e_i = (0, \ldots, 1, \ldots, 0)'$ is the i-th unit (column) vector in $\mathbb{R}^N$. If $X_t \in S$ denotes the state of this Markov chain at time $t \ge 0$, then the corresponding value of ρ_t is $\langle X_t, r \rangle$, where $\langle ., . \rangle$ denotes the inner product in $\mathbb{R}^N$.

A natural process to take as the observation process is Y_t, which can be written

$$Y_t = \int_0^t \langle X_u, g \rangle du + \sigma B_t. \tag{6.5.2}$$

Write $\mathcal{F}_t$ for the right-continuous, complete filtration generated by $\sigma\{X_r, Y_r : r \le t\}$, and $\mathcal{Y}_t$ for the right-continuous, complete filtration $\sigma\{Y_r : r \le t\}$, generated by the observation process. We have the following semimartingale representation result (see Lemma 2.6.18):

$$X_t = X_0 + \int_0^t A_r X_r dr + V_t. \tag{6.5.3}$$

Filtering

We model the above dynamics by supposing that initially we have an "ideal" probability space $(\Omega, \mathcal{F}, P)$ such that under P

1. X is a Markov chain with representation (6.5.3),
2. $\sigma^{-1}Y$ is a standard Brownian motion, independent of X.

Define

$$\Lambda_t = \exp\left\{\int_0^t \langle X_u, g \rangle \sigma^{-1} dY_s - \frac{1}{2} \int_0^t \langle X_u, g \rangle^2 \sigma^{-2} ds\right\},$$

which is also given by

$$\Lambda_t = 1 + \int_0^t \Lambda_s \langle X_u, g \rangle \sigma^{-1} dY_s. \tag{6.5.4}$$

To see this apply the Itô rule to the function $\log \Lambda_t$.

Then Λ_t is an $\mathcal{F}_t$ martingale and $E[\Lambda_t] = 1$.

A new probability measure $\overline{P}$ can be defined by setting $\left.\dfrac{d\overline{P}}{dP}\right|_{\mathcal{F}_t} = \Lambda_t$.

Define the process B_t by the formula

$$dB_t = \sigma^{-1}(Y_t - \langle X_t, g \rangle dt), \quad B_0 = 0.$$

Then Girsanov's theorem 4.3.3 implies that $\{B_t\}$ is a standard Brownian motion process under $\overline{P}$. Therefore, under $\overline{P}$,

$$dY_t = \langle X_t, g \rangle dt + \sigma dB_t. \tag{6.5.5}$$

Note that under $\overline{P}$, the process $\{X_t\}$ still satisfies (6.5.3). Consequently, under $\overline{P}$ the processes $\{X_t\}$ and $\{Y_t\}$ satisfy the real world dynamics (6.5.3) and (6.5.2). However, P is a more convenient measure with which to work. Using a version of Bayes' Theorem (4.1.1),

$$\overline{E}[X_t \mid \mathcal{Y}_t] = \frac{E[\Lambda_t X_t \mid \mathcal{Y}_t]}{E[\Lambda_t \mid \mathcal{Y}_t]}.$$

Write

$$\sigma(X_t) = E[\Lambda_t X_t \mid \mathcal{Y}_t]. \tag{6.5.6}$$

Note that $E[\Lambda_t \mid \mathcal{Y}_t] = \sum_{i=1}^{N} \sigma(\langle X_t, e_i \rangle) = \sigma(\sum_{i=1}^{N} \langle X_t, e_i \rangle) = \sigma(1)$. More simply, $E[\Lambda_t \mid \mathcal{Y}_t] = \langle \sigma(X_t), \mathbf{1} \rangle$, where $\mathbf{1}$ is an N-dimensional vector with all entries equal to 1.

In view of (6.5.4) and (6.5.3) and using the Itô product rule (3.7.15),

$$\Lambda_t X_t = X_0 + \int_0^t \Lambda_s A X_s ds + \int_0^t \Lambda_s \langle X_s, g \rangle X_s \sigma^{-1} dY_s$$

$$= X_0 + \int_0^t \Lambda_s A X_s ds + \int_0^t \Lambda_s G X_s \sigma^{-1} dY_s. \tag{6.5.7}$$

Here G is the diagonal matrix whose entries are $g_1, \dots, g_N$.

Conditioning both sides of (6.5.7) on $\mathcal{Y}_t$ and using the fact that Y_t has independent increments under P (it is Wiener) (see [15] Lemma 3.2 p. 261), we have the following finite dimensional filter for $\sigma(X_t)$:

$$\sigma(X_t) = \sigma(X_0) + \int_0^t A\sigma(X_s)ds + \int_0^t G\sigma(X_s)\sigma^{-2}dY_s. \tag{6.5.8}$$

Note that $\sigma(\rho_t) = \langle \sigma(X_t), r \rangle$.

For $s \leq t$ the smoother for $\langle X_s, e_i \rangle$ is defined as

$$\overline{E}[\langle X_s, e_i \rangle \mid \mathcal{Y}_t],$$

with unnormalized form under P

$$E[\langle X_s, e_i \rangle \Lambda_t \mid \mathcal{Y}_t] \stackrel{\Delta}{=} \sigma_t(\langle X_s, e_i \rangle).$$

However, it is more convenient to work with $\sigma_t(\langle X_s, e_i \rangle X_t)$ (see [10] Chapter 8 for more details) and we have

$$\sigma_t(\langle X_s, e_i \rangle X_t) = \sigma_s(\langle X_s, e_i \rangle X_s) + \int_0^t A\sigma_u(\langle X_s, e_i \rangle X_u)du$$

$$+ \int_0^t G\sigma_u(\langle X_s, e_i \rangle X_u)\sigma^{-2}dY_u.$$

This is a finite dimensional filter for $\sigma_t(\langle X_s, e_i \rangle X_t)$.

Consequently, $\sigma_t(\langle X_s, e_i \rangle) = \langle \sigma_t(\langle X_s, e_i \rangle X_t), \mathbf{1} \rangle$ and

$$\overline{E}[\langle X_s, e_i \rangle \mid \mathcal{Y}_t] = \frac{\sigma_t(\langle X_s, e_i \rangle)}{\sum_{j=1}^{N} \sigma(\langle X_s, e_j \rangle)}.$$

Revising the parameters

In addition to the volatility σ the parameters introduced in the above model are the values r_i, $1 \leq i \leq N$, of the implicit interest rate, and the entries a_{ij}, $1 \leq i, j \leq N$, of the Q-matrix

A. Recall $g_i = r_i - \frac{1}{2}\sigma^2$. Using the expectation maximization (EM) algorithm, it is shown in [10] Chapter 8 that the revised estimates are given by

$$\hat{a}_{ji} = \frac{\sigma(N_t^{ij})}{\sigma(J_t^i)},$$

$$\hat{g}_i = \frac{\sigma(G_t^i)}{\sigma(J_t^i)}.$$

Here N_t^{ij} is the number of jumps of X from e_i to e_j in the time interval $[0, 1]$. $J_t^i = \int_0^t \langle X_u, e_i \rangle du$ is the amount of time X spends in state e_i in the time interval $[0, t]$ and

$$G_t^i = \int_0^t \langle X_u, e_i \rangle \sigma^{-2} dY_u = \int_0^t g_i \langle X_u, e_i \rangle \sigma^{-2} du + \int_0^t \langle X_u, e_i \rangle \sigma^{-2} dB_u,$$

the unnormalized estimates are given by the following linear equations:

$$\sigma(N_t^{ij} X_t) = \int_0^t A\sigma(N_s^{ij} X_s) ds + \int_0^t \langle \sigma(X_s), e_i \rangle a_{ji} e_j ds$$

$$+ \int_0^t G\sigma(N_s^{ij} X_s) \sigma^{-2} dY_s, \tag{6.5.9}$$

$$\sigma(J_t^i X_t) = \int_0^t A\sigma(J_s^i X_s) ds + \int_0^t \langle \sigma(X_s), e_i \rangle e_i ds$$

$$+ \int_0^t G\sigma(J_s^i X_s) \sigma^{-2} dY_s, \tag{6.5.10}$$

$$\sigma(G_t^i X_t) = \int_0^t A\sigma(G_s^i X_s) ds + g_i \int_0^t \langle \sigma(X_s), e_i \rangle e_i \sigma^{-2} ds$$

$$+ \int_0^t (G\sigma(G_s^i X_s) + \langle \sigma(X_s), e_i \rangle e_i) \sigma^{-2} dY_s. \tag{6.5.11}$$

In each case we have

$$\sigma(N_t^{ij}) = \langle \sigma_t(N_t^{ij} X_t), \mathbf{1} \rangle, \quad \sigma(J_t^i) = \langle \sigma_t(J_t^i X_t), \mathbf{1} \rangle, \quad \sigma(G_t^i) = \langle \sigma_t(G_t^i X_t), \mathbf{1} \rangle.$$

Numerical methods

Here we describe numerical approximations to (6.5.8), (6.5.9), (6.5.10), and (6.5.11).
Write $q_t = \sigma(X_t)$. Then (6.5.8) is

$$q_t = \sigma(X_0) + \int_0^t Aq_s ds + \int_0^t Gq_s \sigma^{-2} dY_s.$$

Suppose $h = \frac{t}{n}$. For $0 \le k < n$ a first approximation gives

$$q_{(k+1)h} = q_{kh} + Aq_{kh}.h + Gq_{kh}\sigma^{-2}(Y_{(k+1)h} - Y_{kh}).$$

However, this neglects terms which do not converge to 0 when $h \to 0$. To capture these terms Milshtein [27] noted one should substitute

$$q_{kh} + Aq_{kh}.s + Gq_{kh}\sigma^{-2}(Y_s - Y_{kh})$$

for q_s in the expression

$$q_{(k+1)h} = q_{kh} + \int_{kh}^{(k+1)h} Aq_s ds + \int_{kh}^{(k+1)h} Gq_s \sigma^{-2} dY_s.$$

Neglecting terms which converge to 0 when $h \to 0$, the Milshtein approximation then is

$$q_{(k+1)h} = q_{kh} + Aq_{kh}.h + Gq_{kh}\sigma^{-2}(Y_{(k+1)h} - Y_{kh})$$

$$+ \frac{1}{2}G^2 q_{kh}\sigma^{-4}[(Y_{(k+1)h} - Y_{kh})^2 - \sigma^2 h].$$

A full discussion of the Milshtein scheme, and other more sophisticated schemes, can be found in [22]. Write $n_t^{ij} = \sigma(N_t^{ij} X_t)$. Then (6.5.9) becomes

$$n_t^{ij} = \int_0^t An_s^{ij} ds + \int_0^t \langle q_s, e_i \rangle a_{ji} e_j ds$$

$$+ \int_0^t Gn_s^{ij}\sigma^{-2} dY_s. \tag{6.5.12}$$

The Milshtein form in this case is:

$$n_{(k+1)h}^{ij} = \langle q_{kh}, e_i \rangle a_{ji} e_j h + [I + Ah + q_{kh}.h + G(Y_{(k+1)h} - Y_{kh})\sigma^{-2}$$

$$+ \frac{1}{2}G^2[(Y_{(k+1)h} - Y_{kh})^2 - \sigma^2 h]\sigma^{-4}]n_{kh}^{ij}.$$

Similarly, writing $\tau_t^i = \sigma(J_t^i X_t)$ and discretizing (6.5.10),

$$\tau_{(k+1)h}^i = \langle q_{kh}, e_i \rangle e_i h + [I + Ah + q_{kh}.h + G(Y_{(k+1)h} - Y_{kh})\sigma^{-2}$$

$$+ \frac{1}{2}G^2[(Y_{(k+1)h} - Y_{kh})^2 - \sigma^2 h]\sigma^{-4}]\tau_{kh}^i.$$

Finally, with $\gamma_t^i = \sigma(G_t^i X_t)$ discretizing (6.5.11),

$$\gamma_{(k+1)h}^i = g_i \langle q_{kh}, e_i \rangle e_i \sigma^{-2} h + [I + Ah + q_{kh}.h$$

$$+ G(Y_{(k+1)h} - Y_{kh})\sigma^{-2}$$

$$+ \frac{1}{2}G^2[(Y_{(k+1)h} - Y_{kh})^2 - \sigma^2 h]\sigma^{-4}]\gamma_{kh}^i$$

$$+ \sigma^{-2} \langle q_{kh}, e_i \rangle e_i (Y_{(k+1)h} - Y_{kh})$$

$$+ \frac{1}{2}\sigma^{-4} \langle Gq_{kh}, e_i \rangle [(Y_{(k+1)h} - Y_{kh})^2 - \sigma^2 h]e_i.$$

New estimates for the parameters a_{ji} and g_i, based on the observations of the price up to time $t = nh$, are, therefore,

$$\hat{a}_{ji} = \frac{\langle n_t^{ij}, 1 \rangle}{\langle \tau_t^i, 1 \rangle}, \quad \hat{a}_{ji} = \frac{\langle n_t^{ij}, 1 \rangle}{\langle \tau_t^i, 1 \rangle}.$$

Using the smoothed versions of (6.5.9), (6.5.10) and (6.5.11) (see [10] Chapter 8), and possibly additional data, a second revised estimate for these parameters can be obtained. Iterating this procedure provides a monotonic, increasing sequence of probability densities, so, in terms of maximizing the expectation, the models are improving with each step and the estimation methods are self-tuning.

A genetics model

7.1 Introduction

Consider a population of N independent individuals. At each time $k \in \{0, 1, 2, \ldots\}$ each individual can be in one of n states. The total number, N, of individuals in the population remains constant in time. However, the distribution of the N individuals among the n states changes.

We suppose that initially all random variables are defined on a probability space $(\Omega, \mathcal{F}, P)$. For $1 \le i, j \le n$, p_{ji} is the probability that an individual in the population will jump from state i at time $k - 1$ to state j at time k. That is, we suppose each individual in the population behaves like an independent time-homogeneous Markov chain with transition matrix $P = (p_{ji})$.

Note $\sum_{j=1}^{n} p_{ji} = 1$.

Write $p_j = (p_{1j}, p_{2j}, \ldots, p_{nj})'$ for the j-th column of P.

Write $\Pi(N)$ for the set of all partitions of N into n summands; that is, $z \in \Pi(N)$ if $z = (z_1, z_2, \ldots, z_n)$, where each z_i is a nonnegative integer and $z_1 + z_2 + \cdots + z_n = N$.

Write $X(k) = (X_1(k), X_2(k), \ldots, X_n(k)) \in \Pi(N)$ for the distribution of the population at time k.

It is easily checked that

$$E[X(k) \mid X(k - 1)] = PX(k - 1). \tag{7.1.1}$$

However, the population is sampled by withdrawing (with replacement), at each time k, M individuals from the population and observing to which state they belong. That is, at each time k a sample

$$Y(k) = (Y_1(k), Y_2(k), \ldots, Y_n(k)) \in \Pi(M)$$

is obtained, where $\Pi(M)$ is the set of partitions of M.

Clearly this sequence of samples, $Y(0), Y(1), Y(2), \ldots$ enables us to revise our estimates of the state $X(k)$.

7.2 Recursive estimates

For

$$\alpha = (\alpha_1, \alpha_2, \ldots, \alpha_n) \in R^n \text{ and}$$

$$s = (s_1, s_2, \ldots, s_n) \in \Pi(N),$$

write

$$F(\alpha, s) = \prod_{j=1}^{n} \langle p_j, \alpha \rangle^{s_j},$$

where $\langle , \rangle$ denotes the scalar product in R^n.

For $r = (r_1, r_2, \ldots, r_n) \in \Pi(N)$ write

$$p_{rs} = P(X(k) = r \mid X(k-1) = s).$$

Then p_{rs} is the coefficient of $\alpha_1^{r_1} \alpha_2^{r_2} \ldots \alpha_n^{r_n}$ in $F(\alpha, s)$. That is,

$$p_{rs} = (r_1! r_2! \ldots r_n!) \frac{\partial^N}{\partial \alpha_1^{r_1} \partial \alpha_2^{r_2} \ldots \partial \alpha_n^{r_n}} F(\alpha, s). \tag{7.2.1}$$

For $y = (y_1, y_2, \ldots, y_n) \in \Pi(M)$ write $\begin{pmatrix} M \\ y_1 \ y_2 \ \cdots \ y_n \end{pmatrix}$ for the multinomial coefficient

$\frac{M!}{y_1! y_2! \ldots y_n!}$. This is just the number of ways of selecting y_1 objects from M into state 1, y_2 into state 2 and so on.

Then, under the original probability measure P,

$$P(Y(k) = y \mid X(k) = r) = \begin{pmatrix} M \\ y_1 \ y_2 \ \cdots \ y_n \end{pmatrix} \left(\frac{r_1}{N}\right)^{y_1} \left(\frac{r_2}{N}\right)^{y_2} \cdots \left(\frac{r_n}{N}\right)^{y_n}.$$

Write $\mathcal{G}_k$ for the complete σ-field generated by $X(0), X(1), \ldots, X(k)$ and $Y(0), Y(1), Y(2), \ldots, Y(k-1)$.

$\mathcal{Y}_k$ will denote the complete σ-field generated by $Y(0), Y(1), Y(2), \ldots, Y(k)$. We wish to introduce a new probability measure $\overline{P}$ under which the probability of withdrawing an element in any one of the n states is just $1/n$. For this define factors

$$\gamma_k(Y(k)) = \left(\frac{1}{n}\right)^M \left(\frac{X_1(k)}{N}\right)^{-Y_1(k)} \left(\frac{X_2(k)}{N}\right)^{-Y_2(k)} \cdots \left(\frac{X_n(k)}{N}\right)^{-Y_n(k)},$$

and write

$$\Lambda_k = \prod_{\ell=0}^{k} \gamma_k.$$

A new probability measure can be defined by putting $\left.\frac{d\overline{P}}{dP}\right|_{\mathcal{G}_k} = \Lambda_k.$

Lemma 7.2.1 *For $y \in \Pi(M)$, $r \in \Pi(N)$,*

$$\overline{P}(Y(k) = y \mid \mathcal{G}_k) = \begin{pmatrix} M \\ y_1 \ y_2 \ \cdots \ y_n \end{pmatrix} \left(\frac{1}{n}\right)^M.$$

Proof $\overline{P}(Y(k) = y \mid \mathcal{G}_k) = \overline{E}[I(Y(k) = y) \mid \mathcal{G}_k]$ and by a version of Bayes' Theorem (4.1.1), this is

$$= \frac{E[\Lambda_k I(Y(k) = y) \mid \mathcal{G}_k]}{E[\Lambda_k \mid \mathcal{G}_k]}.$$

Now γ_k is the only factor of Λ_k not $\mathcal{G}_k$-measurable, so this is

$$= \frac{E[\gamma_k I(Y(k) = y) \mid \mathcal{G}_k]}{E[\gamma_k \mid \mathcal{G}_k]}.$$

The denominator $E[\gamma_k \mid \mathcal{G}_k]$ equals

$$\left(\frac{1}{n}\right)^M E\left[\left(\frac{X_1(k)}{N}\right)^{-Y_1(k)} \left(\frac{X_2(k)}{N}\right)^{-Y_2(k)} \cdots \left(\frac{X_n(k)}{N}\right)^{-Y_n(k)} \mid \mathcal{G}_k\right],$$

and the only variables not $\mathcal{G}_k$-measurable are $Y_1(k), \ldots, Y_n(k)$. Consequently, this conditional expectation is

$$= \left(\frac{1}{n}\right)^M \sum_{y \in \Pi(M)} \binom{M}{y_1 \ldots y_n} = 1.$$

The numerator is

$$E[\gamma_k I(Y(k) = y) \mid \mathcal{G}_k] = \binom{M}{y_1\ y_2\ \cdots\ y_n} \left(\frac{1}{n}\right)^M.$$

Consequently,

$$\overline{P}(Y(k) = y \mid \mathcal{G}_k) = \binom{M}{y_1\ y_2\ \cdots\ y_n} \left(\frac{1}{n}\right)^M$$

$$[3pt] = \overline{P}(Y(k) = y).$$

That is, under $\overline{P}$ the n states are i.i.d. with probability $1/n$. ∎

Remark 7.2.2 Under $\overline{P}$, $\overline{P}(X(k) = r \mid X(k-1) = s)$ is still p_{rs} given by (7.2.1). However, as we saw in Lemma 7.2.1,

$$\overline{P}(Y(k) = y \mid \mathcal{G}_k) = \overline{P}(Y(k) = y \mid X(k) = r)$$

$$= \overline{P}(Y(k) = y) = \binom{M}{y_1\ y_2\ \cdots\ y_n} \left(\frac{1}{n}\right)^M.$$

□

To return from $\overline{P}$ to P the inverse density must be introduced. That is, with

$$\overline{\gamma}_k = \gamma_k^{-1} = \left(\frac{1}{n}\right)^{-M} \left(\frac{X_1(k)}{N}\right)^{Y_1(k)} \left(\frac{X_2(k)}{N}\right)^{Y_2(k)} \cdots \left(\frac{X_n(k)}{N}\right)^{Y_n(k)},$$

$$\overline{\Lambda}_k = \Lambda_k^{-1} = \prod_{\ell=0}^k \overline{\gamma}_\ell,$$

the probability P can be defined by putting $\left.\dfrac{dP}{d\overline{P}}\right|_{\mathcal{G}_k} = \overline{\Lambda}_k$.

If $\{\phi_k\}$ is a $\{\mathcal{G}_k\}$ adapted process then Bayes' Theorem (4.1.1) implies

$$E[\phi_k \mid \mathcal{Y}_k] = \frac{\overline{E}[\overline{\Lambda}_k \phi_k \mid \mathcal{Y}_k]}{\overline{E}[\overline{\Lambda}_k \mid \mathcal{Y}_k]}.$$

$\overline{E}[\overline{\Lambda}_k \phi_k \mid \mathcal{Y}_k]$ is, therefore, an unnormalized conditional expectation of ϕ_k given $\mathcal{Y}_k$. The denominator $\overline{E}[\overline{\Lambda}_k \mid \mathcal{Y}_k]$ is a normalizing factor.

For $r \in \Pi(N)$ write $q_r(k) = \overline{E}[\overline{\Lambda}_k I(X(k) = r) \mid \mathcal{Y}_k]$. Note that $\sum_{r \in \Pi(N)} I(X(k) = r) = 1$ so that $\sum_{r \in \Pi(N)} q_r(k) = \overline{E}[\overline{\Lambda}_k \mid \mathcal{Y}_k]$.

We then have the following recursion.

Theorem 7.2.3 *If*
$Y(k) = (Y_1(k), Y_2(k), \ldots, Y_n(k)) = (y_1, y_2, \ldots, y_n) \in \Pi(N)$,

$$q_r(k) = n^{-M} \left(\frac{r_1}{N}\right)^{y_1} \left(\frac{r_1}{N}\right)^{y_2} \cdots \left(\frac{r_n}{N}\right)^{y_n} \sum_{s \in \Pi(N)} p_{rs} q_s(k-1).$$

(Note we take $0^0 = 1$.)

Proof

$$q_r(k) = \overline{E}[\overline{\Lambda}_k I(X(k) = r) \mid \mathcal{Y}_k]$$

$$= \overline{E}[\overline{\Lambda}_k I(X(k) = r) \mid \mathcal{Y}_{k-1}, Y(k) = (y_1, y_2, \ldots, y_n)]$$

$$= \overline{E}[\overline{\Lambda}_{k-1} \overline{\gamma}_k I(X(k) = r) \mid \mathcal{Y}_{k-1}, Y(k) = (y_1, y_2, \ldots, y_n)]$$

$$= n^{-M} \left(\frac{r_1}{N}\right)^{y_1} \left(\frac{r_2}{N}\right)^{y_2} \cdots \left(\frac{r_n}{N}\right)^{y_n} \overline{E}[\overline{\Lambda}_{k-1} I(X(k) = r)$$

$$\times \left(\sum_{s \in \Pi(N)} I(X(k-1) = s)\right) \mid \mathcal{Y}_{k-1}]$$

$$= n^{-M} \left(\frac{r_1}{N}\right)^{y_1} \left(\frac{r_2}{N}\right)^{y_2} \cdots \left(\frac{r_n}{N}\right)^{y_n} \overline{E}[\overline{\Lambda}_{k-1}$$

$$\times \left(\sum_{s \in \Pi(N)} (X(k-1) = s)\right) p_{rs} \mid \mathcal{Y}_{k-1}]$$

$$= n^{-M} \left(\frac{r_1}{N}\right)^{y_1} \left(\frac{r_2}{N}\right)^{y_2} \cdots \left(\frac{r_n}{N}\right)^{y_n} \sum_{s \in \Pi(N)} p_{rs} q_s(k-1).$$

∎

Remarks 7.2.4

$$P(X(k) = r \mid \mathcal{Y}_k) = E[I(X(k) = r) \mid \mathcal{Y}_k]$$

$$= \frac{q_r(k)}{\sum_{s \in \Pi(N)} q_s(k)}.$$

To obtain the expected value of $X(k)$ given the observations $\mathcal{Y}_k$ we consider the vector of values $r = (r_1, r_2, \ldots, r_n)$ for any $r \in \Pi(N)$. Then

$$E[X(k) \mid \mathcal{Y}_k] = \frac{\sum_{r \in \Pi(N)} q_r(k) \cdot r}{\sum_{s \in \Pi(N)} q_r(k)}.$$

Unfortunately this does not have the simple form of (7.1.1).

Also note that the transition probabilities p_{rs} can be re-estimated using the techniques described in Chapter 2 of [10].   □

7.3 Approximate formulae

Unfortunately the recursion for $q_r(k)$ given by Theorem 7.2.3 is not easily evaluated. One approximation would be to use a smaller value N' for N in the summation. To obtain nontrivial partitions of N' into n summands, N' should be greater than n. Substitution of the observed $Y(0)$, $Y(1)$, $Y(2)$, ... then would give a sequence of approximate distributions.

Alternatively, one could replace the martingale "noise" in the dynamics of $X(k)$ by Gaussian noise ([23]). To describe this, first suppose the n states of the individuals in the population are identified with the unit (column) vectors $e_1, \ldots, e_n$, $e_i = (0, \ldots, 1, 0, \ldots, 0)'$ of R^n. Let $X^i(k) \in \{e_1, \ldots, e_n\}$ denote the state of the i-th individual at time k. Then for each i, $1 \leq i \leq N$, $X^i(k)$ behaves like a Markov chain on $(\Omega, \mathcal{F}, P)$ with transition matrix P. Consequently,

$$X^i(k) = PX^i(k-1) + M^i(k), \tag{7.3.1}$$

where $E[M^i(k) \mid \mathcal{G}_{k-1}] = E[M^i(k) \mid X^i(k-1)] = 0$.

Write $p(0) = (p_1(0), \ldots, p_n(0))' = E[X^i(0)]$. Then from (7.3.1)

$$E[X^i(k)] = p(k) = P^k p(0).$$

For (column) vectors $x, y \in R^n$ write $x \otimes y = xy'$ for their Kronecker, or tensor, product, and diag x for the matrix with x on the diagonal.

Then, because $X^i(k)$ is one of the unit vectors $e_1, \ldots, e_n$,

$$X^i(k) \otimes X^i(k) = \text{diag } X^i(k)$$
$$= P \text{ diag } X^i(k-1)P' + M^i(k) \otimes (PX^i(k-1))$$
$$+ (PX^i(k-1)) \otimes M^i(k) + M^i(k) \otimes M^i(k)$$
$$= \text{diag } PX^i(k-1) + \text{diag } M^i(k).$$

Taking the expectation, we have

$$E[M^i(k) \otimes M^i(k)] = \text{diag } Pp(k-1) - P \text{ diag } p(k-1)P'$$
$$= Q(k), \text{ say.}$$

For $i \neq j$ the processes X^i and X^j are independent.

Define

$$X(k) = \frac{\sum_{i=1}^{N} X^i(k)}{N},$$

$$M(k) = \frac{\sum_{i=1}^{N} M^i(k)}{N}.$$

The (vector) process $X(k)$ describes the actual distribution of the population at time k. Its components sum to unity and

$$X(k) = PX(k-1) + M(k). \tag{7.3.2}$$

Also, by independence, $E[M(k) \otimes M(k)]$ is also equal to the matrix $Q(k)$.

The suggestion made in [23] is to replace the martingale increments $M(k)$ in (7.3.2) by independent (vector) Gaussian random variables $W(k)$ of mean 0 and covariance $Q(k)$. Write $\phi_k(w)$ for the normal density on R^n of mean 0 and covariance $Q(k)$.

That is, suppose the signal process $\overline{X}(k)$, taking values in R^n, has dynamics

$$\overline{X}(k) = P\overline{X}(k-1) + W(k).$$

For $y = (y_1, y_2, \ldots, y_n) \in \Pi(M)$ and $x = (x_1, x_2, \ldots, x_n) \in R^n$, $x \neq 0$, define

$$\rho(x, y) = |x|^{-M} |x_1|^{y_1} |x_2|^{y_2} \ldots |x_n|^{y_n};$$

set $\rho(0, y) = 0$ for $y \in \Pi(M)$.

The observation process still gives rise to $Y(0), Y(1), \ldots, Y(k) \in \Pi(M)$ and for $y \in \Pi(M)$, $x \in R^n$ we suppose

$$P(Y(k) = y \mid X(k) = x) = \begin{pmatrix} M \\ y_1 \ y_2 \ \cdots \ y_n \end{pmatrix} \rho(x, y).$$

Starting with the probability $\overline{P}$, now define $\overline{\gamma}_k = n^{-M} \rho(\overline{X}(k), Y(k))$, and $\overline{\Lambda}_k = \prod_{\ell=0}^{k} \overline{\gamma}_\ell$.

Again P can be defined in terms of $\overline{P}$ by setting $\left. \dfrac{dP}{d\overline{P}} \right|_{\mathcal{G}_k} = \overline{\Lambda}_k$.

Suppose $f : R^n \to R$ is any measurable "test" function. Consider

$$E[f(\overline{X}(k)) \mid \mathcal{Y}_k] = \frac{\overline{E}[\overline{\Lambda}_k f(\overline{X}(k)) \mid \mathcal{Y}_k]}{\overline{E}[\overline{\Lambda}_k \mid \mathcal{Y}_k]}.$$

Suppose there is an unnormalized conditional density $q_k(x)$ such that

$$\overline{E}[\overline{\Lambda}_k f(\overline{X}(k)) \mid \mathcal{Y}_k] = \int_{R^n} f(x) q_k(x) dx.$$

The next result gives a recursion for q_k which is the analog of Theorem 7.2.3.

Theorem 7.3.1

$$q_k(z) = n^{-M} \rho(z, y) \int_{R^n} \phi_k(z - Px) q_{k-1}(x) ds.$$

Proof

$$\overline{E}[\overline{\Lambda}_k f(\overline{X}(k)) \mid \mathcal{Y}_k] = \int_{R^n} f(z) q_k(z) \mathrm{d}z$$

$$= n^{-M} \overline{E}[\overline{\Lambda}_{k-1} \rho(\overline{X}(k), Y(k)) f(\overline{X}(k)) \mid \mathcal{Y}_k]$$

$$= n^{-M} \overline{E}[\overline{\Lambda}_{k-1} \rho(P\overline{X}(k-1) + W(k), y)$$

$$\times f(P\overline{X}(k-1) + W(k)) \mid \mathcal{Y}_{k-1}, Y(k) = y]$$

$$= n^{-M} \overline{E}[\overline{\Lambda}_{k-1} \rho(P\overline{X}(k-1) + W(k), y)$$

$$\times f(P\overline{X}(k-1) + W(k)) \mid \mathcal{Y}_{k-1}]$$

$$= n^{-M} \overline{E}[\,\overline{\Lambda}_{k-1} \rho(P\overline{X}(k-1) + W(k), y)$$

$$\times f(Px + w)\phi_k(w) q_{k-1}(x) \mid \mathcal{Y}_{k-1}]$$

$$= n^{-M} \iint \rho(z, y) f(z) \phi_k(z - Px) q_{k-1}(x) \mathrm{d}z \mathrm{d}x.$$

As this identity holds for all such f the result follows. ∎

8

Hidden populations

8.1 Introduction

An important problem in statistical ecology is how to determine the size of an animal population. A large number of techniques for providing an answer are available (see [35]) but the best-known one is the *capture–recapture* method.

A random sample of individuals is captured, tagged or marked in some way, and then released back into the population. After allowing time for the marked and unmarked to mix sufficiently, a second simple random sample is taken and the marked ones are observed.

At epoch ℓ write N_ℓ for the population size, $\tilde{n}_\ell$ for the number of marked and released individuals, $\tilde{\tilde{n}}_k = \sum_{\ell=1}^k \tilde{n}_\ell$ for the total number of captured and marked individuals up to time k, M_ℓ for the sample size, n_ℓ for the number of available marked individuals for sampling and y_ℓ for the number of captured (or recaptured) marked individuals.

We are interested in estimating the size N_ℓ at time ℓ of the population.

All random variables are defined initially on a probability space $(\Omega, \mathcal{F}, P)$. All the filtrations defined here will be assumed to be complete.

Write $\mathcal{G}_k = \sigma(N_\ell, n_\ell, y_\ell, M_\ell \; \ell \leq k)$, and $\mathcal{Y}_k = \sigma(y_\ell \; \ell \leq k)$.

We assume here that

1. The population sizes N_k follow the dynamics:

$$N_k = N_{k-1} + \sigma(N_{k-1})v_k, \qquad (8.1.1)$$

N_0 has distribution π_0 and v_k is a sequence of independent random variables with densities ϕ_k.

2. The n_k are random variables with conditional binomial distributions with parameters $p_k = p(\tilde{\tilde{n}}_k, y_1, \ldots, y_k, \theta)$ and $\tilde{\tilde{n}}_k$. For example,

$$p_1 = \frac{\theta \tilde{n}_1}{\tilde{n}_1} = \theta,$$

$$p_2 = \frac{\theta \tilde{n}_2 + \theta^2 \tilde{n}_1}{\tilde{n}_1 + \tilde{n}_2} = \frac{\theta \tilde{n}_2 + \theta^2 \tilde{n}_1}{\tilde{\tilde{n}}_2}, \ldots,$$

$$p_k = \frac{\sum_{i=1}^k \tilde{n}_i \theta^{k-i+1}}{\tilde{\tilde{n}}_k} = \frac{\tilde{\tilde{n}}_{k-1}}{\tilde{\tilde{n}}_k} \theta p_{k-1} + \theta \frac{\tilde{n}_k}{\tilde{\tilde{n}}_k}. \qquad (8.1.2)$$

$0 < \theta \leq 1$ is a parameter assumed to be known or it is to be estimated. The powers of θ express our belief that as time goes by early marked individuals are becoming less and less available for recapture due to various causes including deaths, emigration, etc.

If the number of captured and marked individuals $\tilde{n}_\ell$ is kept constant (8.1.2) takes the form:

$$p_k(\theta) = \frac{k-1}{k} p_{k-1} + \frac{\theta^k}{k}.$$ (8.1.3)

3. The observed random variable y_k is assumed to have a conditional binomial distribution,

$$P(y_k = m \mid \mathcal{G}_k - \{y_k\}) = \binom{M_k}{m}\left(\frac{n_k}{N_k}\right)^m\left(1 - \frac{n_k}{N_k}\right)^{M_k - m},$$ (8.1.4)

where $\displaystyle \binom{M_k}{m} = \frac{M_k!}{m!(M_k - m)!}.$

8.2 Distribution estimation

Define $\lambda_0 = 1$. For $\ell \geq 1$ and for suitable density functions ψ_ℓ write

$$\lambda_\ell = \frac{\sigma(N_{\ell-1})\psi_\ell(N_\ell)}{\phi_\ell(v_\ell)}\frac{1}{2^{M_\ell + \tilde{n}_\ell}} p_\ell^{-n_\ell}(1 - p_\ell)^{-\tilde{n}_\ell + n_\ell}\left(\frac{n_\ell}{N_\ell}\right)^{-y_\ell}\left(1 - \frac{n_\ell}{N_\ell}\right)^{y_\ell - M_\ell},$$ (8.2.1)

and $\Lambda_k = \prod_{\ell=0}^{k}\lambda_\ell$

Lemma 8.2.1 *The process Λ_k is a $\mathcal{G}$-martingale.*

Proof $E[\Lambda_k \mid \mathcal{G}_{k-1}] = \Lambda_{k-1}E[\lambda_k \mid \mathcal{G}_{k-1}]$. It remains to show that $E[\lambda_k \mid \mathcal{G}_{k-1}] = 1$.

$$E[\lambda_k \mid \mathcal{G}_{k-1}] = E[\frac{\sigma(N_{k-1})\psi_k(N_k)}{\phi_k(v_k)}\frac{1}{2^{M_k + \tilde{n}_k}}\left(\frac{n_k}{N_k}\right)^{-y_k}$$

$$\times \left(1 - \frac{n_k}{N_k}\right)^{y_k - M_k} p_k^{-n_k}(1 - p_k)^{-\tilde{n}_k + n_k} \mid \mathcal{G}_{k-1}]$$

$$= \frac{1}{2^{M_k + \tilde{n}_k}}E[\frac{\sigma(N_{k-1})\psi_k(N_k)}{\phi_k(v_k)} p_k^{-n_k}(1 - p_k)^{-\tilde{n}_k + n_k}E[\left(\frac{n_k}{N_k}\right)^{-y_k}$$

$$\times \left(1 - \frac{n_k}{N_k}\right)^{y_k - M_k} \mid \mathcal{G}_{k-1}, N_k, n_k, M_k] \mid \mathcal{G}_{k-1}].$$

The inner expectation equals 2^{M_k}, so that

$$E[\lambda_k \mid \mathcal{G}_{k-1}] = \frac{1}{2^{\tilde{n}_k}}E[\frac{\sigma(N_{k-1})\psi_k(N_k)}{\phi_k(v_k)} p_k^{-n_k}(1 - p_k)^{-\tilde{n}_k + n_k} \mid \mathcal{G}_{k-1}]$$

$$= \frac{1}{2^{\tilde{n}_k}}E[\frac{\sigma(N_{k-1})\psi_k(N_k)}{\phi_k(v_k)}\sum_{i=0}^{\tilde{n}_k}\binom{\tilde{n}_k}{i} \mid \mathcal{G}_{k-1}]$$

$$= E[\frac{\sigma(N_{k-1})\psi_k(N_{k-1} + \sigma v_k)}{\phi_k(v_k)} \mid \mathcal{G}_{k-1}]$$

$$= \int \frac{\sigma(N_{k-1})\psi_k(N_{k-1} + \sigma v)}{\phi_k(v)}\phi_k(v)dv = \int \psi_k(u)du = 1.$$

∎

A new probability measure $\overline{P}$ can be defined by setting $\dfrac{\mathrm{d}\overline{P}}{\mathrm{d}P}\Big|_{\mathcal{G}_k} = \Lambda_k$. The point here is that:

Lemma 8.2.2 *Under the new probability measure $\overline{P}$, N_k, n_k and y_k are three sequences of independent random variables which are independent of each other. Further, N_k has density ψ_k, n_k has distribution $\mathrm{bin}(\widetilde{n}_k, 1/2)$ and y_k has distribution $\mathrm{bin}(M_k, 1/2)$.*

Proof For any "test" functions f, g and h, using Bayes' Theorem 4.1.1,

$$\overline{E}[f(N_k)g(n_k)h(y_k) \mid \mathcal{G}_{k-1}] = \frac{E[f(N_k)g(n_k)h(y_k)\Lambda_k \mid \mathcal{G}_{k-1}]}{E[\Lambda_k \mid \mathcal{G}_{k-1}]},$$

which equals

$$E[f(N_k)g(n_k)h(y_k)\lambda_k \mid \mathcal{G}_{k-1}]$$
$$= E\Big[f(N_k)g(n_k)\frac{\sigma(N_{k-1})\psi_k(N_k)}{\phi_k(v_k)}\frac{1}{2^{\widetilde{n}_k}}p_k^{-n_k}(1-p_k)^{-\widetilde{n}_k+n_k}$$
$$\times E[h(y_k)\frac{1}{2^{M_k}}\Big(\frac{n_k}{N_k}\Big)^{-y_k}\Big(1 - \frac{n_k}{N_k}\Big)^{y_k-M_k} \mid \mathcal{G}_{k-1}, N_k, n_k, M_k] \mid \mathcal{G}_{k-1}\Big].$$

After cancellation the inner expectation equals

$$\sum_{m=0}^{M_k} h(m)\binom{M_k}{m}\frac{1}{2^{M_k}},$$

which shows that y_k has distribution $\mathrm{bin}(M_k, 1/2)$ independent of N and n. Similarly,

$$E\Big[f(N_k)g(n_k)\frac{\sigma(N_{k-1})\psi_k(N_k)}{\phi_k(v_k)}\frac{1}{2^{\widetilde{n}_k}}p_k^{-n_k}(1-p_k)^{-\widetilde{n}_k+n_k} \mid \mathcal{G}_{k-1}\Big]$$
$$= \overline{E}[g(n_k)]E\Big[f(N_{k-1}+v_k)\frac{\sigma(N_{k-1})\psi_k(N_{k-1}+\sigma v_k)}{\phi_k(v_k)} \mid \mathcal{G}_{k-1}\Big]$$
$$= \overline{E}[g(n_k)]\int f(N_{k-1}+\sigma v)\sigma(N_{k-1})\psi_k(N_{k-1}+\sigma v)\mathrm{d}v$$
$$= \overline{E}[g(n_k)]\int f(u)\psi_k(u)\mathrm{d}u$$
$$= \overline{E}[g(n_k)]\overline{E}[f(N_k)].$$

That is, under $\overline{P}$ the three processes are independent sequences of random variables with the desired distributions. ∎

Using this fact we derive a recursive equation for the unnormalized conditional distribution of N_k given $\mathcal{Y}_k$.

For any "test" function f consider

$$E[f(N_k) \mid \mathcal{Y}_k] = \frac{\overline{E}[f(N_k)\Lambda_k^{-1} \mid \mathcal{Y}_k]}{\overline{E}[\Lambda_k^{-1} \mid \mathcal{Y}_k]} \overset{\triangle}{=} \frac{\int f(z)q_k(z)\mathrm{d}z}{\overline{E}[\Lambda_k^{-1} \mid \mathcal{Y}_k]}. \tag{8.2.2}$$

The denominator of (8.2.2) being a normalizing factor, we focus only on the numerator.

In view of Lemma 8.2.2,

$$\overline{E}[f(N_k)\Lambda_k^{-1} \mid \mathcal{Y}_k] = \overline{E}[f(N_k)\Lambda_{k-1}^{-1}\lambda_k^{-1} \mid \mathcal{Y}_k]$$

$$= 2^{\widetilde{n}_k + M_k}\overline{E}\left[\sum_{i=0}^{\widetilde{n}_k}\int f(z)\left(\frac{i}{z}\right)^{y_k}\left(1-\frac{i}{z}\right)^{M_k-y_k}\frac{\phi_k\left(\frac{z-N_{k-1}}{\sigma(N_{k-1})}\right)}{\sigma(N_{k-1})\psi_k(z)}\psi_k(z)\right.$$

$$\left. \times p_k^i(1-p_k)^{\widetilde{n}_k-i}dz\frac{1}{2^{\widetilde{n}_k}}\binom{\widetilde{n}_k}{i}\Lambda_{k-1}^{-1}\mid \mathcal{Y}_k\right]$$

$$= 2^{M_k}\sum_{i=0}^{\widetilde{n}_k}\int\int f(z)\left(\frac{i}{z}\right)^{y_k}\left(1-\frac{i}{z}\right)^{M_k-y_k}\frac{\phi_k\left(\frac{z-u}{\sigma(u)}\right)}{\sigma(u)}$$

$$\times p_k^i(1-p_k)^{\widetilde{n}_k-i}\binom{\widetilde{n}_k}{i}q_{k-1}(u)dzdu.$$

Comparing this last expression with the right hand side of (8.2.2) we have:

Theorem 8.2.3 *The unnormalized conditional probability density function of the hidden Markov model given by (8.1.1), (8.1.2), and (8.1.4) follow the recursions*

$$q_k(z) = \sum_{i=0}^{\widetilde{n}_k}B(y_k, z, i)\int \Phi_k(z, u)q_{k-1}(u)du. \qquad (8.2.3)$$

Here $\quad B_k(y, z, i) = 2^{M_k}\left(\frac{i}{z}\right)^{y_k}\left(1-\frac{i}{z}\right)^{M_k-y_k}\binom{\widetilde{n}_k}{i}p_k^i(1-p_k)^{\widetilde{n}_k-i}\quad$ *and* $\quad \Phi_k(z, u) =$

$\dfrac{\phi_k\left(\frac{z-u}{\sigma(u)}\right)}{\sigma(u)}$. *(Note we take* $0^0 = 1$.*)*

Remarks 8.2.4

1. The normalized conditional density of N_k is given by $\dfrac{q_k(z)}{\int q_k(u)du}$.

2. The initial (normalized) probability density of N_0, prior to sampling, is $\pi_0(.)$, so $q_0(z) = \pi_0(z)$. Using the notation in Theorem 8.2.3,

$$q_1(z) = \sum_{i=0}^{\widetilde{n}_1}B_1(y, z, i)\int \Phi_1(z, u)\pi_0(u)du, \qquad (8.2.4)$$

and further estimates follow from (8.2.3).

3. If the distribution of N_0 is a delta function concentrated at some number A, (8.2.4) becomes

$$q_1(z) = \Phi_1(z, A)\sum_{i=0}^{\widetilde{n}_1}B_1(y, z, i). \qquad (8.2.5)$$

□

8.3 Parameter estimation

Our model is function of the parameter p_k, the proportion of the accessible marked individuals at epoch k. Suppose p_k has dynamics given by (8.1.2). We also assume that θ will take values in some measurable space (Θ, β, γ). We now derive a recursive joint conditional unnormalized distribution for N_k and θ. We keep working under the probability measure $\overline{P}$.

Lemma 8.3.1 *Write*

$$q_k(z, \theta)dzd\theta = \overline{E}[I(N_k \in dz, \theta \in d\theta)\Lambda_k^{-1} \mid \mathcal{Y}_k].$$

Then

$$q_k(z, \theta) = \sum_{i=0}^{\widetilde{\widetilde{n}}_k} B_k(y, z, i, \theta) \int \Phi_k(z, u)q_{k-1}(u, \theta)\,du. \qquad (8.3.1)$$

Here $B_k(y, z, i, \theta) = 2^{M_k}\left(\dfrac{i}{z}\right)^{y_k}\left(1 - \dfrac{i}{z}\right)^{M_k - y_k} p_k(\theta)^i\left(1 - p_k(\theta)\right)^{\widetilde{\widetilde{n}}_k - i}\dbinom{\widetilde{\widetilde{n}}_k}{i},$ *and*

$$\Phi_k(z, u) = \frac{\phi_k\left(\dfrac{z - u}{\sigma(u)}\right)}{\sigma(u)}.$$

Proof   Let f, g be integrable test functions.

$$\overline{E}[f(N_k)g(\theta)\Lambda_k^{-1} \mid \mathcal{Y}_k] = \iint f(z)g(v)q_k(z, v)\,dzd\gamma(v). \qquad (8.3.2)$$

Using the independence assumption under $\overline{P}$ the left hand side of (8.3.2) is

$$= \overline{E}[f(N_k)g(\theta)\Lambda_{k-1}^{-1}\lambda_k^{-1} \mid \mathcal{Y}_k]$$

$$= 2^{M_k}\overline{E}[\sum_{i=0}^{\widetilde{\widetilde{n}}_k} \int\int f(z)g(v)\left(\frac{i}{z}\right)^{y_k}\left(1 - \frac{i}{z}\right)^{M_k - y_k}\frac{\phi_k\left(\dfrac{z - N_{k-1}}{\sigma(N_{k-1})}\right)}{\sigma(N_{k-1})\psi_k(z)}\psi_k(z)$$

$$\times p_k(v)^i\left(1 - p_k(v)\right)^{\widetilde{\widetilde{n}}_k - i}dzd\gamma(v)\dbinom{\widetilde{\widetilde{n}}_k}{i}\Lambda_{k-1}^{-1} \mid \mathcal{Y}_k]$$

$$= 2^{M_k}\sum_{i=0}^{\widetilde{\widetilde{n}}_k}\int\int\int f(z)g(v)\left(\frac{i}{z}\right)^{y_k}\left(1 - \frac{i}{z}\right)^{M_k - y_k}\frac{\phi_k\left(\dfrac{z - u}{\sigma(u)}\right)}{\sigma(u)}$$

$$\times p_k(v)^i\left(1 - p_k(v)\right)^{\widetilde{\widetilde{n}}_k - i}\dbinom{\widetilde{\widetilde{n}}_k}{i}q_{k-1}(u, v)dzdud\gamma(v).$$

Comparing this last expression with the right hand side of (8.3.2) gives (8.3.1).   ∎

If at time 1, θ has density $h(\theta)$, then

$$q_1(z, \theta) = \sum_{i=0}^{\widetilde{\widetilde{n}}_k} B_1(y, z, i, \theta)h(\theta) \int \Phi_1(z, u)\pi_0(u)du,$$

and further updates are given by Lemma 8.3.1.

If no dynamics enter the population size and N_k has density $\phi_k(.)$ independent of N_ℓ, $\ell < k$, the recursion in Lemma 8.3.1 simplifies to:

$$q_k(z, \theta) = \phi_k(z)q_{k-1}(z, \theta) \sum_{i=0}^{\widetilde{\widetilde{n}}_k} B_k(y, z, i, \theta). \qquad (8.3.3)$$

Maximum posterior estimators

Quantity (8.2.4) (or 8.2.5) is a function of the unknown population size and could be maximized with respect to z, yielding a critical value $\widehat{N}_1$, which is the maximum posterior estimate of N at epoch 1 given y_1. Similar maximizations at later times will provide MAP estimators for the population size at these times.

8.4 Pathwise estimation

We now derive a recursive equation, which *does not involve any integration*, for the unnormalized density of the whole path up to epoch k.

Write $q_k(z_0, \ldots, z_k)dz_0 \ldots dz_k \stackrel{\triangle}{=} \overline{E}[I(z_0 \in dz_0) \ldots I(z_k \in dz_k)\Lambda_k^{-1} \mid \mathcal{Y}_k]$.

Theorem 8.4.1 *Using the notation in Theorem 8.2.3,*

$$q_k(z_0, \ldots, z_k) = \sum_{i=0}^{\widetilde{\widetilde{n}}_k} B_k(y, z, i)\Phi_k(z_{k-1}, z_k)q_{k-1}(z_0, \ldots, z_{k-1}). \qquad (8.4.1)$$

Proof Let $f_0, \ldots, f_k$ be "test functions". Then

$$\overline{E}[f_0(N_0) \ldots f_k(N_k)\Lambda_k^{-1} \mid \mathcal{Y}_k] = \int f_0(z_0) \ldots f_k(z_k)q_k(z_0, \ldots, z_k)dz_0 \ldots dz_k, \qquad (8.4.2)$$

and

$$\overline{E}[f_0(N_0) \ldots f_k(N_k)\Lambda_k^{-1} \mid \mathcal{Y}_k] = \overline{E}[f_0(N_0) \ldots f_k(N_k)\Lambda_{k-1}^{-1}\lambda_k^{-1} \mid \mathcal{Y}_k]$$

$$= 2^{M_k}\overline{E}[f_0(N_0) \ldots f_k(N_{k-1})\Lambda_{k-1}^{-1} \int f_k(z_k)\left(\frac{n}{z_k}\right)^{y_k}\left(1 - \frac{n}{z_k}\right)^{M_k - y_k}$$

$$\times \frac{\phi_k\left(\dfrac{z_k - N_{k-1}}{\sigma(N_{k-1})}\right)}{\sigma(N_{k-1})}dz_k \mid \mathcal{Y}_k]$$

$$= 2^{M_k}\int \int f_0(z_0) \ldots f_k(z_k)\left(\frac{n}{z_k}\right)^{y_k}\left(1 - \frac{n}{z_k}\right)^{M_k - y_k}$$

$$\times \frac{\phi_k\left(\dfrac{z_k - z_{k-1}}{\sigma(z_{k-1})}\right)}{\sigma(z_{k-1})}q_{k-1}(z_0, \ldots, z_{k-1})dz_0 \ldots dz_k.$$

Comparing the last expression with (8.4.2) yields at once (8.4.1).   ∎

Again we have $q_0(z) = \pi_0(z)$, $q_1(z_0, z_1) = \Phi_1(z_0, z_1)\pi_0(z_0) \sum_{i=0}^{\widetilde{\widetilde{n}}_1} B_1(y, z, i,)$, and further estimates follow from (8.4.1). However, no integration is needed in subsequent recursions.

Maximum posterior estimators

Expression (8.4.1) is a function of the path $(z_0, \ldots, z_k)$ and could be maximized yielding a critical path $(\widehat{N}_0, \ldots, \widehat{N}_k)$. Since no integration is involved here one could substitute, at time k say, the sequence of critical values $\widehat{N}_0, \ldots, \widehat{N}_{k-1}$ and then maximize $q_k(\widehat{N}_0, \ldots, \widehat{N}_{k-1}, z_k)$ with respect to the variable z_k to obtain an estimate for N_k.

8.5 A Markov chain model

Suppose that on probability space $(\Omega, \mathcal{F}, \overline{P})$ are given three sequences of independent random variables, N_k, n_k, and y_k. For $k \in \mathbb{N}$, N_k is uniformly distributed over some finite set $S = \{s_1, \ldots, s_L\} \subset \mathbb{N} - \{0\}$, n_k has a binomial distribution with parameters $(\widetilde{n}_k, 1/2)$ and y_k has a binomial distribution with parameters $(M_k, 1/2)$, where $M_k \in \mathbb{N} - \{0\}$ is given.

We wish to define a new probability measure P such that y_k has a binomial distribution with parameters $(M_k, n_k/N_k)$, N_k is a Markov chain with state space S and stochastic matrix $C = \{c_{ij}\} = P[N_{k+1} = s_i \mid N_k = s_j]$, n_k are random variables with conditional binomial distributions with parameters $(p_k, \widetilde{n}_k)$.

Define the $\mathcal{G}$-predictable sequences $\alpha_\ell^i = \sum_{j=1}^{L} I(N_{\ell-1} = s_j)c_{ij}$, for $i = 1, \ldots, L$. In vector notation this is $\alpha_\ell(N_{\ell-1}) = C'I(N_{\ell-1})$, where $I(N_{\ell-1}) = (I(N_{\ell-1} = s_1), \ldots, I(N_{\ell-1} = s_L))$.

Now write

$$\lambda_\ell = 2^{M_\ell + \widetilde{n}_\ell} \, p_\ell^{n_\ell} (1 - p_\ell)^{\widetilde{n}_\ell - n_\ell} \left(\frac{n_\ell}{N_\ell}\right)^{y_\ell} \left(1 - \frac{n_\ell}{N_\ell}\right)^{M_\ell - y_\ell} \prod_{i=1}^{L} (L\alpha_\ell^i)^{I(N_\ell = s_i)}, \qquad (8.5.1)$$

$\Lambda_k = \prod_{\ell=0}^{k} \lambda_\ell$.

Lemma 8.5.1 *The process Λ_k is a $\mathcal{G}$-martingale.*

Proof $E[\Lambda_k \mid \mathcal{G}_{k-1}] = \Lambda_{k-1} E[\lambda_k \mid \mathcal{G}_{k-1}]$, so we must show that $E[\lambda_k \mid \mathcal{G}_{k-1}] = 1$.

$$E[\lambda_k \mid \mathcal{G}_{k-1}] = E[\frac{\phi_k(N_k)}{\phi_k(v_k)} \frac{1}{2^{M_k}} \left(\frac{n}{N_k}\right)^{-y_k} \left(1 - \frac{n}{N_k}\right)^{y_k - M_k} \mid \mathcal{G}_{k-1}]$$

$$= \frac{1}{2^{M_k}} E[\frac{\phi_k(N_k)}{\phi_k(v_k)} E[\left(\frac{n}{N_k}\right)^{-y_k} \left(1 - \frac{n}{N_k}\right)^{y_k - M_k} \mid \mathcal{G}_{k-1}, N_k, M_k] \mid \mathcal{G}_{k-1}]$$

$$= \frac{1}{2^{M_k}} E[\frac{\phi_k(N_k)}{\phi_k(v_k)} \sum_{m=0}^{M_k} \frac{M_k!}{m!(M_k - m)!} \mid \mathcal{G}_{k-1}]$$

$$= \int \frac{\phi_k(N_{k-1} + v)}{\phi_k(v)} \phi_k(v) dv = 1,$$

and a new probability measure P can be defined by setting $\left.\dfrac{dP}{d\overline{P}}\right|_{\mathcal{G}_k} = \Lambda_k$. ∎

Lemma 8.5.2 *Under the probability measure P the above processes obey the desired dynamics, i.e. N_k is a Markov chain with state space S and stochastic matrix $C = \{c_{ij}\}$, y_k and n_k are random variables with conditional binomial distributions with parameters $(M_k, n_k/N_k)$ and $(p_k, \widetilde{n}_k)$ respectively.*

Proof We give a proof only for the first statement regarding N_k.

$$P[N_k = s_j \mid \mathcal{G}_{k-1}] = E[I(N_k = s_j) \mid \mathcal{G}_{k-1}]$$

$$= \frac{\overline{E}[I(N_k = s_j)\Lambda_k \mid \mathcal{G}_{k-1}]}{\overline{E}[\Lambda_k \mid \mathcal{G}_{k-1}]}$$

$$= \overline{E}[I(N_k = s_j)\lambda_k \mid \mathcal{G}_{k-1}]$$

$$= \overline{E}[I(N_k = s_j)2^{M_k + \widetilde{n}_k} p_\ell^{n_k}(1 - p_k)^{\widetilde{n}_k - n_k} \left(\frac{n_k}{s_j}\right)^{y_k}$$

$$\times \left(1 - \frac{n_k}{s_j}\right)^{M_k - y_k} L\alpha_k^j \mid \mathcal{G}_{k-1}]$$

$$= L\alpha_k^j 2^{M_k + \widetilde{n}_k} \overline{E}[I(N_k = s_j) p_\ell^{n_k}(1 - p_k)^{\widetilde{n}_k - n_k} \left(\frac{n_k}{s_j}\right)^{y_k}$$

$$\times \left(1 - \frac{n_k}{s_j}\right)^{M_k - y_k} \mid \mathcal{G}_{k-1}]$$

$$= \alpha_k^j L 2^{M_k + \widetilde{n}_k} \frac{1}{L} \sum_{n=0}^{\widetilde{n}_k} \sum_{m=0}^{M_k} \binom{M_k}{m} \left(\frac{n}{s_i}\right)^m \left(1 - \frac{n}{s_i}\right)^{M_k - m} \frac{1}{2^{M_k}}$$

$$\times \binom{\widetilde{n}_k}{n} p_k^n (1 - p_k)^{\widetilde{n}_k - n} \frac{1}{2^{\widetilde{n}_k}}$$

$$= \alpha_k^j = P[N_k = s_j \mid N_{k-1}].$$

■

Working under the probability measure $\overline{P}$, we derive recursive equations for the unnormalized conditional probability distribution of N_k. Write

$$P[N_k = s_i \mid \mathcal{Y}_k] = E[I(N_k = s_i) \mid \mathcal{Y}_k]$$

$$= \frac{\overline{E}[I(N_k = s_i)\Lambda_k \mid \mathcal{Y}_k]}{\overline{E}[\Lambda_k \mid \mathcal{Y}_k]},$$

and $q_k^{s_i} = \overline{E}[I(N_k = s_i)\Lambda_k \mid \mathcal{Y}_k]$.

Theorem 8.5.3

$$q_k^{s_i} = \sum_{n=0}^{\widetilde{n}_k} B_k(y, s_i, n) \sum_{j=1}^{L} c_{ij} q_{k-1}^{s_j}. \tag{8.5.2}$$

Here $B_k(y, s_i, n) = 2^{M_k} \left(\frac{n}{s_i}\right)^{y_k} \left(1 - \frac{n}{s_i}\right)^{M_k - y_k} \binom{\widetilde{n}_k}{n} p_k^n (1 - p_k)^{\widetilde{n}_k - n}.$

If at time 0, $q_0 = \pi = (\pi_1, \ldots, \pi_L)$,

$$q_1^{s_i} = \sum_{n=0}^{\widetilde{n}_1} B_1(y, s_i, n) \sum_{j=1}^{L} c_{ij} \pi_j.$$

$$(8.5.3)$$

If $N_0 = s_\alpha$ with probability 1,

$$q_1^{s_i} = c_{i\alpha} \sum_{n=0}^{\widetilde{\widetilde{n}}_1} B_1(y, s_i, n), \tag{8.5.4}$$

and further updates are given by (8.5.2).

MAP estimators of $N_1, \ldots, N_k$ are provided by

$$\widehat{N}_1 = \mathrm{argmax}\{q_1^{s_1}, q_1^{s_2}, \ldots, q_1^{s_L}\}, \ldots,$$
$$\widehat{N}_k = \mathrm{argmax}\{q_k^{s_1}, q_k^{s_2}, \ldots, q_k^{s_L}\}.$$

8.6 Recursive parameter estimation

The previous model is function of the parameters p_k and $C = c_{ij}$. Let $p_k = p_k(\theta_1)$ and $C = C(\theta_2) = c_{ij}(\theta_2)$ and $\theta = (\theta_1, \theta_2)$. Suppose θ belongs to some measurable space (Θ, β, γ).

Working again under the probability measure $\overline{P}$, write

$$q_k^{s_i}(\theta)\mathrm{d}\theta = \overline{E}[I(N_k = s_i)I(\theta \in \mathrm{d}\theta)\Lambda_k \mid \mathcal{Y}_k]. \tag{8.6.1}$$

Lemma 8.6.1

$$q_k^{s_i}(\theta) = \sum_{n=0}^{\widetilde{\widetilde{n}}_k} B_k(y, s_i, n, \theta) \sum_{j=1}^{L} c_{ij}(\theta)q_{k-1}^{s_j}(\theta). \tag{8.6.2}$$

Here
$$B_k(y, s_i, n, \theta) = 2^{M_k}\left(\frac{n}{s_i}\right)^{y_k}\left(1 - \frac{n}{s_i}\right)^{M_k - y_k}\binom{\widetilde{\widetilde{n}}_k}{n} p_k^n(\theta)(1 - p_k(\theta))^{\widetilde{\widetilde{n}}_k - n}.$$

If θ_1 has density $h(.)$ and θ_2 has density $g(.)$,

$$q_1^{s_i}(\theta) = \sum_{n=0}^{\widetilde{\widetilde{n}}_1} B_1(y, s_i, n, \theta) \sum_{j=1}^{L} c_{ij}(\theta_2)g(\theta_2)\pi_j. \tag{8.6.3}$$

8.7 A tags loss model

In this section we propose a model where the marks or tags are not permanent. In this situation the marking is done using double tagging where each individual is marked with two tags. For simplicity we assume that the two tags on each individual are nondistinguishable and that individuals retain or lose their tags independently.

We start again with a probability space $(\Omega, \mathcal{F}, \overline{P})$ on which are given two sequences of independent random variables N_k and y_k. For $k \in \mathbb{N}$, N_k is uniformly distributed over some finite set $S = \{s_1, \ldots, s_L\} \subset \mathbb{N} - \{0\}$, and y_k has a trinomial distribution with parameters $(M_k, 1/3)$, where $M_k \in \mathbb{N} - \{0\}$ is given.

At any epoch ℓ each individual in the population is in any of three states, namely unmarked, marked with only one tag, and marked with two tags, which states we shall call 0, 1, 2 respectively. We suppose that each individual behaves like an independent time homogeneous Markov chain with transition matrix $\{p_{ij}\}$.

At each time ℓ the population size N_ℓ is distributed or partitioned into three groups $N_\ell(2)$, $N_\ell(1)$, and $N_\ell(0) = N_\ell - N_\ell(2) - N_\ell(1)$ among the three states, and we would like to define the set of all such partitions as the states of a three-dimensional Markov chain $(N_\ell(0), N_\ell(1), N_\ell(2))$. Recall that at each epoch ℓ, $0 \leq N_\ell(2), N_\ell(1) \leq \tilde{\tilde{n}}_\ell$.

Write

$$p_{(i_0,i_1,i_2),(j_0,j_1,j_2)} = P[(N_k(0), N_k(1), N_k(2)) = (i_0, i_1, i_2)$$
$$| (N_{k-1}(0), N_{k-1}(1), N_{k-1}(2)) = (j_0, j_1, j_2)],$$

and for any real numbers x_0, x_1, x_2 define the function

$$F(x_0, x_1, x_2, j_0, j_1, j_2) = (\sum_{\ell=0}^{2} p_{\ell 0} x_\ell)^{j_0} (\sum_{\ell=0}^{2} p_{\ell 1} x_\ell)^{j_1} (\sum_{\ell=0}^{2} p_{\ell 2} x_\ell)^{j_2}.$$

Then $p_{(i_0,i_1,i_2),(j_0,j_1,j_2)}$ is the coefficient of $x_0^{i_0} x_1^{i_1} x_2^{i_2}$ in $F(x_0, x_1, x_2, j_0, j_1, j_2)$.

We wish to define a new probability measure P such that y_k has a conditional trinomial distribution with parameters $\left(M_k, N_k(0)/N_k, N_k(1)/N_k, N_k(2)/N_k\right)$, N_k is a Markov chain with state space S and stochastic matrix $C = \{c_{ij}\}$. The Markov chain $(N_\ell(0), N_\ell(1), N_\ell(2))$ is the same under both probability measures.

Define again the $\mathcal{G}$-predictable sequences

$$\alpha_\ell^i = \sum_{j=1}^{L} I(N_{\ell-1} = s_j) c_{ij}, \quad i = 1, \dots, L.$$

Now write

$$\lambda_\ell = 3^{M_\ell} \left(\frac{N_\ell(0)}{N_\ell}\right)^{y_\ell(0)} \left(\frac{N_\ell(1)}{N_\ell}\right)^{y_\ell(1)} \left(\frac{N_\ell(2)}{N_\ell}\right)^{y_\ell(2)} \prod_{i=1}^{L} (L\alpha_\ell^i)^{I(N_\ell=s_i)}, \tag{8.7.1}$$

$$\Lambda_k = \prod_{\ell=0}^{k} \lambda_\ell.$$

The process Λ_k is a $\mathcal{G}$-martingale and a new probability measure P can be defined by setting $\left.\dfrac{dP}{d\overline{P}}\right|_{\mathcal{G}_k} = \Lambda_k$. It can be checked that under P the above processes have the desired distributions. Working under the probability measure $\overline{P}$, we derive recursive equations for the unnormalized conditional joint probability distribution of N_k and $(N_k(0), N_k(1), N_k(2))$. Write

$$P[N_k = s_i, (N_k(0), N_k(1), N_k(2)) = (i_0, i_1, i_2) \mid \mathcal{Y}_k]$$
$$= E[I(N_k = s_i)I[(N_k(0), N_k(1), N_k(2)) = (i_0, i_1, i_2)] \mid \mathcal{Y}_k]$$
$$= \frac{\overline{E}[I(N_k = s_i)I[(N_k(0), N_k(1), N_k(2)) = (i_0, i_1, i_2)]\Lambda_k \mid \mathcal{Y}_k]}{\overline{E}[\Lambda_k \mid \mathcal{Y}_k]},$$

and

$$q_k(s_i, i_1, i_2) = \overline{E}[I(N_k = s_i, N_k(2) = i_2, N_k(1) = i_1, N_k(0) = s_i - i_1 - i_2)\Lambda_k \mid \mathcal{Y}_k].$$

It can be shown that $q_k(s_i, i_1, i_2)$ is given by the following recursions.

Theorem 8.7.1

$$q_k(s_i, i_1, i_2) = 3^{M_k} \left(\frac{s_i - i_1 - i_2}{s_i} \right)^{y_k(0)} \left(\frac{i_1}{s_i} \right)^{y_k(1)} \left(\frac{i_2}{s_i} \right)^{y_k(2)} \sum_{j=1}^{L} c_{ij}$$

$$\times \sum_{j_1+j_2=0}^{\widetilde{\widetilde{n}}_{k-1}} p_{(s_i-i_1-i_2,i_1,i_2)(s_j-j_1-j_2,j_1,j_2)} q_{k-1}(s_j, j_1, j_2). \qquad (8.7.2)$$

The expected value of N_k given the observations $\mathcal{Y}_k$ is given by

$$E[N_k \mid \mathcal{Y}_k] = \frac{\sum_{i=1}^{L} s_i \sum_{i_1+i_2=0}^{\widetilde{n}_k} q_k(s_i, i_1, i_2)}{\sum_{i=1}^{L} \sum_{i_1+i_2=0}^{\widetilde{n}_k} q_k(s_i, i_1, i_2)}.$$

Another way of looking at the problem is by considering only the subpopulation of tagged individuals in the definition of the Markov chain $(N_k(0), N_k(1), N_k(2))$. In this case the state space is the set of all the partitions of the totality of tagged individuals into three groups: the ones with two tags, the ones with one tag and the ones who lost both tags. Hence we write the total number of tagged individuals as $\widetilde{\widetilde{n}}_k = \widetilde{\widetilde{n}} = N_k(2) + N_k(1) + N_k(0)$. Note that, when sampling, we cannot observe directly members belonging to the group of individuals who lost their two tags as they become undistinguishable from the unmarked ones in the sample.

Now we assume that under $\overline{P}$ the observation process is multinomial with parameters $(M_k, 1/4)$ and under P it is (conditional) multinomial with parameters $\left(M_k, N_k(0)/N_k, N_k(1)/N_k, N_k(2)/N_k, N_k(u)/N_k \right)$. Here $N_k(u)$ is the number of unmarked individuals in the population. Again note that $N_k(u)$ is not $N_k(0)$. Given $y_k(1)$, $y_k(2)$, the unobserved component $y_k(0)$, under the probability measure $\overline{P}$, is binomial with parameters $(M_k - y_k(1) - y_k(2), 1/2)$.

Write

$$P[N_k = s_i, (N_k(0), N_k(1), N_k(2)) = (i_0, i_1, i_2) \mid \mathcal{Y}_k]$$

$$= E[I(N_k = s_i)I((N_k(0), N_k(1), N_k(2)) = (i_0, i_1, i_2)) \mid \mathcal{Y}_k]$$

$$= \frac{\overline{E}[I(N_k = s_i)I((N_k(0), N_k(1), N_k(2)) = (i_0, i_1, i_2))\Lambda_k \mid \mathcal{Y}_k]}{\overline{E}[\Lambda_k \mid \mathcal{Y}_k]},$$

and

$$q_k(s_i, i_0, i_1, i_2) = \overline{E}[I(N_k = s_i)I((N_k(0), N_k(1), N_k(2)) = (i_0, i_1, i_2))\Lambda_k \mid \mathcal{Y}_k]$$

$$= \overline{E}[I(N_k = s_i)I((N_k(0), N_k(1), N_k(2)) = (i_0, i_1, i_2))4^{M_k}$$

$$\times \left(\frac{i_0}{s_i} \right)^{y_k(0)} \left(\frac{i_1}{s_i} \right)^{y_k(1)} \left(\frac{i_2}{s_i} \right)^{y_k(2)} \left(\frac{i_u}{s_i} \right)^{y_k(u)}$$

$$\times L \sum_{j=1}^{L} I(N_{k-1} = s_j)c_{ij}\Lambda_{k-1} \mid \mathcal{Y}_k]$$

$$
= 4^{M_k} \left(\frac{i_1}{S_i}\right)^{y_k(1)} \left(\frac{i_2}{S_i}\right)^{y_k(2)} \overline{E}\left[\left(\frac{i_0}{S_i}\right)^{y_k(0)}\right.
$$

$$
\times \left(\frac{S_i - i_2 - i_1 - i_0}{S_i}\right)^{M_k - y_k(2) - y_k(1) - y_k(0)}
$$

$$
\times \sum_{j=1}^{L} I(N_{k-1} = s_j) c_{ij} I((N_k(0), N_k(1), N_k(2)) = (i_0, i_1, i_2))
$$

$$
\times \sum_{j_0 + j_1 + j_2 = \widetilde{\widetilde{n}}} I((N_{k-1}, N_{k-1}(1), N_{k-1}(2)) = (j_0, j_1, j_2)) \Lambda_{k-1} \mid \mathcal{Y}_k \right]
$$

$$
= 4^{M_k} \left(\frac{i_1}{S_i}\right)^{y_k(1)} \left(\frac{i_2}{S_i}\right)^{y_k(2)} \overline{E}\left[\sum_{m=0}^{M_k - y_k(2) - y_k(1)} \left(\frac{i_0}{S_i}\right)^{m}\right.
$$

$$
\times \left(\frac{S_i - i_2 - i_1 - i_0}{S_i}\right)^{M_k - y_k(2) - y_k(1) - m}
$$

$$
\times \binom{M_k - y_k(2) - y_k(1)}{m} \left(\frac{1}{2}\right)^{M_k - y_k(2) - y_k(1)} \sum_{j=1}^{L} I(N_{k-1} = s_j) c_{ij}
$$

$$
\times \sum_{j_0 + j_1 + j_2 = \widetilde{\widetilde{n}}} I((N_{k-1}, N_{k-1}(1), N_{k-1}(2)) = (j_0, j_1, j_2))
$$

$$
\times p_{(i_0, i_1, i_2)(j_0, j_1, j_2)} \Lambda_{k-1} \mid \mathcal{Y}_k \right].
$$

Using the definition of q we have:

Theorem 8.7.2

$$
q_k(s_i, i_0, i_1, i_2) = \sum_{m=0}^{M_k - y_k(2) - y_k(1)} B_k(m, i_0, i_1, i_2) \sum_{j=1}^{L} c_{ij}
$$

$$
\sum_{j_0 + j_1 + j_2 = \widetilde{\widetilde{n}}} p_{(i_0, i_1, i_2)(j_0, j_1, j_2)} q_{k-1}(s_j, j_0, j_1, j_2).
$$

Here

$$
B_k(m, i_0, i_1, i_2) = 2^{M_k + y_k(1) + y_k(2)} \left(\frac{i_1}{S_i}\right)^{y_k(1)} \left(\frac{i_2}{S_i}\right)^{y_k(2)} \left(\frac{i_0}{S_i}\right)^{m}
$$

$$
\times \left(\frac{S_i - i_2 - i_1 - i_0}{S_i}\right)^{M_k - y_k(2) - y_k(1) - m} \times \binom{M_k - y_k(2) - y_k(1)}{m}.
$$

8.8 Gaussian noise approximation

An approximate but simpler form of the recursion in Lemma 8.7.2 is to use a suggestion proposed by [23] where the martingale increment "noise" present in the representation of a Markov chain is replaced by Gaussian noise. To this effect, let's identify, as it is explained in [10], the three states 0, 1, 2 with the standard unit (column) vectors e_1, e_2, e_3 of $\mathbb{R}^3$. Write $X_k^n \in \{e_1, e_2, e_3\}$ for the state of the n-th individual at time k, $1 \leq n \leq \widetilde{\widetilde{n}}$. Then each individual behaves like a Markov chain on $(\Omega, \mathcal{F}, P)$ with transition matrix P.

Define $X_k = \dfrac{1}{\overset{\approx}{n}} \overset{\overset{\approx}{n}}{\underset{n=1}{\sum}} X_k^n$. Then

$$X_k = P X_{k-1} + M_k, \tag{8.8.1}$$

where M_k is a martingale increment. The suggestion made in [23] is to replace the martingale increment M_k in (8.8.1) by an independent Gaussian random variable v_k of mean 0 and covariance matrix $E[M_k M_k']$ whose density is denoted by ϕ_k.

That is, the signal process x_k, taking values in $\mathbb{R}^3$, has dynamics

$$x_k = P x_{k-1} + v_k. \tag{8.8.2}$$

We assume that under $\overline{P}$ the observation process is multinomial with parameters $(M_k, 1/4)$, x_k has density ϕ_k and N_k is uniformly distributed over the set $(s_1, \ldots, s_L)$. Under the "real world" probability measure P, N_k is a Markov chain with transition matrix C, x_k has dynamics (8.8.2) and y_k has conditional probability distribution given by

$$P[y_k = y_k(2) + y_k(1) + y_k(0) + y_k(u)$$

$$\mid M_k, N_k = s_i, x_k = (x_0, x_1, x_2), N_k(u) = s_i - x_2 - x_1 - x_0]$$

$$= \binom{M_k}{y_k(2),\, y_k(1),\, y_k(0),\, y_k(u)} \left(\frac{x_0}{s_i}\right)^{y_k(0)} \left(\frac{x_1}{s_i}\right)^{y_k(1)} \left(\frac{x_2}{s_i}\right)^{y_k(2)}$$

$$\times \left(\frac{s_i - x_2 - x_1 - x_0}{s_i}\right)^{M_k - y_k(2) - y_k(1) - y_k(0)}.$$

P is defined in terms of $\overline{P}$ using the $\mathcal{G}$–martingale

$$\Lambda_k = \prod_{k=0}^{k} 4^{M_k} \left(\frac{x_k(0)}{N_k}\right)^{y_k(0)} \left(\frac{x_k(1)}{N_k}\right)^{y_k(1)} \left(\frac{x_k(2)}{N_k}\right)^{y_k(2)}$$

$$\times \left(\frac{N_k - x_k(2) - x_k(1) - x_k(0)}{N_k}\right)^{M_k - y_k(2) - y_k(1) - y_k(0)} \prod_{i=1}^{L} (L\alpha_k^i)^{I(N_k = s_i)}.$$

The next theorem is the analog of Theorem 8.7.2.

Theorem 8.8.1 *The unnormalized joint conditional probability distribution of N_k and x_k* $\overline{E}[I(N_k = s_i)I(x_k \in dx)\Lambda_k \mid \mathcal{Y}_k] := q_k^{s_i}(x)dx$ *is given recursively as follows:*

$$q_k^{s_i}(x) = \sum_{m=0}^{M_k - y_k(2) - y_k(1)} B_k(m, x_0, x_1, x_2) \sum_{j=1}^{L} c_{ij} \int \phi_k(x - Pu) q_{k-1}^{s_j}(u) du,$$

using the notation in Theorem 8.7.2.

References

[1] Arnold, L. *Stochastic Differential Equations: Theory and Applications.* John Wiley & Sons (1974).

[2] Bensoussan, A. *Stochastic Control of Partially Observed Systems*, Cambridge University Press (1992).

[3] Baum, L. E. and Petrie, T. Statistical inference for probabilistic functions of finite state markov chains, *Ann. Inst. Statistical Mathematics* **37** (1966) 1554–1563.

[4] Billingsley, P. *Probability and Measure.* Third edn. Wiley Series in Probability and Mathematical Statistics (1995).

[5] Brémaud, P. *Markov Chains, Gibbs Fields, Monte Carlo Simulation and Queues.* Text in Applied Mathematics 31. Springer (1999).

[6] Chung, K. L. and Williams, R. J. *Introduction to Stochastic Integration.* Second edn. Birkhauser (1990).

[7] Chung, K. L. *A Course in Probability Theory.* Academic Press (1974).

[8] Davis, M. H. A. Martingales of Wiener and Poisson processes. *J. London Math. Soc.* (2) **13** (1976) 336–338.

[9] Dempster, A. P., Laird, N. M., and Rubin, D. B. Maximum likelihood from incomplete data via the EM algorithm, *J. Royal Statistical Society, B* **39** (1977) 1–38.

[10] Elliott, R. J., Aggoun L., and Moore, J. B. *Hidden Markov Models: Estimation and Control.* Applications of Mathematics, Vol. 29. Springer-Verlag (1995).

[11] Elliott, R. J. *Stochastic Calculus and Applications.* Applications of Mathematics Vol. 18. Springer-Verlag (1982).

[12] Elliott, R. J. and Moore, J. B. A martingale Kronecker lemma and parameter estimation for linear systems. *IEEE Trans. Auto. Control* **43**, No. 9 (1998).

[13] Elliott, R. J. and Krishnamurthy, Vikrum. New finite dimensional filters for parameter estimation of discrete-time linear Gaussian models. *IEEE Trans. Auto. Control* **44** (1998) 938–951.

[14] Elliott, R. J. and Krishnamurthy, Vikrum. Exact finite dimensional filters for maximum likelihood parameter estimation of continuous-time linear Gaussian systems. *SIAM J. Control Optim.* **35**, No. 6 (1997) 1908–1923.

[15] Hajek, B. and Wong, E. *Stochastic Processes in Engineering Systems.* Springer-Verlag (1985).

[16] Ikeda, N. and Watanabe, S. *Stochastic Differential Equations and Diffusion Processes.* Second edn. North-Holland Publishing Company (1989).

[17] Itô, K. Stochastic integral. *Proc. Imperial Acad. Tokyo* **20** (1944) 519–524.

[18] Jacod, J. *Calcul Stochastique et Problemes de Martingales*, Lecture Notes in Math. Vol. 714, Springer-Verlag (1979).

[19] Jacod, J. and Shiryayev, A. N. *Limit Theorems for Stochastic Processes.* Springer (1987).

[20] Jazwinski, A. H. *Stochastic Processes and Filtering Theory.* Academic Press (1970).

[21] Karatzas, Ioannis and Shreve, S. E. *Brownian Motion and Stochastic Calculus*. Graduate Texts in Mathematics, Vol. 113. Springer-Verlag (1988).

[22] Kloeden, P. E. and Platen, E. *Numerical Solution of Stochastic Differential Equations*. Springer-Verlag (1992).

[23] Krichagina, N. V., Lipster, R. S., and Rubinovich, E. Y. Kalman filter for Markov processes. In *Steklov Seminar* (1984). Eds. N. V. Krylov, R. S. Lipster and A. A. Novikov, Optimization Software Inc. (1985) 197–213.

[24] Kunita, H. and Watanabe, S. On square integrable martingales, *Nagoya Math. J.* **30** (1967) 209–245.

[25] Lipster, R. S. and Shiryayev, A. N. (1977). *Statistics of Random Processes 2*. Springer-Verlag (1977).

[26] Loève, M. *Probability Theory I*. Fourth edn. Springer-Verlag (1977).

[27] Milshtein, G. N. Approximate integration of stochastic differential equations. *Theory of Prob. Appl.* **19** (1974) 562–577.

[28] Meyer, P. A. *Probability and Potentials*. Blaisdell Publishing Company (1966).

[29] Neveu, J. *Discrete Parameter Martingales*. North-Holland (1975).

[30] Oksendal, B. *Stochastic Differential Equations, an Introduction with Applications*. Fourth edn. Universitext. Springer (1995).

[31] Papoulis, A. *Probability, Random Variables and Stochastic Processes*. McGraw Hill (1984).

[32] Revuz, Daniel and Yor, Marc. *Continuous Martingales and Brownian Motion*. Third edn. Grundlehren der mathematischen Wissenschaften, Vol. 293. Springer-Verlag (1999).

[33] Rogers, L. C. G. and Satchell, S. E. Estimating variance from high, low and closing prices, *Ann. Appl. Probability* **1** (1991) 504–512.

[34] Rogers, L. C. G. and Williams, David. *Diffusions, Markov Processes, and Martingales*. Vol. 1: Foundations. Second edn. John Wiley & Sons (1994).

[35] Seber, G. A. F. *The Estimation of Animal Abundance and Related Parameters*. Second edn. Edward Arnold (1982).

[36] Shiryayev, A. N. *Probability Theory*. Springer-Verlag (1984).

[37] Wu, C. F. J. On the convergence properties of the EM algorithm, *Ann. Statistics* **11** (1983) 95–103.

Index

Printed in the United States
By Bookmasters